T0202745

Pro Deep Learning with TensorFlow 2.0

A Mathematical Approach to Advanced Artificial Intelligence in Python

Second Edition

Santanu Pattanayak

Apress®

Pro Deep Learning with TensorFlow 2.0: A Mathematical Approach to Advanced Artificial Intelligence in Python

Santanu Pattanayak
Prestige Ozone
Bangalore, Karnataka, India

ISBN-13 (pbk): 978-1-4842-8930-3   ISBN-13 (electronic): 978-1-4842-8931-0
https://doi.org/10.1007/978-1-4842-8931-0

Managing Director, Apress Media LLC: Welmoed Spahr
Acquisitions Editor: Celestin Suresh John
Development Editor: Laura Berendson
Coordinating Editor: Aditee Mirashi

Cover designed by eStudioCalamar

Cover image by Pixabay (www.pixabay.com)

Distributed to the book trade worldwide by Apress Media, LLC, 1 New York Plaza, New York, NY 10004, U.S.A. Phone 1-800-SPRINGER, fax (201) 348-4505, e-mail orders-ny@springer-sbm.com, or visit www.springeronline.com. Apress Media, LLC is a California LLC and the sole member (owner) is Springer Science + Business Media Finance Inc (SSBM Finance Inc). SSBM Finance Inc is a **Delaware** corporation.

For information on translations, please e-mail booktranslations@springernature.com; for reprint, paperback, or audio rights, please e-mail bookpermissions@springernature.com.

Apress titles may be purchased in bulk for academic, corporate, or promotional use. eBook versions and licenses are also available for most titles. For more information, reference our Print and eBook Bulk Sales web page at http://www.apress.com/bulk-sales.

Any source code or other supplementary material referenced by the author in this book is available to readers on GitHub (https://github.com/Apress). For more detailed information, please visit http://www.apress.com/source-code.

Printed on acid-free paper

To my wife, Sonia.

Table of Contents

About the Author

Santanu Pattanayak works as a Senior Staff Machine Learning Specialist at Qualcomm Corp R&D and is the author of *Quantum Machine Learning with Python*, published by Apress. He has more than 16 years of experience, having worked at GE, Capgemini, and IBM before joining Qualcomm. He graduated with a degree in electrical engineering from Jadavpur University, Kolkata, and is an avid math enthusiast. Santanu has a master's degree in data science from the Indian Institute of Technology (IIT), Hyderabad. Currently, he resides in Bangalore with his wife.

About the Technical Reviewer

Manohar Swamynathan is a data science practitioner and an avid programmer, with over 14+ years of experience in various data science–related areas that include data warehousing, business intelligence (BI), analytical tool development, ad hoc analysis, predictive modeling, data science product development, consulting, formulating strategies, and executing analytics programs. He's had a career covering the life cycle of data across different domains, such as US mortgage banking, retail/ecommerce, insurance, and industrial IoT. He has a bachelor's degree with a specialization in physics, mathematics, and computers and a master's degree in project management. He's currently living in Bengaluru, the Silicon Valley of India.

Introduction

Pro Deep Learning with TensorFlow 2.0 is a practical and mathematical guide to deep learning using TensorFlow. Deep learning is a branch of machine learning where you model the world in terms of a hierarchy of concepts. This pattern of learning is similar to the way a human brain learns, and it allows computers to model complex concepts that often go unnoticed in other traditional methods of modeling. Hence, in the modern computing paradigm, deep learning plays a vital role in modeling complex real-world problems, especially by leveraging the massive amount of unstructured data available today.

Because of the complexities involved in a deep-learning model, many times it is treated as a black box by people using it. However, to derive the maximum benefit from this branch of machine learning, one needs to uncover the hidden mystery by looking at the science and mathematics associated with it. In this book, great care has been taken to explain the concepts and techniques associated with deep learning from a mathematical as well as a scientific viewpoint. Also, the first chapter is totally dedicated toward building the mathematical base required to comprehend deep-learning concepts with ease. TensorFlow has been chosen as the deep-learning package because of its flexibility for research purposes and its ease of use. Another reason for choosing TensorFlow is its capability to load models with ease in a live production environment using its serving capabilities.

In summary, *Pro Deep Learning with TensorFlow 2.0* provides practical, hands-on expertise so you can learn deep learning from scratch and deploy meaningful deep-learning solutions. This book will allow you to get up to speed quickly using TensorFlow and to optimize different deep-learning architectures. All the practical aspects of deep learning that are relevant in any industry are emphasized in this book. You will be able to use the prototypes demonstrated to build new deep-learning applications. The code presented in the book is available in the form of iPython notebooks and scripts that allow you to try out examples and extend them in interesting ways. You will be equipped with the mathematical foundation and scientific knowledge to pursue research in this field and give back to the community.

Who This Book Is For

This book is for data scientists and machine-learning professionals looking at deep-learning solutions to solve complex business problems.

This book is for software developers working on deep-learning solutions through TensorFlow.

This book is for graduate students and open source enthusiasts with a constant desire to learn.

What You'll Learn

The chapters covered in this book are as follows:

Chapter 1—Mathematical Foundations: In this chapter, all the relevant mathematical concepts from linear algebra, probability, calculus, optimization, and machine-learning formulation are discussed in detail to lay the mathematical foundation required for deep learning. The various concepts are explained with a focus on their use in the fields of machine learning and deep learning.

Chapter 2—Introduction to Deep-Learning Concepts and TensorFlow: This chapter introduces the world of deep learning and discusses its evolution over the years. The key building blocks of neural networks, along with several methods of learning, such as the Perceptron learning rule and backpropagation methods, are discussed in detail. Also, this chapter introduces the paradigm of TensorFlow coding so that readers are accustomed to the basic syntax before moving on to more involved implementations in TensorFlow.

Chapter 3—Convolutional Neural Networks: This chapter deals with convolutional neural networks used for image processing. Image processing is a computer vision issue that has seen a huge boost in performance in the areas of object recognition and detection, object classification, localization, and segmentation using convolutional neural networks. The chapter starts by illustrating the operation of convolution in detail and then moves on to the working principles of a convolutional neural network. Much emphasis is given to the building blocks of a convolutional neural network to give the reader the tools needed to experiment and extend their networks in interesting ways. Further, backpropagation through convolutional and pooling layers is discussed in detail so that the reader has a holistic view of the training process of convolutional networks. Also covered in this chapter are the properties of equivariance and translation invariance, which are central to the success of convolutional neural networks.

Chapter 4—Natural Language Processing: This chapter deals with natural language processing using deep learning. It starts with different vector space models for text processing; word-to-vector embedding models, such as the continuous bag of words method and Skip-grams; and then moves to much more advanced topics that involve recurrent neural networks (RNNs), LSTM, bidirectional RNN, and GRU. Language modeling as well as neural machine translation is covered in detail in this chapter to help the reader utilize these networks in real-world problems involving the same. Also, the mechanism of backpropagation in cases of RNNs and LSTM as well vanishing-gradient problems are discussed in much detail. This chapter also puts emphasis on the attention mechanism and how it fits into the powerful transformer architecture that has changed the world of natural language processing.

Chapter 5—Unsupervised Learning with Restricted Boltzmann Machines and Autoencoders: In this chapter, you will learn about unsupervised methods in deep learning that use restricted Boltzmann machines (RBMs) and autoencoders. Also, the chapter will touch upon Bayesian inference and Markov chain Monte Carlo (MCMC) methods, such as the Metropolis algorithm and Gibbs sampling, since the RBM training process requires some knowledge of sampling. Further, this chapter introduces the concepts of contrastive divergence in the context of a customized version of Gibbs sampling for enabling practical training of RBMs. We will further discuss how RBMs can be used for collaborative filtering in recommender systems as well as their use in unsupervised pretraining of deep-belief networks (DBNs).

In the second part of the chapter, various kinds of autoencoders are covered, such as sparse encoders, denoising autoencoders, variational autoencoders, and so forth. Also, the reader will learn about how internal features learned from the autoencoders can be utilized for dimensionality reduction as well as for supervised learning. Finally, the chapter ends with a little brief on data preprocessing techniques, such as PCA whitening and ZCA whitening.

Chapter 6—Advanced Neural Networks: In this chapter, the reader will learn about some of the advanced neural networks, such as fully convolutional neural networks, R-CNN, Fast R-CNN, Faster, U-Net, and so forth, that deal with semantic segmentation of images, object detection, and localization. This chapter also introduces the readers to traditional image segmentation methods so that they can combine the best of both worlds as appropriate. In the second half of the chapter, the reader will learn about the generative adversarial network (GAN), a new schema of generative model used for producing synthetic data like the data produced by a given distribution. GAN has usages and potential in several fields, such as in image generation, image inpainting, abstract reasoning, semantic segmentation, video generation, style transfer from one domain to another, and text-to-image generation applications, among others. In addition to standard GAN, the book introduces the theoretical concepts as well as the practical deployment intricacies associated with cycle consistency GAN that is used for domain translational problems.

To summarize, the key learnings the reader can expect from this book are as follows:

- Understand full-stack deep learning using TensorFlow and gain a solid mathematical foundation for deep learning.

- Deploy complex deep-learning solutions in production using TensorFlow.

- Carry out research on deep learning and perform experiments using TensorFlow.

Source Code

All source code used in this book can be downloaded from `github.com/apress/pro-deep-learning-tensorflow2`.

Mathematical Foundations

Deep learning is a branch of machine learning that uses many layers of artificial neurons stacked one on top of the other for identifying complex features within the input data and solving complex real-world problems. It can be used for both supervised and unsupervised machine-learning tasks. Deep learning is currently used in areas such as computer vision, video analytics, pattern recognition, anomaly detection, text processing, sentiment analysis, and recommender system, among other things. Also, it has widespread use in robotics, self-driving car mechanisms, and artificial intelligence systems in general.

Mathematics is at the heart of any machine-learning algorithm. A strong grasp of the core concepts of mathematics goes a long way in enabling one to select the right algorithms for a specific machine-learning problem, keeping in mind the end objectives. Also, it enables one to tune machine-learning/deep-learning models better and understand the possible reasons for an algorithm not performing as desired. Deep learning being a branch of machine learning demands as much expertise in mathematics, if not more, than that required for other machine-learning tasks. Mathematics as a subject is vast, but there are a few specific topics that machine-learning or deep-learning professionals and/or enthusiasts should be aware of to extract the most out of this wonderful domain of machine learning, deep learning, and artificial intelligence. Illustrated in Figure 1-1 are the different branches of mathematics along with their importance in the field of machine learning and deep learning. We will discuss the relevant concepts in each of the following branches in this chapter:

- Linear algebra

- Calculus

- Probability

- Optimization and formulation of machine-learning algorithms

© Santanu Pattanayak 2023
S. Pattanayak, *Pro Deep Learning with TensorFlow 2.0*, https://doi.org/10.1007/978-1-4842-8931-0_1

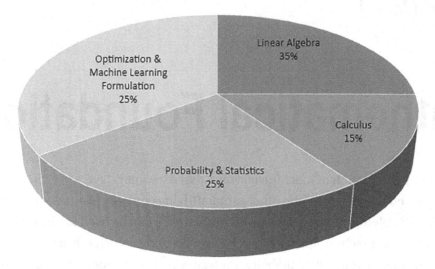

Figure 1-1. *Importance of mathematics topics for machine learning and data science*

Note   Readers who are already familiar with these topics can choose to skip this chapter or have a casual glance through the content.

Linear Algebra

Linear algebra is a branch of mathematics that deals with vectors and their transformation from one vector space to another vector space. Since in machine learning and deep learning we deal with multidimensional data and their manipulation, linear algebra plays a crucial role in almost every machine-learning and deep-learning algorithm. Illustrated in Figure 1-2 is a three-dimensional vector space where v_1, v_2, and v_3 are vectors and P is a 2-D plane within the three-dimensional vector space.

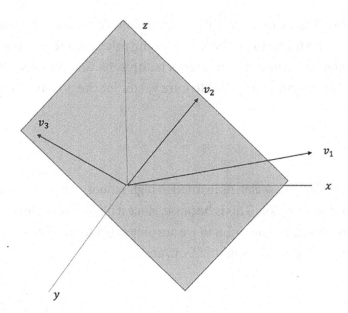

Figure 1-2. *Three-dimensional vector space with vectors and a vector plane*

Vector

An array of numbers, either continuous or discrete, is called a vector, and the space consisting of vectors is called a vector space. Vector space dimensions can be finite or infinite, but most machine-learning or data-science problems deal with fixed-length vectors, for example, the velocity of a car moving in the plane with velocities *Vx* and *Vy* in the *x* and *y* direction respectively (see Figure 1-3).

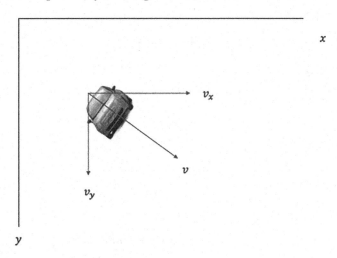

Figure 1-3. *Car moving in the x-y vector plane with velocity components Vx and Vy*

In machine learning, we deal with multidimensional data, so vectors become very crucial. Let's say we are trying to predict the housing prices in a region based on the area of the house, number of bedrooms, number of bathrooms, and population density of the locality. All these features form an input feature vector for the housing price prediction problem.

Scalar

A one-dimensional vector is a scalar. As learned in high school, a scalar is a quantity that has only magnitude and no direction. This is because, since it has only one direction along which it can move, its direction is immaterial, and we are only concerned about the magnitude.

Examples: height of a child, weight of fruit, etc.

Matrix

A matrix is a two-dimensional array of numbers arranged in rows and columns. The size of the matrix is determined by its row length and column length. If a matrix A has m rows and n columns, it can be represented as a rectangular object (see Figure 1-4a) having $m \times n$ elements, and it can be denoted as $A_{m \times n}$.

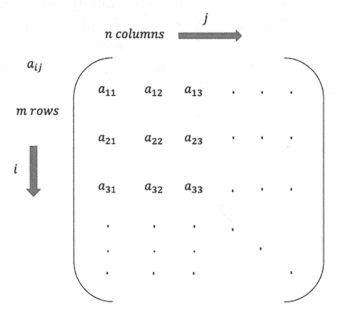

$m \times n\ matrix$

Figure 1-4a. *Structure of a matrix*

A few vectors belonging to the same vector space form a matrix.

For example, an image in grayscale is stored in a matrix form. The size of the image determines the image matrix size, and each matrix cell holds a value from 0 to 255 representing the pixel intensity. Illustrated in Figure 1-4b is a grayscale image followed by its matrix representation.

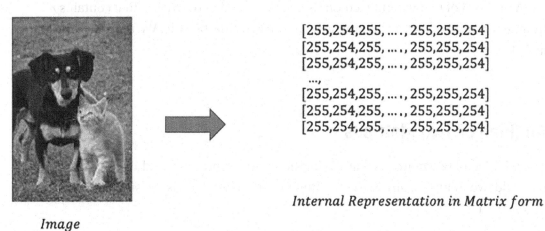

[255,254,255, ... , 255,255,254]
[255,254,255, ... , 255,255,254]
[255,254,255, ... , 255,255,254]

...,

[255,254,255, ... , 255,255,254]
[255,254,255, ... , 255,255,254]
[255,254,255, ... , 255,255,254]

Internal Representation in Matrix form

Image

Figure 1-4b. *Structure of a matrix*

Tensor

A tensor is a multidimensional array of numbers. In fact, vectors and matrices can be treated as 1-D and 2-D tensors. In deep learning, tensors are mostly used for storing and processing data. For example, an image in RGB is stored in a three-dimensional tensor, where along one dimension we have the horizontal axis and along the other dimension we have the vertical axis and where the third dimension corresponds to the three color channels, namely, red, green, and blue. Another example is the four-dimensional tensors used in feeding images through mini-batches in a convolutional neural network. Along the first dimension, we have the image number in the batch, and along the second dimension, we have the color channels, and the third and fourth dimensions correspond to pixel location in the horizontal and vertical directions.

Matrix Operations and Manipulations

Most deep-learning computational activities are done through basic matrix operations, such as multiplication, addition, subtraction, transposition, and so forth. Hence, it makes sense to review the basic matrix operations.

A matrix A of m rows and n columns can be considered a matrix that contains n number of column vectors of dimension m stacked side-by-side. We represent the matrix as follows:

$$A_{m \times n} \in \mathbb{R}^{m \times n}$$

Addition of Two Matrices

The addition of two matrices A and B implies their element-wise addition. We can only add two matrices, provided their dimensions match. If C is the sum of matrices A and B, then

$$c_{ij} = a_{ij} + b_{ij} \quad \forall \; i \in \{1,2,..m\}, \forall \; j \in \{1,2,..n\}$$

$$where \; a_{ij} \in A, b_{ij} \in B, c_{ij} \in C$$

Example: $A = \begin{bmatrix} 1 & 2 \\ 3 & 4 \end{bmatrix} B = \begin{bmatrix} 5 & 6 \\ 7 & 8 \end{bmatrix}$ then $A + B = \begin{bmatrix} 1+5 & 2+6 \\ 3+7 & 4+8 \end{bmatrix} = \begin{bmatrix} 6 & 8 \\ 10 & 12 \end{bmatrix}$

Subtraction of Two Matrices

The subtraction of two matrices A and B implies their element-wise subtraction. We can only subtract two matrices provided their dimensions match.

If C is the matrix representing $A - B$, then

$$c_{ij} = a_{ij} - b_{ij} \forall \; i \in \{1,2,..m\}, \forall \; j \in \{1,2,..n\}$$

$$where \; a_{ij} \in A, b_{ij} \in B, c_{ij} \in C$$

Example: $A = \begin{bmatrix} 1 & 2 \\ 3 & 4 \end{bmatrix} B = \begin{bmatrix} 5 & 6 \\ 7 & 8 \end{bmatrix}$ then $A - B = \begin{bmatrix} 1-5 & 2-6 \\ 3-7 & 4-8 \end{bmatrix} = \begin{bmatrix} -4 & -4 \\ -4 & -4 \end{bmatrix}$

Product of Two Matrices

For two matrices $A \in R^{m \times n}$ and $B \in R^{p \times q}$ to be multipliable, n should be equal to p. The resulting matrix is $C \in R^{m \times q}$. The elements of C can be expressed as follows:

$$c_{ij} = \sum_{k=1}^{n} a_{ik} b_{kj} \ \forall \ i \in \{1,2,..m\}, \forall \ j \in \{1,2,..q\}$$

For example, the matrix multiplication of the two matrices $A, B \in R^{2 \times 2}$ can be computed as seen here:

$$A = \begin{bmatrix} 1 & 2 \\ 3 & 4 \end{bmatrix} B = \begin{bmatrix} 5 & 6 \\ 7 & 8 \end{bmatrix}$$

$$c_{11} = \begin{bmatrix} 1 & 2 \end{bmatrix} \begin{bmatrix} 5 \\ 7 \end{bmatrix} = 1 \times 5 + 2 \times 7 = 19 \ c_{12} = \begin{bmatrix} 1 & 2 \end{bmatrix} \begin{bmatrix} 6 \\ 8 \end{bmatrix} = 1 \times 6 + 2 \times 8 = 22$$

$$c_{21} = \begin{bmatrix} 3 & 4 \end{bmatrix} \begin{bmatrix} 5 \\ 7 \end{bmatrix} = 3 \times 5 + 4 \times 7 = 43 \ c_{22} = \begin{bmatrix} 3 & 4 \end{bmatrix} \begin{bmatrix} 6 \\ 8 \end{bmatrix} = 3 \times 6 + 4 \times 8 = 50$$

$$C = \begin{bmatrix} c_{11} & c_{12} \\ c_{21} & c_{22} \end{bmatrix} = \begin{bmatrix} 19 & 22 \\ 43 & 50 \end{bmatrix}$$

Transpose of a Matrix

The transpose of a matrix $A \in R^{m \times n}$ is generally represented by $A^T \in R^{n \times m}$ and is obtained by transposing the column vectors as row vectors.

$$a'_{ji} = a_{ij} \ \ \forall \ i \in \{1,2,..m\}, \forall \ j \in \{1,2,..n\}$$

$$where \ a'_{ji} \in A^T \ and \ a_{ij} \in A$$

Example: Example : $A = \begin{bmatrix} 1 & 2 \\ 3 & 4 \end{bmatrix}$ then $A^T = \begin{bmatrix} 1 & 3 \\ 2 & 4 \end{bmatrix}$

The transpose of the product of two matrices A and B is the product of the transposes of matrices A and B in the reverse order; i.e., $(AB)^T = B^T A^T$.

For example, if we take two matrices $A = \begin{bmatrix} 19 & 22 \\ 43 & 50 \end{bmatrix}$ and $B = \begin{bmatrix} 5 & 6 \\ 7 & 8 \end{bmatrix}$, then

$$(AB) = \begin{bmatrix} 19 & 22 \\ 43 & 50 \end{bmatrix}\begin{bmatrix} 5 & 6 \\ 7 & 8 \end{bmatrix} = \begin{bmatrix} 95 & 132 \\ 301 & 400 \end{bmatrix} \text{ and hence } (AB)^T = \begin{bmatrix} 95 & 301 \\ 132 & 400 \end{bmatrix}$$

Now, $A^T = \begin{bmatrix} 19 & 43 \\ 22 & 50 \end{bmatrix}$ and $B^T = \begin{bmatrix} 5 & 7 \\ 6 & 8 \end{bmatrix}$

$$B^T A^T = \begin{bmatrix} 5 & 7 \\ 6 & 8 \end{bmatrix}\begin{bmatrix} 19 & 43 \\ 22 & 50 \end{bmatrix} = \begin{bmatrix} 95 & 301 \\ 132 & 400 \end{bmatrix}$$

Hence, the equality $(AB)^T = B^T A^T$ holds.

Dot Product of Two Vectors

Any vector of dimension n can be represented as a matrix $v \in R^{n \times 1}$. Let us denote two n-dimensional vectors $v_1 \in R^{n \times 1}$ and $v_2 \in R^{n \times 1}$.

$$v_1 = \begin{bmatrix} v_{11} \\ v_{12} \\ . \\ . \\ . \\ v_{1n} \end{bmatrix} \quad v_2 = \begin{bmatrix} v_{21} \\ v_{22} \\ . \\ . \\ . \\ v_{2n} \end{bmatrix}$$

The dot product of two vectors is the sum of the product of corresponding components—i.e., components along the same dimension—and can be expressed as follows:

$$v_1.v_2 = v_1{}^T v_2 = v_2{}^T v_1 = v_{11}v_{21} + v_{12}v_{22} + . . + v_{1n}v_{2n} = \sum_{k=1}^{n} v_{1k}v_{2k}$$

Example: $v_1 = \begin{bmatrix} 1 \\ 2 \\ 3 \end{bmatrix} v_2 = \begin{bmatrix} 3 \\ 5 \\ -1 \end{bmatrix} v_1.v_2 = v_1{}^T v_2 = 1 \times 3 + 2 \times 5 - 3 \times 1 = 10$

Matrix Working on a Vector

When a matrix is multiplied by a vector, the result is another vector. Let's say $A \in R^{m \times n}$ is multiplied by the vector $x \in R^{n \times 1}$. The result would produce a vector $b \in R^{m \times 1}$:

$$A = \begin{bmatrix} c_1^{(1)} c_1^{(2)} \dots c_1^{(n)} \\ c_2^{(1)} c_2^{(2)} \dots c_2^{(n)} \\ \cdot \\ \cdot \\ \cdot \\ c_m^{(1)} c_m^{(2)} \dots c_m^{(n)} \end{bmatrix} \quad x = \begin{bmatrix} x_1 \\ x_2 \\ \cdot \\ \cdot \\ \cdot \\ x_n \end{bmatrix}$$

A consists of n column vectors $c^{(i)} \in R^{m \times 1}$ $\quad \forall\, i \in \{1, 2, \dots, n\}$.

$$A = \begin{bmatrix} c^{(1)} c^{(2)} c^{(3)} \dots c^{(n)} \end{bmatrix}$$

$$b = Ax = \begin{bmatrix} c^{(1)} c^{(2)} c^{(3)} \dots c^{(n)} \end{bmatrix} \begin{bmatrix} x_1 \\ x_2 \\ \cdot \\ \cdot \\ \cdot \\ x_n \end{bmatrix} = x_1 c^{(1)} + x_2 c^{(2)} + \dots + x_n c^{(n)}$$

As we can see, the product is nothing but the linear combination of the column vectors of matrix A, with the components of vector x being the linear coefficients.

The new vector b formed through the multiplication has the same dimension as that of the column vectors of A and stays in the same column space. This is such a beautiful fact; no matter how we combine the column vectors, we can never leave the space spanned by the column vectors.

Now, let's work on an example.

$$A = \begin{bmatrix} 1 & 2 & 3 \\ 4 & 5 & 6 \end{bmatrix} \quad x = \begin{bmatrix} 2 \\ 2 \\ 3 \end{bmatrix} \quad b = Ax = 2\begin{bmatrix} 1 \\ 4 \end{bmatrix} + 2\begin{bmatrix} 2 \\ 5 \end{bmatrix} + 3\begin{bmatrix} 3 \\ 6 \end{bmatrix} = \begin{bmatrix} 15 \\ 36 \end{bmatrix}$$

We can see both the column vectors of A and $b \in R^{2 \times 1}$.

Linear Independence of Vectors

A vector is said to be linearly dependent on other vectors if it can be expressed as the linear combination of other vectors.

If $v_1 = 5v_2 + 7v_3$, then v_1, v_2, and v_3 are not linearly independent since at least one of them can be expressed as the sum of other vectors. In general, a set of n vectors v_1, $v_2, .. , v_n$ where $v_i \in R^{m \times 1} \, \forall \, i \in \{1, 2, .. \, n\}$ is said to be linearly independent if and only if $a_1 v_1 + a_2 v_2 + a_3 v_3 + . . . + a_n v_n = 0$ implies each of $a_i = 0 \quad \forall \, i \in \{1, 2, ...n\}$.

If $a_1 v_1 + a_2 v_2 + a_3 v_3 + . . . + a_n v_n = 0$ and not all $a_i = 0$, then the vectors are not linearly independent.

Given a set of vectors, the following method can be used to check whether they are linearly independent or not.

$a_1 v_1 + a_2 v_2 + a_3 v_3 + . . . + a_n v_n = 0$ can be written as follows:

$$\begin{bmatrix} v_1 \, v_2 ... v_n \end{bmatrix} \begin{bmatrix} a_1 \\ . \\ a_n \end{bmatrix} = 0$$

Solving for $[a_1 \, a_2 .... a_n]^T$, if the only solution we get is the zero vector, then the set of vectors $v_1, v_2, ... v_n$ is said to be linearly independent.

If a set of n vectors $v_i \in R^{n \times 1}$ is linearly independent, then those vectors span the whole n-dimensional space. In other words, by taking linear combinations of the n vectors, one can produce all possible vectors in the n-dimensional space. If the n vectors are not linearly independent, they span only a subspace within the n-dimensional space.

To illustrate this fact, let us take vectors in three-dimensional space, as illustrated in Figure 1-5.

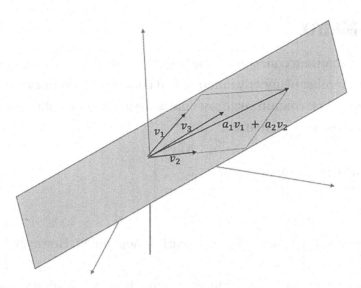

Figure 1-5. *A two-dimensional subspace spanned by v_1 and v_2 in a three-dimensional vector space*

If we have a vector $v_1 = [1 \ \ 2 \ \ 3]^T$, we can span only one dimension in the three-dimensional space because all the vectors that can be formed with this vector would have the same direction as that of v_1, with the magnitude being determined by the scaler multiplier. In other words, each vector would be of the form $a_1 v_1$.

Now, let's take another vector $v_2 = [\ 5 \ \ 9 \ \ 7 \ ]^T$, whose direction is not the same as that of v_1. So, the span of the two vectors $Span(v_1, v_2)$ is nothing but the linear combination of v_1 and v_2. With these two vectors, we can form any vector of the form $av_1 + bv_2$ that lies in the plane of the two vectors. Basically, we will span a two-dimensional subspace within the three-dimensional space. The same is illustrated in the following diagram.

Let's us add another vector $v_3 = [\ 4 \ \ 8 \ \ 1]^T$ to our vector set. Now, if we consider the $Span(v_1, v_2, v_3)$, we can form any vector in the three-dimensional plane. You take any three-dimensional vector you wish, and it can be expressed as a linear combination of the preceding three vectors.

These three vectors form a basis for the three-dimensional space. Any three linearly independent vectors would form a basis for the three-dimensional space. The same can be generalized for any n-dimensional space.

If we had taken a vector v_3, which is a linear combination of v_1 and v_2, then it wouldn't have been possible to span the whole three-dimensional space. We would have been confined to the two-dimensional subspace spanned by v_1 and v_2.

Rank of a Matrix

One of the most important concepts in linear algebra is the rank of a matrix. The rank of a matrix is the number of linearly independent column vectors or row vectors. The number of independent column vectors would always be equal to the number of independent row vectors for a matrix.

Example: Consider the matrix $A = \begin{bmatrix} 1 & 3 & 4 \\ 2 & 5 & 7 \\ 3 & 7 & 10 \end{bmatrix}$

The column vectors $\begin{bmatrix} 1 \\ 2 \\ 3 \end{bmatrix}$ and $\begin{bmatrix} 3 \\ 5 \\ 7 \end{bmatrix}$ are linearly independent. However, $\begin{bmatrix} 4 \\ 7 \\ 10 \end{bmatrix}$ is not linearly independent since it's the linear combination of the other two column vectors; i.e., $\begin{bmatrix} 4 \\ 7 \\ 10 \end{bmatrix} = \begin{bmatrix} 1 \\ 2 \\ 3 \end{bmatrix} + \begin{bmatrix} 3 \\ 5 \\ 7 \end{bmatrix}$. Hence, the rank of the matrix is 2 since it has two linearly independent column vectors.

As the rank of the matrix is 2, the column vectors of the matrix can span only a two-dimensional subspace inside the three-dimensional vector space. The two-dimensional subspace is the one that can be formed by taking the linear combination of $\begin{bmatrix} 1 \\ 2 \\ 3 \end{bmatrix}$ and $\begin{bmatrix} 3 \\ 5 \\ 7 \end{bmatrix}$.

The following are a few important notes:

- A square matrix $A \in R^{n \times n}$ is said to be full rank if the rank of A is n. A square matrix of rank n implies that all the n column vectors and even the n row vectors for that matter are linearly independent, and hence it would be possible to span the whole n-dimensional space by taking the linear combination of the n column vectors of the matrix A.

- If a square matrix $A \in R^{n \times n}$ is not full rank, then it is a singular matrix; i.e., all its column vectors or row vectors are not linearly independent. A singular matrix has an undefined matrix inverse and zero determinant.

Identity Matrix or Operator

A matrix $I \in R^{n \times n}$ is said to be an identity matrix or operator if any vector or matrix when multiplied by I remains unchanged. A 3×3 identity matrix is given by

$$I = \begin{bmatrix} 1 & 0 & 0 \\ 0 & 1 & 0 \\ 0 & 0 & 1 \end{bmatrix}$$

Let's say we take the vector $v = [2\ \ 3\ \ 4]^T$

$$Iv = \begin{bmatrix} 1 & 0 & 0 \\ 0 & 1 & 0 \\ 0 & 0 & 1 \end{bmatrix}\begin{bmatrix} 2 \\ 3 \\ 4 \end{bmatrix} = \begin{bmatrix} 2 \\ 3 \\ 4 \end{bmatrix}$$

Similarly, let's say we have a matrix $A = \begin{bmatrix} 1 & 2 & 3 \\ 4 & 5 & 6 \\ 7 & 8 & 9 \end{bmatrix}$

The matrices AI and IA are both equal to matrix A. Hence, the matrix multiplication is commutative when one of the matrices is an identity matrix.

Determinant of a Matrix

A determinant of a square matrix A is a number and is denoted by $det(A)$. It can be interpreted in several ways. For a matrix $A \in R^{n \times n}$, the determinant denotes the n-dimensional volume enclosed by the n row vectors of the matrix. For the determinant to be nonzero, all the column vectors or the row vectors of A should be linearly independent. If the n row vectors or column vectors are not linearly independent, then they don't span the whole n-dimensional space, but rather a subspace of dimension less than n, and hence the n-dimensional volume is zero. For a matrix $A \in R^{2 \times 2}$, the determinant is expressed as follows:

$$A = \begin{bmatrix} a_{11} & a_{12} \\ a_{21} & a_{22} \end{bmatrix} \in R^{2 \times 2}$$

$$det(A) = \begin{vmatrix} a_{11} & a_{12} \\ a_{21} & a_{22} \end{vmatrix} = a_{11}a_{22} - a_{12}a_{21}$$

Similarly, for a matrix $B \in R^{3 \times 3}$, the determinant of the matrix is given by the following:

$$B = \begin{bmatrix} a_{11} & a_{12} & a_{13} \\ a_{21} & a_{22} & a_{23} \\ a_{31} & a_{32} & a_{33} \end{bmatrix} \in \mathbb{R}^{3 \times 3}$$

$$B \in \begin{bmatrix} a_{11} & a_{12} & a_{13} \\ a_{21} & a_{22} & a_{23} \\ a_{31} & a_{32} & a_{33} \end{bmatrix} \in R^{3 \times 3}$$

$$det(B) = a_{11} \begin{vmatrix} a_{22} & a_{23} \\ a_{32} & a_{33} \end{vmatrix} - a_{12} \begin{vmatrix} a_{21} & a_{23} \\ a_{31} & a_{33} \end{vmatrix} + a_{13} \begin{vmatrix} a_{21} & a_{22} \\ a_{31} & a_{32} \end{vmatrix}$$

where $det(B) = a_{11} \begin{vmatrix} a_{22} & a_{23} \\ a_{32} & a_{33} \end{vmatrix} - a_{12} \begin{vmatrix} a_{21} & a_{23} \\ a_{31} & a_{33} \end{vmatrix} + a_{13} \begin{vmatrix} a_{21} & a_{22} \\ a_{31} & a_{32} \end{vmatrix}$

The method for determinant computation can be generalized to matrices in $R^{n \times n}$. Treating B as an n-dimensional matrix, its determinant can be expressed as follows:

$$det(B) = \begin{bmatrix} a_{11} \times \\ & a_{22} & a_{23} \\ & a_{32} & a_{33} \end{bmatrix} \begin{bmatrix} a_{12 \times} \\ a_{21} & & a_{23} \\ a_{31} & & a_{33} \end{bmatrix} + \begin{bmatrix} & & a_{13 \times} \\ a_{21} & a_{22} \\ a_{31} & a_{32} \end{bmatrix}$$

For example, the determinant of the matrix $A = \begin{bmatrix} 6 & 1 & 1 \\ 4 & -2 & 5 \\ 2 & 8 & 7 \end{bmatrix}$ can be computed as follows:

$$det(A) = \begin{bmatrix} 6 \times \\ & -2 & 5 \\ & 8 & 7 \end{bmatrix} \begin{bmatrix} 1 \times \\ 4 & & 5 \\ 2 & & 7 \end{bmatrix} + \begin{bmatrix} 1 \times \\ 4 & -2 \\ 2 & 8 \end{bmatrix}$$

$$= 6 \times \begin{vmatrix} -2 & 5 \\ 8 & 7 \end{vmatrix} - 1 \times \begin{vmatrix} 4 & 5 \\ 2 & 7 \end{vmatrix} + 1 \times \begin{vmatrix} 4 & -2 \\ 2 & 8 \end{vmatrix} = 6(-14-40) - 1(28-10) + 1(32+4)$$

$$= 6 \times (-54) - 1(18) + 36 = -306$$

Interpretation of Determinant

As stated earlier, the absolute value of the determinant of a matrix determines the volume enclosed by the row vectors acting as edges.

For a matrix $A \in R^{2 \times 2}$, it denotes the area of the parallelogram with the two-row vector acting as edges.

For a matrix $A = \begin{bmatrix} a & b \\ c & d \end{bmatrix}$, the $det(A)$ is equal to the area of the parallelogram with vectors $u = [a\ b]^T$ and $v = [\ c\ d]^T$ as edges.

Area of the parallelogram = $|u||v|\sin\theta$ where θ is the angle between u and v (see Figure 1-6).

$$= \sqrt{a^2 + b^2}\sqrt{c^2 + d^2}\ \frac{(ad - bc)}{\sqrt{a^2 + b^2}\sqrt{c^2 + d^2}} = (ad - bc)$$

Parallelogram

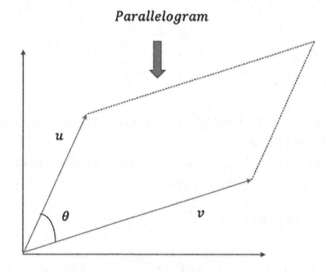

Figure 1-6. *Parallelogram formed by two vectors*

Similarly, for a matrix $B \in R^{3 \times 3}$, the determinant is the volume of the parallelepiped with the three-row vectors as edges.

Inverse of a Matrix

An inverse of a square matrix $A \in R^{n \times n}$ is denoted by A^{-1} and produces the identity matrix $I \in R^{n \times n}$ when multiplied by A.

$$AA^{-1} = A^{-1}A = I$$

Not all square matrices have inverses for A. The formula for computing the inverse of A is as follows:

$$A^{-1} = \frac{adjoint(A)}{\det(A)} = \frac{(cofactor\ matrix\ of\ A)^T}{\det(A)}$$

If a square matrix $A \in R^{n \times n}$ is singular—i.e., if A doesn't have n independent column or row vectors—then the inverse of A doesn't exist. This is because for a singular matrix $det(A) = 0$ and hence the inverse becomes undefined.

$$A = \begin{bmatrix} a & b & c \\ d & e & f \\ g & h & i \end{bmatrix}$$

Let the elements of A be represented by a_{ij}, where i represents the row number and j the column number for an element.

Then, the cofactor for $a_{ij} = (-1)^{i+j}d_{ij}$, where d_{ij} is the determinant of the matrix formed by deleting the row i and the column j from A.

The cofactor for the element $a = (-1)^{1+1}\begin{vmatrix} e & f \\ h & i \end{vmatrix} = ei - fh$.

Similarly, the cofactor for element $b = (-1)^{1+2}\begin{vmatrix} d & f \\ g & i \end{vmatrix} = -(di - fg)$.

Once the cofactor matrix is formed, the transpose of the cofactor matrix would give us $adjoint(A)$. The $adjoint(A)$ divided by the $det(A)$ gives A^{-1}.

For example, the inverse matrix of $A = \begin{bmatrix} 4 & 3 \\ 3 & 2 \end{bmatrix}$ can be computed as follows:

Cofactor matrix of $A = \begin{bmatrix} 1(2) & -1(3) \\ -1(3) & 1(4) \end{bmatrix} = \begin{bmatrix} 2 & -3 \\ -3 & 4 \end{bmatrix}$

$$det(A) = \begin{vmatrix} 4 & 3 \\ 3 & 2 \end{vmatrix} = 8 - 9 = -1$$

Therefore, $A^{-1} = \dfrac{(cofactor\ matrix\ of\ A)^T}{det(A)} = \dfrac{\begin{bmatrix} 2 & -3 \\ -3 & 4 \end{bmatrix}^T}{-1} = \begin{bmatrix} -2 & 3 \\ 3 & -4 \end{bmatrix}.$

The following are a few rules for inverses of a matrix:

- $(AB)^{-1} = B^{-1}A^{-1}$

- $I^{-1} = I$, where I is the identity matrix.

Norm of a Vector

The norm of a vector is a measure of its magnitude. There are several kinds of such norms. The most familiar is the Euclidean norm, defined next. It is also known as the l^2 norm.

For a vector $x \in R^{n \times 1}$, the l^2 norm is as follows:

$$\|x_2\| \left(|x_1|^2 + |x_2|^2 + \ldots + |x_n|^2 \right)^{1/2} = (x.x)^{1/2} = (x^T x)^{1/2}$$

Similarly, the l^1 norm is the sum of the absolute values of the vector components.

$$\|x\|_1 = |x_1| + |x_2| + \ldots + |x_n|$$

In general, the l^p norm of a vector can be defined as follows when $1 < p < \infty$:

$$\left(|x_1|^p + |x_2|^p + \ldots + |x_n|^p \right)^{1/p}$$

When $p \rightarrow \infty$ then the norm is called Supremum norm and is defined as follows:

$$\lim_{p \to \infty} \|x_p\| = \lim_{p \to \infty} \left(|x_1|^p + |x_2|^p + \ldots + |x_n|^p \right)^{1/p}$$

$$= max(x_1, x_2, \ldots, x_n)$$

In Figure 1-7, the unit norm curves have been plotted for l^1, l^2 and Supremum norm.

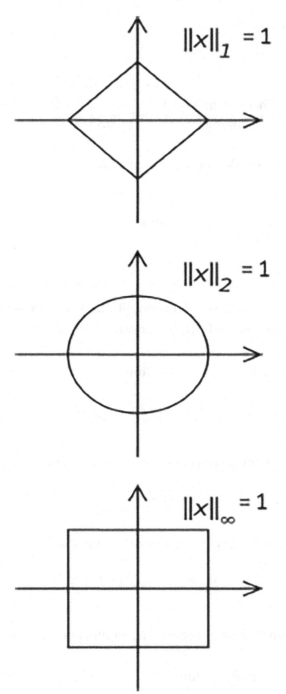

Figure 1-7. *Unit l^1, l^2 and Supremum norms of vectors $R^{2 \times 1}$*

Generally, for machine learning, we use both l^2 and l^1 norms for several purposes. For instance, the least square cost function that we use in linear regression is the l^2 norm of the error vector; i.e., the difference between the actual target-value vector and the predicted target-value vector. Similarly, very often we would have to use regularization for our model, with the result that the model doesn't fit the training data very well and fails to generalize to new data. To achieve regularization, we generally add the square of either the l^2 norm or the l^1 norm of the parameter vector for the model as a penalty in the cost function for the model. When the l^2 norm of the parameter vector is used for regularization, it is generally known as Ridge regularization, whereas when the l^1 norm is used instead, it is known as Lasso regularization.

Pseudo-Inverse of a Matrix

If we have a problem $Ax = b$ where $A \in R^{n \times n}$ and $b \in R^{n \times 1}$ are provided and we are required to solve for $x \in R^{n \times 1}$, we can solve for x as $x = A^{-1}b$ provided A is not singular and its inverse exists.

However, if $A \in R^{m \times n}$—i.e., if A is a rectangular matrix and $m > n$—then A^{-1} doesn't exist, and hence we can't solve for x by the preceding approach. In such cases, we can get an optimal solution, as $x^* = (A^TA)^{-1}A^Tb$. The matrix $(A^TA)^{-1}A^T$ is called the pseudo-inverse since it acts as an inverse to provide the optimal solution. This pseudo-inverse would come up in least square techniques, such as linear regression.

Unit Vector in the Direction of a Specific Vector

Unit vector in the direction of the specific vector is the vector divided by its magnitude or norm. For a Euclidian space, also called an l^2 space, the unit vector in the direction of the vector $x = [3 \ 4]^T$ is as follows:

$$\frac{x}{\|x\|_2} = \frac{x}{\left(x^Tx\right)^{1/2}} = \frac{[3\,4]^T}{5} = [0.6\,0.8]^T$$

Projection of a Vector in the Direction of Another Vector

Projection of a vector v_1 in the direction of v_2 is the dot product of v_1 with the unit vector in the direction of v_2. $\|v_{12}\| = v_1{}^T u_2$, where $\|v_{12}\|$ is the projection of v_1 onto v_2 and u_2 is the unit vector in the direction of v_2.

Since $u_2 = \dfrac{v_2}{v_{22}}$ as per the definition of a unit vector, the projection can also be

expressed as $\|v_{12}\| = v_1{}^T u_2 = v_1{}^T \dfrac{v_2}{v_{22}} = v_1{}^T \dfrac{v_2}{\left(v_2{}^T v_2\right)^{1/2}}$

For example, the projection of the vector $[1\ 1]^T$ in the direction of vector $[3\ 4]^T$ is the dot product of $[1\ 1]^T$ with the unit vector in the direction of $[3\ 4]^T$, i.e., $[0.6\ 0.8]^T$ as computed earlier.

The required projection $= \begin{bmatrix} 1 & 1 \end{bmatrix}^T \begin{bmatrix} 0.6 \\ 0.8 \end{bmatrix} = 1 \times 0.6 + 1 \times 0.8 = 1.4$.

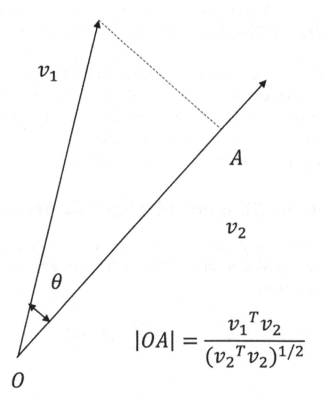

$$|OA| = \frac{v_1{}^T v_2}{(v_2{}^T v_2)^{1/2}}$$

Figure 1-8. *The length of the projection of the vector v_1 onto v_2*

In Figure 1-8, the length of the line segment OA gives the length of the projection of the vector v_1 onto v_2.

Eigen Vectors

Here we come to one of the most important concepts in linear algebra—Eigen vectors and Eigen values. Eigen values and Eigen vectors come up in several areas of machine learning. For example, the principal components in principal-component analysis are the Eigen vectors of the covariance matrix, while the Eigen values are the covariances along the principal components. Similarly, in Google's page-rank algorithm, the vector of the page-rank score is nothing but an Eigen vector of the page transition-probability matrix corresponding to the Eigen value of 1.

A matrix works on a vector as an operator. The operation of the matrix on the vector is to transform the vector into another vector whose dimensions might or might not be same as the original vector based on the matrix dimension.

When a matrix $A \in R^{n \times n}$ works on a vector $x \in R^{n \times 1}$, we again get back a vector $Ax \in R^{n \times 1}$. Generally, the magnitude as well as the direction of the new vector is different from that of the original vector. If in such a scenario the newly generated vector has the same direction or exactly the opposite direction as that of the original vector, then any vector in such a direction is called an Eigen vector. The magnitude by which the vector gets stretched is called the Eigen value (see Figure 1-9):

$$Ax = \lambda x$$

where A is the matrix operator operating on the vector v by multiplication, which is also the Eigen vector, and λ is the Eigen value.

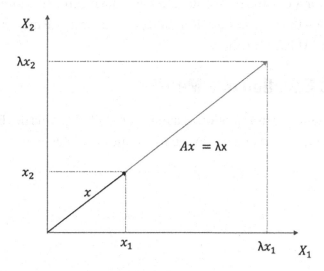

Figure 1-9. *Eigen vector unaffected by the matrix transformation A*

Figure 1-10. *The famous Mona Lisa image has a transformation applied to the vector space of pixel location*

As we can see from Figure 1-10, the pixels along the horizontal axis represented by a vector have changed direction when a transformation to the image space is applied, while the pixel vector along the horizontal direction hasn't changed direction. Hence, the pixel vector along the horizontal axis is an Eigen vector to the matrix transformation being applied to the *Mona Lisa* image.

Characteristic Equation of a Matrix

The roots of the characteristic equation of a matrix $A \in R^{n \times n}$ give us the Eigen values of the matrix. There would be n Eigen values corresponding to n Eigen vectors for a square matrix of order n.

For an Eigen vector $v \in R^{n \times 1}$ corresponding to an Eigen value of λ, we have the following:

$$Av = \lambda v$$

$$=> (A - \lambda I)v = 0$$

Now, v being an Eigen vector is nonzero, and hence $(A - \lambda I)$ must be singular for the preceding to hold true.

For $(A - \lambda I)$ to be singular, $det(A - \lambda I) = 0$, which is the characteristic equation for matrix A. The roots of the characteristic equation give us the Eigen values. Substituting the Eigen values in the $Av = \lambda v$ equation and then solving for v gives the Eigen vector corresponding to the Eigen value.

For example, the Eigen values and Eigen vectors of the matrix

$$A = \begin{bmatrix} 0 & 1 \\ -2 & -3 \end{bmatrix}$$

can be computed as seen next.

The characteristic equation for the matrix A is $det(A - \lambda I) = 0$.

$$\begin{vmatrix} -\lambda & 1 \\ -2 & -3-\lambda \end{vmatrix} = 0 \quad => \lambda^2 + 3\lambda + 2 = 0 => \lambda = -2, -1$$

The two Eigen values are -2 and -1.

Let the Eigen vector corresponding to the Eigen value of -2 be $u = [a \ \ b]^T$.

$$\begin{bmatrix} 0 & 1 \\ -2 & -3 \end{bmatrix} \begin{bmatrix} a \\ b \end{bmatrix} = -2 \begin{bmatrix} a \\ b \end{bmatrix}$$

This gives us the following two equations:

$$0a + 1b = -2a => 2a + b = 0 \quad - \quad (1)$$

$$-2a - 3b = -2b => 2a + b = 0 - \quad (2)$$

Both the equations are the same; i.e., $2a + b = 0 => \dfrac{a}{b} = \dfrac{1}{-2}$.

Let $a = k_1$ and $b = -2k_1$, where k_1 is a constant.

Therefore, the Eigen vector corresponding to the Eigen value -2 is $u = k_1 \begin{bmatrix} 1 \\ -2 \end{bmatrix}$.

Using the same process, the Eigen vector v corresponding to the Eigen value of -1 is $v = k_2 \begin{bmatrix} 1 \\ -1 \end{bmatrix}$.

One thing to note is that Eigen vectors and Eigen values are always related to a specific operator (in the preceding case, matrix A is the operator) working on a vector space. Eigen values and Eigen vectors are not specific to any vector space.

Functions can be treated as vectors. Let's say we have a function $f(x) = e^{ax}$.

Each of the infinite values of x would be a dimension, and the value of $f(x)$ evaluated at those values would be the vector component along that dimension. So, what we would get is an infinite vector space.

Now, let's look at the differentiator operator.

$$\frac{dy}{dx}\left(f(x)\right) = \frac{dy}{dx}\left(e^{ax}\right) = ae^{ax}$$

Here, $\dfrac{dy}{dx}$ is the operator and e^{ax} is an Eigen function with respect to the operator, while a is the corresponding Eigen value.

As expressed earlier, the applications of Eigen vectors and Eigen values are profound and far reaching in almost any domain, and this is true for machine learning as well. To get an idea of how Eigen vectors have influenced modern applications, we will look at the Google page-ranking algorithm in a simplistic setting.

Let us look at the page-ranking algorithm for a simple website that has three pages—A, B, and C—as illustrated in Figure 1-11.

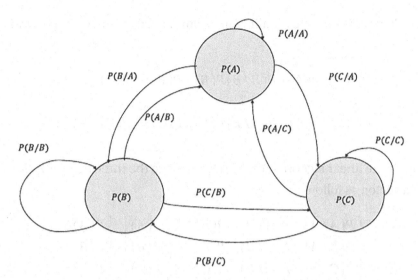

Figure 1-11. _Transition-probability diagram for three pages A, B, and C_

In a web setting, one can jump from one page to another page given that the original page has a link to the next page. Also, a page can self-reference and have a link to itself. So, if a user goes from page A to B because page A references page B, the event can be denoted by B/A. $P(B/A)$ can be computed by the total number of visits to page B from page A divided by the total number of visits to page A. The transition probabilities for all page combinations can be computed similarly. Since the probabilities are computed by normalizing count, the individual probabilities for pages would carry the essence of the importance of the pages.

In the steady state, the probabilities of each page would become constant. We need to compute the steady-state probability of each page based on the transition probabilities.

For the probability of any page to remain constant at steady state, probability mass going out should be equal to probability mass coming in, and each of them—when summed up with probability mass that stays in a page—should equal the probability of the page. In that light, if we consider the equilibrium equation around page A, the probability mass going out of A is $P(B/A)P(A) + P(C/A)P(A)$ whereas the probability mass coming into A is $P(A/B)P(B) + P(A/C)P(C)$. The probability mass $P(A/A)P(A)$ remains at A itself. Hence, at equilibrium, the sum of probability mass coming from outside—i.e., $P(A/B)P(B) + P(A/C)P(C)$—and probability mass remaining at A, i.e., $P(A/A)P(A)$, should equal $P(A)$, as expressed here:

$$P(A/A)P(A)+P(A/B)P(B)+P(A/C)P(C)=P(A) \qquad (1)$$

Similarly, if we consider the equilibrium around pages B and C, the following holds true:

$$P(B/A)P(A)+P(B/B)P(B)+P(B/C)P(C)=P(B) \tag{2}$$

$$P(C/A)P(A)+P(C/B)P(B)+P(C/C)P(C)=P(C) \tag{3}$$

Now comes the linear algebra part. We can arrange the three equations into a matrix working on a vector, as follows:

$$\begin{bmatrix} P(A/A) & P(A/B) & P(A/C) \\ P(B/A) & P(B/B) & P(B/C) \\ P(C/A) & P(C/B) & P(C/C) \end{bmatrix} \begin{bmatrix} P(A) \\ P(B) \\ P(C) \end{bmatrix} = \begin{bmatrix} P(A) \\ P(B) \\ P(C) \end{bmatrix}$$

The transition-probability matrix works on the page-probability vector to produce again the page-probability vector. The page-probability vector, as we can see, is nothing but an Eigen vector to the page-transition-probability matrix, and the corresponding Eigen value for the same is 1.

So, computing the Eigen vector corresponding to the Eigen value of 1 would give us the page-probability vector, which in turn can be used to rank the pages. Several page-ranking algorithms of reputed search engines work on the same principle. Of course, the actual algorithms of the search engines have several modifications to this naïve model, but the underlying concept is the same. The probability vector can be determined through methods such as power iteration, as discussed in the next section.

Power Iteration Method for Computing Eigen Vector

The power iteration method is an iteration technique used to compute the Eigen vector of a matrix corresponding to the Eigen value of largest magnitude.

Let $A \in R^{n \times n}$ and then let that the n Eigen values in order of magnitude are $\lambda_1 > \lambda_2 > \lambda_3 > \ldots > \lambda_n$ and the corresponding Eigen vectors are $v_1 > v_2 > v_3 > \ldots > v_n$.

Power iteration starts with a random vector v, which should have some component in the direction of the Eigen vector corresponding to the largest Eigen value, i.e., v_1.

The approximate Eigen vector in any iteration is given by the following:

$$v^{(k+1)} = \frac{Av^{(k)}}{Av^{(k)}}$$

After a sufficient number of iterations, $v^{(k+1)}$ converges to v_1. In every iteration, we multiply the matrix A by the vector obtained from the prior step. If we remove the normalizing of the vector to convert it to a unit vector in the iterative method, we have $v^{(k+1)} = A^k v$.

Let the initial vector v be represented as a combination of the Eigen vectors: $v = k_1 v_1 + k_2 v_2 + \ . \ . + k_n v_n$ where $k_i \ \forall \ i \ \in \{1, 2, 3, ..n\}$ are constants.

$$
\begin{aligned}
v^{(k+1)} = A^k v &= A^k \left(k_1 v_1 + k_2 v_2 + .. + k_n v_n \right) \\
&= k_1 A^k v_1 + k_2 A^k v_2 + .. + k_n A^k v_n \\
&= k_1 \lambda_1^{\ k} v_1 + k_2 \lambda_2^{\ k} v_2 + .. + k_n \lambda_n^{\ k} v_n \\
&= \lambda_1^{\ k} \left(k_1 v_1 + k_2 \left(\frac{\lambda_2}{\lambda_1} \right)^k v_2 + .. + k_n \left(\frac{\lambda_n}{\lambda_1} \right)^k v_n \right)
\end{aligned}
$$

Now, when k is sufficiently large—i.e., (—all the terms except the first will vanish since

$$\left(\frac{\lambda_i}{\lambda_1} \right)^k \to 0 \ \forall \ i \in \{2, 3, ..n\}$$

Therefore, $v^{(k+1)} = \lambda_1^{\ k} k_1 v_1$, which gives us the Eigen vector corresponding to the Eigen value of largest magnitude. The rate of convergence depends on the magnitude of the second largest Eigen value in comparison with the largest Eigen value. The method converges slowly if the second largest Eigen value is close in magnitude to the largest one.

Note   In this chapter, I have touched upon the basics of linear algebra so that readers who are not familiar with this subject have some starting point. However, I would suggest the reader to take up linear algebra in more detail in his or her spare time. Renowned Professor Gilbert Strang's book *Linear Algebra and Its Applications* is a wonderful way to get started.

Calculus

In its very simplest form, calculus is a branch of mathematics that deals with differentials and integrals of functions. Having a good understanding of calculus is important for machine learning for several reasons:

- Different machine-learning models are expressed as functions of several variables.

- To build a machine-learning model, we generally compute a cost function for the model based on the data and model parameters, and through optimization of the cost function, we derive the model parameters that best explain the given data.

Differentiation

Differentiation of a function generally means the rate of change of a quantity represented by a function with respect to another quantity on which the function is dependent on.

Let's say a particle moves in a one-dimensional plane—i.e., a straight line—and its distance at any specific time is defined by the function $f(t) = 5t^2$.

The velocity of the particle at any specific time would be given by the derivative of the function with respect to time t.

The derivative of the function is defined as $\dfrac{df(t)}{dt}$ and is generally expressed by the following formulae based on whichever is convenient:

$$\frac{df}{dt} = \lim_{h \to 0} \frac{f(t+h) - f(t)}{h}$$

or

$$\frac{df}{dt} = \lim_{h \to 0} \frac{f(t+h) - f(t-h)}{2h}$$

When we deal with a function that is dependent on multiple variables, the derivative of the function with respect to each of the variables keeping the others fixed is called a partial derivative, and the vector of partial derivatives is called the gradient of the function.

Let's say the price z of a house is dependent on two variables: square feet area of the house x and the number of bedrooms y.

$$z = f(x,y)$$

The partial derivative of z with respect to x is represented by the following:

$$\frac{\partial z}{\partial x} = \lim_{h \to 0} \frac{f(x+h,y) - f(x,y)}{h}$$

Similarly, the partial derivative of z with respect to y is as follows:

$$\frac{\partial z}{\partial y} = \lim_{h \to 0} \frac{f(x,y+h) - f(x,y)}{h}$$

Bear in mind that in a partial derivate, except the variable with respect to which the derivate is being taken, are held constant.

Gradient of a Function

For a function with two variables $z = f(x,y)$, the vector of partial derivatives $\left[\frac{\partial z}{\partial x} \ \frac{\partial z}{\partial y} \right]^T$ is called the gradient of the function and is denoted by ∇z. The same can be generalized for a function with n variables. A multivariate function $f(x_1, x_2,.., x_n)$ can also be expressed as $f(x)$, where $x = [x_1, x_2.... x_n]^T \in R^{n \times 1}$. The gradient vector for the multivariate function $f(x)$ with respect to x can be expressed as $\nabla f = \left[\frac{\partial f}{\partial x_1} \ \frac{\partial f}{\partial x_2} ...... \frac{\partial f}{\partial x_n} \right]^T$.

For example, the gradient of a function with three variables $f(x, y, z) = x + y^2 + z^3$ is given by the following:

$$\nabla f = \begin{bmatrix} 1 & 2y & 3z^2 \end{bmatrix}^T$$

The gradient and the partial derivatives are important in machine-learning algorithms when we try to maximize or minimize cost functions with respect to the model parameters, since at the maxima and minima the gradient vector of a function is zero. At the maxima and minima of a function, the gradient vector of the function should be a zero vector.

Successive Partial Derivatives

We can have successive partial derivatives of a function with respect to multiple variables. For example, for a function $z = f(x, y)$

$$\frac{\partial}{\partial y}\left(\frac{\partial z}{\partial x}\right) = \frac{\partial^2 z}{\partial y \partial x}$$

This is the partial derivative of z with respect to x first and then with respect to y. Similarly,

$$\frac{\partial}{\partial x}\left(\frac{\partial z}{\partial y}\right) = \frac{\partial^2 z}{\partial x \partial y}$$

If the second derivatives are continuous, the order of partial derivatives doesn't matter and

$$\frac{\partial^2 z}{\partial x \partial y} = \frac{\partial^2 z}{\partial y \partial x}.$$

Hessian Matrix of a Function

The Hessian of a multivariate function is a matrix of second-order partial derivatives. For a function $f(x, y, z)$, the Hessian is defined as follows:

$$Hf = \begin{bmatrix} \dfrac{\delta^2 f}{\delta x^2} & \dfrac{\delta^2 f}{\delta x \delta y} & \dfrac{\delta^2 f}{\delta x \delta z} \\[2ex] \dfrac{\delta^2 f}{\delta y \delta x} & \dfrac{\delta^2 f}{\delta y^2} & \dfrac{\delta^2 f}{\delta y \delta z} \\[2ex] \dfrac{\delta^2 f}{\delta z \delta x} & \dfrac{\delta^2 f}{\delta z \delta y} & \dfrac{\delta^2 f}{\delta z^2} \end{bmatrix}$$

The Hessian is useful in the optimization problems that we come across so frequently in the machine-learning domain. For instance, in minimizing a cost function to arrive at a set of model parameters, the Hessian is used to get better estimates for the next set of parameter values, especially if the cost function is nonlinear in nature. Nonlinear optimization techniques, such as Newton's method, Broyden-Fletcher-Goldfarb-Shanno (BFGS), and its variants, use the Hessian for minimizing cost functions.

Maxima and Minima of Functions

Evaluating the maxima and minima of functions has tremendous applications in machine learning. Building machine-learning models relies on minimizing cost functions or maximizing likelihood functions, entropy, and so on in both supervised and unsupervised learning.

Rules for Maxima and Minima for a Univariate Function

- The derivative of $f(x)$ with respect to x would be zero at maxima and minima.

- The second derivative of $f(x)$, which is nothing but the derivative of the first derivative represented by $\dfrac{d^2 f(x)}{dx^2}$, needs to be investigated at the point where the first derivative is zero. If the second derivative is less than zero, then it's a point of maxima, while if it is greater than zero, it's a point of minima. If the second derivative turns out to be zero as well, then the point is called a point of inflection.

Let's take a very simple function, $y = f(x) = x^2$. If we take the derivative of the function w.r.t x and set it to zero, we get $\dfrac{dy}{dx} = 2x = 0$, which gives us $x = 0$. Also, the second derivative $\dfrac{d^2 y}{dx^2} = 2$. Hence, for all values of x, including $x = 0$, the second derivative is greater than zero, and hence $x = 0$ is the minima point for the function $f(x)$.

Let's try the same exercise for $y = g(x) = x^3$.

$\dfrac{dy}{dx} = 3x^2 = 0$ gives us $x = 0$. The second derivative $\dfrac{d^2 y}{dx^2} = 6x$, and if we evaluate it at $x = 0$ we get 0. So, $x = 0$ is neither the minima nor the maxima point for the function $g(x)$. Points at which the second derivative is zero are called points of inflection. At points of inflection, the sign of the curvature changes.

The points at which the derivative of a univariate function is zero or the gradient vector for a multivariate function is a zero vector are called stationary points. They may or may not be points of maxima or minima.

Illustrated in Figure 1-12 are different kinds of stationary points, i.e., maxima, minima, and points of inflection.

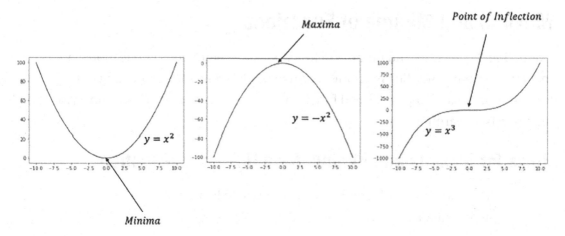

Figure 1-12. *Different types of stationary points—maxima, minima, point of inflection*

Maxima and minima for a multivariate function are a bit more complicated. Let's proceed with an example, and then we will define the rules. We look at a multivariate function with two variables:

$$f(x,y) = x^2 y^3 + 3y + x + 5$$

To determine the stationary points, the gradient vector needs to be zero.

$$\left[\frac{\partial f}{\partial x} \quad \frac{\partial f}{\partial y} \right]^T = \begin{bmatrix} 0 \\ 0 \end{bmatrix}$$

Setting $\frac{\partial f}{\partial x}$ and $\frac{\partial f}{\partial y}$ to zero, we get the following:

$$\frac{\partial f}{\partial x} = 2xy^3 + 1 = 0, \quad \frac{\partial f}{\partial x} = 3x^2 y^2 + 3 = 0$$

We need to compute the Hessian as well:

$$\frac{\partial^2 f}{\partial x^2} = f_{xx} = 2y^3,$$

$$\frac{\partial^2 f}{\partial y^2} = f_{yy} = 6x^2 y,$$

$$\frac{\partial^2 f}{\partial x \partial y} = f_{xy} = 6xy^2$$

$$\frac{\partial^2 f}{\partial y \partial x} = f_{yx} = 6xy^2$$

For functions with continuous second derivatives, $f_{xy} = f_{yx}$.
Let's say the gradient is zero at $(x = a, y = b)$:

- If $f_{xx}f_{yy} - (f_{xy})^2 < 0$ at $(x = a, y = b)$ then $(x = a, y = b)$ is a saddle point.

- If $f_{xx}f_{yy} - (f_{xy})^2 > 0$ at $(x = a, y = b)$ then $(x = a, y = b)$ is an extremum point; i.e., maxima or minima exists.

  a. If $f_{xx} < 0$ and $f_{yy} < 0$ at $(x = a, y = b)$ then $f(x, y)$ has the maximum at $(x = a, y = b)$.

  b. If $f_{xx} > 0$ and $f_{yy} > 0$ at $(x = a, y = b)$ then $f(x, y)$ has the minimum at $(x = a, y = b)$.

- If $f_{xx}f_{yy} - (f_{xy})^2 = 0$ then more advanced methods are required to classify the stationary point correctly.

For a function with n variables, the following are the guidelines for checking for the maxima, minima, and saddle points of a function:

- Computing the gradient and setting it to zero vector would give us the list of stationary points.

- For a stationary point $x_0 \in R^{n \times 1}$, if the Hessian matrix of the function at x_0 has both positive and negative Eigen values, then x_0 is a saddle point. If the Eigen values of the Hessian matrix are all positive, then the stationarity point is a local minima where as if the Eigen values are all negative, then the stationarity point is a local maxima.

Local Minima and Global Minima

Functions can have multiple minima at which the gradient is zero, each of which is called a local minima point. The local minima at which the function has the minimum value is called the global minima. The same applies for maxima. Maxima and minima of a function are derived by optimization methods. Since closed-form solutions are not always available or are computationally intractable, the minima and maxima are most often derived through iterative approaches, such as gradient descent, gradient ascent, and so forth. In the iterative way of deriving minima and maxima, the optimization method may get stuck in a local minima or maxima and be unable to reach the global minima or maxima. In iterative methods, the algorithm utilizes the gradient of the function at a point to get to a more optimal point. When traversing a series of points in this fashion, once a point with a zero gradient is encountered, the algorithm stops assuming the desired minima or maxima is reached. This works well when there is a global minima or maxima for the function. Also, the optimization can get stuck at a saddle point too. In all such cases, we would have a suboptimal model.

Illustrated in Figure 1-13 are global and local minima as well as global and local maxima of a function.

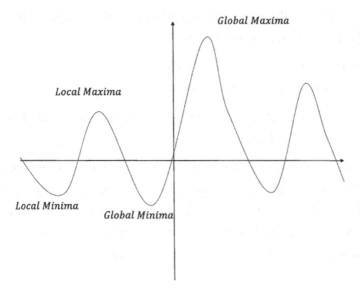

Figure 1-13. *Local and global minima/maxima*

Positive Semi-definite and Positive Definite

A square matrix $A \in R^{n \times 1}$ is positive semi-definite if for any nonzero vector $x \in R^{n \times 1}$ the expression $x^T A x > 0$. The matrix A is positive definite if the expression $x^T A x > 0$. All the Eigen values for a positive semi-definite matrix should be nonnegative, whereas for a positive definite matrix, the Eigen values should be positive. For example, if we consider A as the 2×2 identity matrix—i.e., $\begin{bmatrix} 1 & 0 \\ 0 & 1 \end{bmatrix}$—then it is positive definite since both of its Eigen values, i.e., 1,1, are positive. Also, if we compute $x^T A x$, where $x = [x_1 \ x_2]^T$, we get $x^T A x = x_1^2 + x_2^2$, which is always greater than zero for nonzero vector x, which confirms that A is a positive definite matrix.

Convex Set

A set of points is called convex if, given any two points x and y belonging to the set, all points joining the straight line from x to y also belong to the set. In Figure 1-14, a convex set and a non-convex set are illustrated.

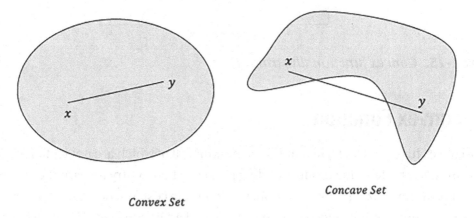

Concave Set

Convex Set

Figure 1-14. *Convex and non-convex set*

Convex Function

A function $f(x)$ defined on a convex set D, where $x \in R^{n \times 1}$ and D is the domain, is said to be convex if the straight line joining any two points in the function lies above or on the graph of the function. Mathematically, this can be expressed as the following:

$$f(tx + (1-t)y) \leq tf(x) + (1-t)f(y) \forall \ x,y \in D, \ \forall \ t \in [0,1]$$

35

For a convex function that is twice continuously differentiable, the Hessian matrix of the function evaluated at each point in the domain D of the function should be positive semi-definite; i.e., for any vector $x \in R^{n \times 1}$,

$$x^T H x \geq 0$$

A convex function has the local minima as its global minima. Bear in mind that there can be more than one global minima, but the value of the function would be same at each of the global minima for a convex function.

In Figure 1-15, a convex function $f(x)$ is illustrated. As we can see, the $f(x)$ clearly obeys the property of convex functions stated earlier.

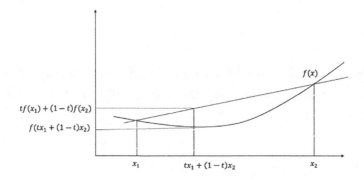

Figure 1-15. *Convex function illustration*

Non-convex Function

A non-convex function can have many local minima, all of which are not global minima.

In any machine-learning model building process where we try to learn the model parameters by minimizing a cost function, we prefer the cost function to be convex, since with a proper optimization technique we would attain the global minima for sure. For a non-convex cost function, there is a high chance that the optimization technique will get stuck at a local minima or a saddle point, and hence it might not attain its global minima.

Multivariate Convex and Non-convex Functions Examples

Since we would be dealing with high-dimensional functions in deep learning, it makes sense to look at convex and non-convex functions with two variables.

$f(x, \ y) = 2x^2 + 3y^2 - 5$ is a convex function with minima at $x = 0$, $y = 0$, and the minimum value of $f(x, y)$ at $(x = 0, y = 0)$ is -5. The function is plotted in Figure 1-16 for reference.

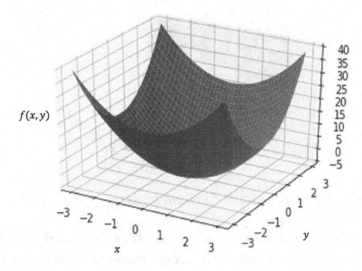

Figure 1-16. *Plot of the convex function $2x^2 + 3y^2 - 5$*

Now, let's consider the function $f(x, y) = \log x/y \ \ \forall \ x > 0, y > 0$.

The preceding is a non-convex function, and the easiest way to verify this is to look at the Hessian matrix for the following function:

$$\text{Hessian } H = \begin{bmatrix} -\dfrac{1}{x^2} & 0 \\ 0 & \dfrac{1}{y^2} \end{bmatrix}$$

The Eigen values of the Hessian are $-\dfrac{1}{x^2}$ and $\dfrac{1}{y^2}$. $-\dfrac{1}{x^2}$ would always be negative for real x. Hence, the Hessian is not positive semi-definite, making the function non-convex.

We can see in Figure 1-17 that the plot of the function $log(x/y)$ looks non-convex.

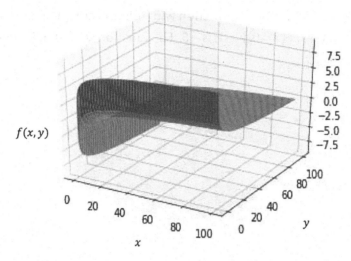

Figure 1-17.   *Plot of the non-convex function log(x/y)*

Linear regression through least squares and logistic regression through log-loss cost functions (binary cross-entropy) are all convex optimization problems, and hence the model parameters learned through optimization are a global minima solution. Similarly, in SVM, the cost function that we optimize is convex.

Whenever there are hidden layers or latent factors involved in any model, the cost function tends to be non-convex in nature. A neural network with hidden layers gives non-convex cost or error surface irrespective of whether we are solving regression or classification problems.

Similarly, in K-means, clustering the introduction of clusters makes the cost function to optimize a non-convex cost function. Intelligent methods need to be adopted for non-convex cost functions so that we achieve some local minima that are good enough if it is not possible to reach the global minima.

Parameter initialization becomes very important when dealing with a non-convex problem. The closer the initialized parameters are to the global minima or to some acceptable local minima, the better. For k means, one method of ensuring that the solution is not suboptimal is to run the k-means algorithm several times with different randomly initialized model parameters, i.e., the cluster centroids. We may then take the one that reduces the sum of the intra-cluster variances the most. For neural networks, one needs to use advanced gradient-descent methods involving momentum parameters to come out of the local minima and move forward. We will get to gradient-based optimization methods for neural networks in more detail later in this book.

Taylor Series

Any function can be expressed as an infinite sum by considering the value of the function and its derivatives at a specific point. Such an expansion of the function is called Taylor series expansion. The Taylor series expansion of a univariate function around a point x can be expressed as follows:

$$f(x+h) = f(x) + hf'(x) + \frac{1}{2!}h^2 f''(x) + \frac{1}{3!}h^3 f'''(x) + . . . + \frac{1}{n!}h^n f^n(x) + ..$$

where $f^n(x)$ is the *nth* derivative of the function $f(x)$ and n ! denotes the factorial of the number n. The term h has the same dimension as that of x, and both h, x are scalars.

- If $f(x)$ is a constant function, then all the derivatives are zero and $f(x + h)$ and $f(x)$ are same.

- If the function is linear around the neighborhood of x, then for any point $(x + h)$ that lies in the region of the linearity,

- If the function is quadratic around the neighborhood of x, then for any point that lies in the quadratic zone, $f(x + h) = f(x) + hf'(x)$.

- Taylor series expansion becomes very important in iterative methods such as gradient-descent methods and Newton's methods for optimization as well as in numerical methods for integration and differentiation.

Taylor series expansion for multivariate functions around a point $x \in R^{n \times 1}$ can be expressed as $f(x + \Delta x) = f(x) + \Delta x^T \nabla f(x) + \frac{1}{2}\Delta x^T \nabla^2 f(x)\Delta x +$ higher-order terms

where $\nabla f(x)$ is the gradient vector and $\nabla^2 f(x)$ is the Hessian matrix for the function $f(x)$.

Generally, for practical purposes, we don't go beyond second-order Taylor series expansion in machine-learning applications since in numerical methods they are hard to compute. Even for second-order expansion, computing the Hessian is cost-intensive, and hence several second-order optimization methods rely on computing the approximate Hessians from gradients instead of evaluating them directly. Please note that the third-order derivatives object $\nabla^3 f(x)$ would be a three-dimensional tensor.

Probability

Before we go on to probability, it is important to know what a random experiment and a sample space are.

In many types of work, be it in a research laboratory or an otherwise, repeated experimentation under almost identical conditions is a standard practice. For example, a medical researcher may be interested in the effect of a drug that is to be launched, or an agronomist might want to study the effect of chemical fertilizer on the yield of a specific crop. The only way to get information about these interests is to conduct experiments. At times, we might not need to perform experiments, as the experiments are conducted by nature and we just need to collect the data.

Each experiment would result in an outcome. Suppose the outcome of the experiments cannot be predicted with absolute certainty. However, before we conduct the experiments, suppose we know the set of all possible outcomes. If such experiments can be repeated under almost the same conditions, then the experiment is called a random experiment, and the set of all possible outcomes is called the sample space.

Do note that sample space is only the set of outcomes we are interested in. A throw of dice can have several outcomes. One is the set of outcomes that deals with the face on which the dice lands. The other possible set of outcomes can be the velocity with which the dice hits the floor. If we are only interested in the face on which the dice lands, then our sample space is $\Omega = \{1, 2, 3, 4, 5, 6\}$, i.e., the face number of the dice.

Let's continue with the experiment of throwing the dice and noting down the face on which it lands as the outcome. Suppose we conduct the experiment n times and face 1 turns up m times. Then, from the experiment, we can say that the probability of the event of the dice face's being 1 is equal to the number of experiments in which the dice face that turned up was 1 divided by the total number of experiments conducted; i.e.,

$P(x=1) = \dfrac{m}{n}$, where x denotes the number on the face of the dice.

Let's suppose we are told that a dice is fair. What is the probability of the number coming up as 1?

Well, given that the dice is fair and that we have no other information, most of us would believe in the near symmetry of the dice such that it would produce 100 faces with number 1 if we were to throw the dice 600 times. This would give us the probability as $\dfrac{1}{6}$.

Now, let's say we have gathered some 1000 data points about the numbers on the dice head during recent rolls of the dice. Here are the statistics:

$$1 \rightarrow 200 \; times$$

$$2 \rightarrow 100 \; times$$

$$3 \rightarrow 100 \; times$$

$$4 \rightarrow 100 \; times$$

$$5 \rightarrow 300 \; times$$

$$6 \rightarrow 200 \; times$$

In this case, you would come up with the probability of the dice face's being 1 as $P(x = 1) = 200/1000 = 0.2$. The dice is either not symmetric or is biased.

Unions, Intersection, and Conditional Probability

$P(A \cup B)$ = Probability of the event A or event B or both
   $P(A \cap B)$ = Probability of event A and event B
   $P(A/B)$) = Probability of event A given that B has already occurred.

$$P(A \cap B) = P(A/B)P(B) = P(B/A)P(A)$$

From now on, we will drop the notation of A intersection B as $A \cap B$ and will denote it as AB for ease of notation.

$$P(A - B) = P(A) - P(AB)$$

All the preceding proofs become easy when we look at the Venn diagram of the two events A and B as represented in Figure 1-18.

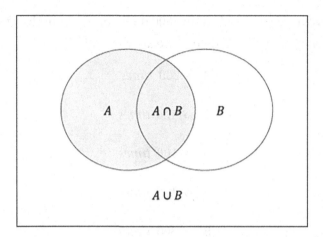

Figure 1-18. *Venn diagram of two events A and B showing the union and intersection of the two events*

Let's say there are n occurrences of an experiment in which A has occurred n_1 times, B has occurred n_2 times, and A and B together have occurred m times.

Let's represent this with a Venn diagram.

$P(A \cup B)$ can be represented by the sum of the probability of the three disjointed events: $(A - B)$, $(B - A)$, and AB.

$$P(AUB) = P(A - B) + P(B - A) + P(AB)$$
$$= P(A) - P(AB) + P(B) - P(AB) + P(AB)$$
$$= P(A) + P(B) - P(AB)$$

$P(A/B)$ is the probability of A given that B has already occurred. Given than B has already happened in n_2 ways, the event A is restricted to the event AB that can occur in m different ways. So, the probability of A given B can be expressed as follows:

$$P(A/B) = \frac{m}{n_2}$$

Now, $\dfrac{m}{n_2}$ can be written as $\dfrac{\frac{m}{n}}{\frac{n_2}{n}} = \dfrac{P(AB)}{P(B)}$.

Hence, $P(A/B) = P(AB)/P(B) = > P(AB) = P(B)P(A/B)$.

Similarly, if we consider $P(B/A)$, then the relation $P(AB) = P(A)P(B/A)$ also holds true.

Chain Rule of Probability for Intersection of Event

The product rule of intersection as just discussed for two events can be extended to n events.

If A_1, A_2, A_3, ... A_n are the set of n events, then the joint probability of these events can be expressed as follows:

$$P\left(A_1 A_2 A_3 \ldots A_n\right) = P\left(A_1\right) P\left(A_2 / A_1\right) P\left(A_3 / A_1 A_2\right) \ldots\ P\left(A_n / A_1 A_2 .. A_{(n-1)}\right)$$

$$= P\left(A_1\right) \prod_{i=2}^{n} P\left(A_i / A_1 A_2 A_3 \ldots A_{(n-1)}\right)$$

Mutually Exclusive Events

Two events A and B are said to be mutually exclusive if they do not co-occur. In other words, A and B are mutually exclusive if $P(AB) = 0$. For mutually exclusive events, $P(A\ U\ B) = P(A) + P(B)$.

In general, the probability of the union of n mutually exclusive events can be written as the sum of their probabilities:

$$P\left(A_1 \cup A_2\ .. \cup A_n\right) = P\left(A_1\right) + P\left(A_2\right)) + .. \ P\left(A_n\right) = \sum_{i=1}^{n} P\left(A_i\right)$$

Independence of Events

Two events A and B are said to be independent if the probability of their intersection is equal to the product of their individual probabilities, i.e.,

$$P\left(AB\right) = P\left(A\right) P\left(B\right)$$

This is possible because the conditional probability of A given B is the same as the probability of A; i.e.,

$$P\left(A / B\right) = P\left(A\right)$$

This means that A is as likely to happen in the set of all the events as it is in the domain of B.

Similarly, $P(B/A) = P(B)$ in order for events A and B to be independent.

When two events are independent, neither of the events is influenced by the fact the other event has happened.

Conditional Independence of Events

Two events A and B are conditionally independent given a third event C if the probability of co-occurrence of A and B given C can be written as follows:

$$P(AB/C) = P(A/C)P(B/C)$$

By the factorization property, $P(AB/C) = P(A/C)P(B/AC)$.

By combining the preceding equations, we see that $P(B/AC) = P(B/C)$ as well.

Do note that the conditional independence of events A and B doesn't guarantee that A and B are independent too. The conditional independence of events property is used a lot in machine-learning areas where the likelihood function is decomposed into simpler form through the conditional independence assumption. Also, a class of network models known as Bayesian networks uses conditional independence as one of several factors to simplify the network.

Bayes Rule

Now that we have a basic understanding of elementary probability, let's discuss a very important theorem called the Bayes rule. We take two events A and B to illustrate the theorem, but it can be generalized for any number of events.

We take $P(AB) = P(A)P(B/A)$ from the product rule of probability. (1)

Similarly, $P(AB) = P(B)P(A/B)$. (2)

Combining (1) and (2), we get

$$P(A)P(B/A) = P(B)P(A/B)$$

$$=> P(A/B) = P(A)P(B/A)/P(B)$$

The preceding deduced rule is called the Bayes rule, and it would come handy in many areas of machine learning, such as in computing posterior distribution from likelihood, using Markov chain models, maximizing a posterior algorithm, and so forth.

Probability Mass Function

The probability mass function (pmf) of a random variable is a function that gives the probability of each discrete value that the random variable can take up. The sum of the probabilities must add up to 1.

For instance, in a throw of a fair dice, let the number on the dice face be the random variable X.

Then, the pmf can be defined as follows:

$$P(X=i)=\frac{1}{6}i\in\{1,2,3,4,5,6\}$$

Probability Density Function

The probability density function (pdf) gives the probability density of a continuous random variable at each value in its domain. Since it's a continuous variable, the integral of the probability density function over its domain must be equal to 1.

Let X be a random variable with domain D. $P(x)$ denotes it's a probability density function, so that

$$\int_D P(x)\ dx=1$$

For example, the probability density function of a continuous random variable that can take up values from 0 to 1 is given by $P(x) = 2x$ where $x \in [0, 1]$. Let's validate whether it is a probability density function.

For $P(x)$ to be a probability density function, $\int_{x=0}^{1} P(x)dx$ should be 1.

$\int_{x=0}^{1} P(x)dx = \int_{x=0}^{1} 2xdx = \left[x^2\right]_0^1 = 1$. Hence, $P(x)$ is a probability density function.

One thing to be noted is that the integral computes the area under the curve, and since $P(x)$ is a probability density function (pdf), the area under the curve for a probability curve should be equal to 1.

Expectation of a Random Variable

Expectation of a random variable is nothing but the mean of the random variable. Let's say the random variable X takes n discrete values, $x_1, x_2, \ldots x_n$, with probabilities $p_1, p_2, \ldots p_n$. In other words, X is a discrete random variable with pmf $P(X = x_i) = p_i$. Then, the expectation of the random variable X is given by

$$E[X] = x_1 p_1 + x_2 p_2 + \ldots + x_n p_n = \sum_{i=1}^{n} x_i p_i$$

If X is a continuous random variable with a probability density function of $P(x)$, the expectation of X is given by

$$E[X] = \int_D x P(x) \ dx$$

where D is the domain of $P(x)$.

Variance of a Random Variable

Variance of a random variable measures the variability in the random variable. It is the mean (expectation) of the squared deviations of the random variable from its mean (or expectation).

Let X be a random variable with mean $\mu = E[X]$

$$Var[X] = E[(X - \mu)^2] \text{ where } \mu = E[X]$$

If X is a discrete random variable that takes n discrete values with a pmf given by $P(X = x_i) = p_i$, the variance of X can be expressed as follows:

$$Var[X] = E\left[(X - \mu)^2 \right]$$

$$= \sum_{i=1}^{n} (x_i - \mu)^2 p_i$$

If X is a continuous random variable having a probability density function of $P(x)$, then $Var[X]$ can be expressed as follows:

$$Var[X] = \int_D (x - \mu)^2 \, P(x)dx$$

where D is the domain of $P(x)$.

Skewness and Kurtosis

Skewness and Kurtosis are higher-order moment statistics for a random variable. Skewness measures the symmetry in a probability distribution, whereas Kurtosis measures whether the tails of the probability distribution are heavy or not. Skewness is a third-order moment and is expressed as follows:

$$Skew(X) = \frac{E\left[(X - \mu)^3\right]}{\left(Var[X]\right)^{3/2}}$$

A perfectly symmetrical probability distribution has a skewness of 0, as shown in the Figure 1-19. A positive value of skewness means that the bulk of the data is toward the left, as illustrated in Figure 1-20, while a negative value of skewness means the bulk of the data is toward the right, as illustrated in Figure 1-21.

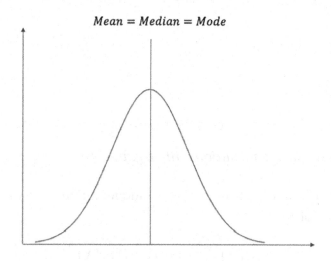

Symmetrical Probability Distribution

Figure 1-19. *Symmetric probability distribution*

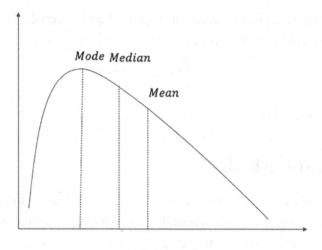

Positive Skewness

Figure 1-20. *Probability distribution with positive skewness*

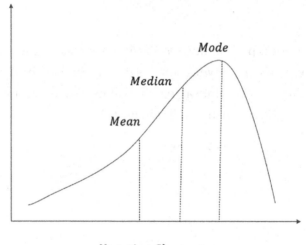

Negative Skewness

Figure 1-21. *Probability distribution with negative skewness*

Kurtosis is a fourth-order statistic, and for a random variable X with a mean of μ, it can be expressed as follows:

$$Kurt(X) = E\left[\left[X - \mu\right]^4\right] / \left(Var[X]\right)^2$$

Higher Kurtosis leads to heavier tails for a probability distribution, as we can see in Figure 1-23. The Kurtosis for a normal distribution (see Figure 1-22) is 3. However, to measure the Kurtosis of other distributions in terms of a normal distribution, one generally refers to excess Kurtosis, which is the actual Kurtosis minus the Kurtosis for a normal distribution—i.e., 3.

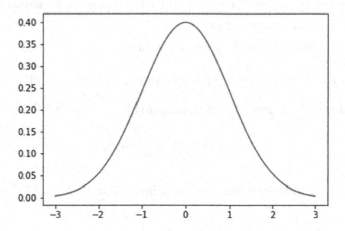

Figure 1-22. *Standard normal distribution with Kurtosis = 3*

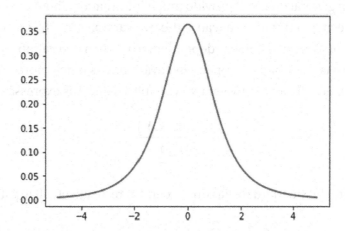

Figure 1-23. *Student's T distribution with Kurtosis = ∞*

Covariance

The covariance between two random variables X and Y is a measure of their joint variability. The covariance is positive if higher values of X correspond to higher values of Y and lower values of X correspond to lower values of Y. On the other hand, if higher values of X correspond to lower values of Y and lower values of X correspond to higher values of Y, then the covariance is negative.

The formula for covariance of X and Y is as follows:

$$cov(X, Y) = E[X - u_x][Y - u_y] \text{ where } u_x = E[X], \ u_y = E[Y]$$

On simplification of the preceding formula, an alternate is as follows:

$$cov(X, Y) = E[XY] - u_x u_y$$

If two variables are independent, their covariance is zero since $E[XY] = E[X]E[Y] = u_x u_y$

Correlation Coefficient

The covariance in general does not provide much information about the degree of association between two variables, because the two variables maybe on very different scales. Getting a measure of the linear dependence between two variables' correlation coefficients, which is a normalized version of covariance, is much more useful.

The correlation coefficient between two variables X and Y is expressed as follows:

$$\rho = \frac{cov(X, Y)}{\sigma_x \sigma_y}$$

where σ_x and σ_y are the standard deviation of X and Y, respectively. The value of ρ lies between -1 and $+1$.

Figure 1-24 illustrates both positive and negative correlations between two variables X and Y.

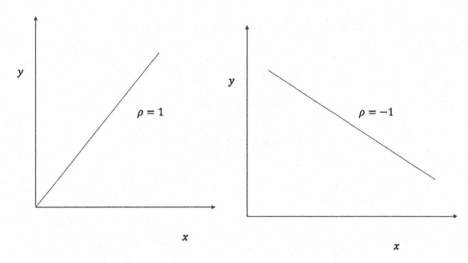

Figure 1-24. *Plot of variables with correlation coefficients of +1 and -1*

Some Common Probability Distribution

In this section, we will go through some of the common probability distributions that are frequently used in the machine-learning and deep-learning domains.

Uniform Distribution

The probability density function for a uniform distribution is constant. For a continuous random variable that takes up values between a and b($b > a$), the probability density function is expressed as follows:

$$P(X = x) = f(x) = \begin{cases} 1/(b-a) \text{ for } x \in [a,b] \\ 0 \text{ elsewhere} \end{cases}$$

Illustrated in Figure 1-25 is the probability density curve for a uniform distribution. The different statistics for a uniform distribution are outlined here:

$$E[X] = \frac{(b+a)}{2}$$

$$Median[X] = \frac{(b+a)}{2}$$

$$Mode[X] = All\ points\ in\ the\ interval\ a\ to\ b$$

$$Var[X] = (b - a)^2 / 12$$

$$Skew[X] = 0$$

$$Excessive\ Kurt[X] = -6/5$$

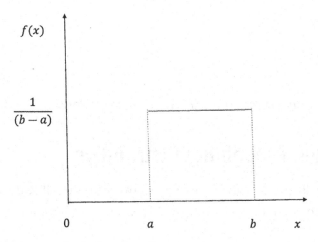

Figure 1-25. *Uniform probability distribution*

Please note that the excess Kurtosis is the actual Kurtosis minus 3, 3 being the actual Kurtosis for a normal distribution. Hence, the excess Kurtosis is the relative Kurtosis with respect to a normal distribution.

Normal Distribution

This is probably the most important scenario for probability distribution in the real world. In a normal distribution, the maximum probability density is at the mean of the distribution, and the density falls symmetrically and exponentially to the square of the distance from the mean. The probability density function of a normal distribution can be expressed as follows:

$$P(X = x) = \frac{1}{\sqrt{2\pi}\ \sigma} e^{\frac{-(x-\mu)^2}{2\sigma^2}} \qquad -\infty < x < +\infty$$

where μ is the mean and σ^2 is the variance of the random variable X. Illustrated in Figure 1-26 is the probability density function of a univariate normal distribution.

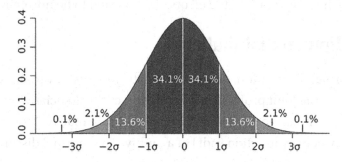

Figure 1-26. *Normal probability distribution*

As shown in Figure 1-26, 68.2% of the data in a normal distribution falls within one standard deviation (+1/-1σ) of the mean, and around 95.4% of the data is expected to fall within +2/-2σ of the mean. The important statistics for a normal distribution are outlined here:

$$E[X] = \mu$$

$$Median[X] = \mu$$

$$Mode[X] = \mu$$

$$Var[X] = \sigma^2$$

$$Skew[X] = 0$$

$$Excess\ Kurt[X] = 0$$

Any normal distribution can be transformed into a standard normal distribution by using the following transformation:

$$z = \frac{(x - \mu)}{\sigma}$$

The mean and standard deviation for the standard normal random variable z are 0 and 1, respectively. The standard normal distribution is used a lot in statistical inference tests. Similarly, in linear regression, the errors are assumed to be normally distributed.

Multivariate Normal Distribution

A multivariate normal distribution, or Gaussian distribution in n variables denoted by vector $x \in R^{n \times 1}$, is the joint probability distribution of the associated variables parameterized by the mean vector $\mu \in R^{n \times 1}$ and covariance matrix $\Sigma \in R^{n \times n}$.

The probability density function (pdf) of a multivariate normal distribution is as follows:

$$P(x/\mu;\Sigma) = \frac{1}{(2\pi)^{n/2}|\Sigma|^{-1/2}} e^{-\frac{1}{2}(x-\mu)^T \Sigma^{-1}(x-\mu)}$$

where $x = [x_1 x_2 ... x_n]^T$

$$-\infty < x_i < +\infty \forall i \in \{1,2,3,..n\}$$

Illustrated in Figure 1-27 is the probability density function of a multivariate normal distribution. A multivariate normal distribution, or Gaussian distribution, has several applications in machine learning. For instance, for multivariate input data that has correlation, the input features are often assumed to follow multivariate normal distribution, and based on the probability density function, points with low probability density are tagged as anomalies. Also, multivariate normal distributions are widely used in a mixture of Gaussian models wherein a data point with multiple features is assumed to belong to several multivariate normal distributions with different probabilities. Mixtures of Gaussians are used in several areas, such as clustering, anomaly detection, hidden Markov models, and so on.

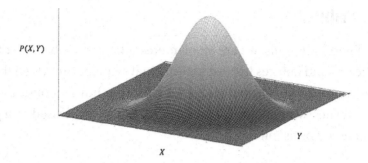

P(X,Y)

X

Y

Figure 1-27. *Multivariate normal distribution in two variables*

Bernoulli Distribution

An experiment in which the two outcomes are mutually exclusive and exhaustive (the sum of probability of the two outcomes is 1) is called a Bernoulli trail.

A Bernoulli trail follows a Bernoulli distribution. Let's say in a Bernoulli trail the two outcomes are success and failure. If the probability of success is p then, since these two events exhaust the sample space, the probability of failure is $1 - p$. Let $x = 1$ denote success. Thus, the probability of success or failure can be denoted as follows:

$$P(X=x)=f(x)=p^x(1-p)^{(1-x)} \quad x \in \{0,1\}$$

The preceding expression for $P(X = x)$ denotes the probability mass function of a Bernoulli distribution. The expectation and variance of the probability mass function are as follows:

$$E[X]=p$$

$$Var[X]=p(1-p)$$

The Bernoulli distribution can be extended to multiclass events that are mutually exclusive and exhaustive. Any two-class classification problem can be modeled as a Bernoulli trail. For instance, the logistic regression likelihood function is based on a Bernoulli distribution for each training data point, with the probability p being given by the sigmoid function.

Binomial Distribution

In a sequence of Bernoulli trails, we are often interested in the probability of the total number of successes and failures instead of the actual sequence in which they occur. If in a sequence of n successive Bernoulli trails x denotes the number of successes, then the probability of x successes out of n Bernoulli trails can be expressed by a probability mass function denoted by as follows:

$$P(X=x)=\binom{n}{x}p^x(1-p)^{(n-x)}\; x \in\{0,1,2\ldots,n\}$$

where p is the probability of success.

The expectation and variance of the distribution are as follows:

$$E[X]=np$$

$$Var[X]=np(1-p)$$

Illustrated in Figure 1-28 is the probability mass function of a binomial distribution with $n = 4$ and $p = 0.3$.

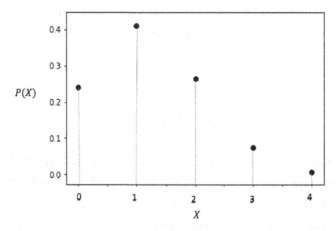

Figure 1-28. *Probability mass function of a binomial distribution with n=4 and p = 0.3*

Poisson Distribution

Whenever the rate of some quantity is of concern, like the number of defects in a 1000-product lot, the number of alpha particles emitted by a radioactive substance in the previous four-hour duration, and so on, Poisson distribution is generally the best way to represent such phenomenon. The probability mass function for Poisson distribution is as follows:

$$P(X = x) = \frac{e^{-\lambda}\lambda^x}{x!} \; where \; x \in \{0,1,2,\ldots\ldots\infty\}$$

$$E[X] = \lambda$$

$$Var[X] = \lambda$$

Illustrated in Figure 1-29 is the probability mass function of a Poisson distribution with mean of $\lambda = 15$.

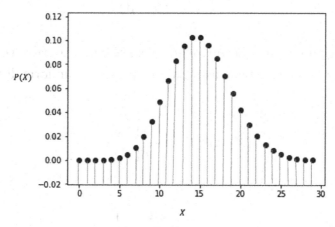

Figure 1-29. *Probability mass function of a Poisson distribution with mean = 15*

Beta Distribution

Beta distribution comes up several times in machine learning especially in Bayesian analysis for defining Priors over parameters as we will see later. It is characterized by two parameters α and β, and the probability density function of a variable X following gamma distribution can be represented as follows:

$$f(X=x;\alpha;b)=\frac{x^{\alpha-1}(1-x)^{\beta-1}}{B(\alpha,\beta)};\alpha>0,\beta>0$$

The domain of X is from 0 to 1, and hence it helps us define distribution over parameters $\theta \in [0,1]$. In Bayesian analysis, it is used to define distribution over the probabilities as probabilities lie between 0 and 1.

The function $B(\alpha,\beta)$ is a normalization factor in the density function and is given by the following:

$$B(\alpha,\beta)=\frac{\tau(\alpha)\tau(\beta)}{\tau(\alpha+\beta)}$$

τ is called the gamma function and for any positive integer n $\tau(n)=(n-1)!$

When $\alpha=1; \beta=1$, then the beta distribution represents a uniform distribution as follows:

$$(X=x;\alpha;b)=\frac{x^{1-1}(1-x)^{1-1}}{B(1,1)}$$

$$=\frac{x^{0}(1-x)^{0}}{1}=1$$

The probability density function for different values of α, β is illustrated in Figure 1-30.

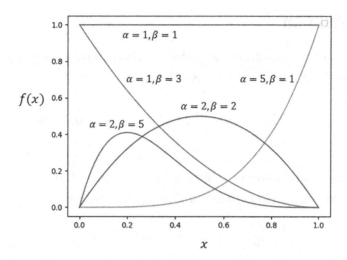

Figure 1-30. *Probability density function of beta distribution for different alpha and beta values*

Dirichlet Distribution

The Dirichlet distribution is the multivariate equivalent of the beta distribution and is characterized by a vector of parameters $\alpha = [\alpha_1, \alpha_2, ...\alpha_k]$ where k is the number of variables involved in the Dirichlet distribution.

If $X_1, X_2, ...X_k$ are the k random variables such that $\sum_{i=1}^{k} x_i = 1$, then the Dirichlet distribution is defined as follows:

$$f\left(X_1 = x_1, X_2 = x_2, .., X_k = x_k; \alpha\right) = \frac{1}{B(\alpha)} \prod_{i=1}^{k} x_i^{\alpha_i - 1}$$

$B(\alpha)$ much like in the beta distribution can be expressed in terms of Gamma function as follows:

$$B(\alpha) = \frac{\prod_{i=1}^{k} \tau(\alpha_i)}{\tau\left(\sum_{i=1}^{k} \alpha_i\right)}$$

In Bayesian statistics, the Dirichlet distribution is generally used to express a prior probability distribution over the probabilities of a multinomial distribution. As we know for a multinomial distribution, the probabilities of the different events $\sum_{i=1}^{k} x_i$ should sum to 1 which is the constraint on the random variables for a Dirichlet distribution.

Gamma Distribution

The gamma distribution of a random variable X is characterized by two parameters: α and β. Its probability density function is as follows:

$$f(X = x;\alpha,\beta) = \frac{1}{\tau(\alpha)}\beta^{\alpha}x^{\alpha-1}e^{-\beta x} \quad ;x > 0\,\alpha,\beta > 0$$

The gamma distribution is often used to model time to failure of machine, time to next earthquake, and time to loan defaults, among other things.

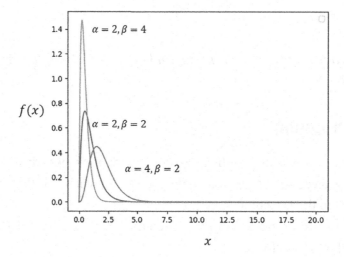

Figure 1-31. *Probability density function of gamma distribution for different alpha and beta values*

The gamma probability density function corresponding to different pairs of α, β is illustrated in Figure 1-31.

Likelihood Function

Likelihood is the probability of the observed data given the parameters that generate the underlying data. Let's suppose we observe n observations $x_1, x_2,...x_n$ and assume that the observations are independent and identically normally distributed with mean μ and variance σ^2.

The likelihood function in this case would be as follows:

$$P\left(Data\,/\,Model\ parameters\right)=P\left(x_1,x_2,.....x_n\,/\,\mu,\sigma^2\right)$$

Since the observations are independent, we can factorize the likelihood as follows:

$$P\left(Data\,/\,Model\ parameters\right)=\prod_{i=1}^{n}P\left(x_i\,/\,\mu,\sigma^2\right)$$

Each of the $x_i\sim Normal(\mu,\sigma^2)$; hence, the likelihood can be further expanded as follows:

$$P\left(Data\,/\,Model\,parameters\right)=\prod_{i=1}^{n}\frac{1}{\sqrt{2\pi}\sigma}e^{\frac{-(x_i-\mu)^2}{2\sigma^2}}$$

Maximum Likelihood Estimate

Maximum likelihood estimate (MLE) is a technique for estimating the parameters of a distribution or model. This is achieved by deriving the parameters that would maximize the likelihood function—i.e., maximize the probability of observing the data given the parameters of the model. Let's work through an example to understand maximum likelihood estimates.

Suppose Adam tosses a coin ten times and observes seven heads and three tails. Also, assume that the tosses are independent and identical. What would be the maximum likelihood estimate for the probability of heads for the given coin?

Each toss of a coin is a Bernoulli trial, with the probability of heads being, let us say, p, which is an unknown parameter that we want to estimate. Also, let the event that a toss produces heads be denoted by 1 and tails by 0.

The likelihood function can be represented as follows:

$$P\left(Data\,/\,parameter\right)=L\left(p\right)=P\left(x_1,x_2,.....x_{10}\,/\,p\right)$$

$$=\prod_{i=1}^{10}P\left(x_i\,/\,p\right)$$

$$=p^7\left(1-p\right)^3$$

Just for clarification, let us see how the likelihood L came to be $p^7(1-p)^3$.

For each head, the probability from the Bernoulli distribution is $P(x_i = 1/\ p) = p^1(1-p)^0 = p$. Similarly, for each tails, the probability is $P(x_i = 0/\ p) = p^0(1-p)^1 = 1-p$. As we have seven heads and three tails, we get the likelihood $L(p)$ to be $p^7(1-p)^3$.

To maximize the likelihood L, we need to take the derivate of L with respect to p and set it to 0.

Now, instead of maximizing the likelihood $L(p)$, we can maximize the logarithm of the likelihood—i.e., $logL(p)$. Since logarithmic is a monotonically increasing function, the parameter value that maximizes $L(p)$ would also maximize $logL(p)$. Taking the derivative of the log of the likelihood is mathematically more convenient than taking the derivative of the product form of the original likelihood.

$$logL(p) = 7logp + 3log(1-p)$$

Taking the derivative of both sides and setting it to zero look as follows:

$$\frac{dLog(L(p))}{dp} = \frac{7}{p} - \frac{3}{1-p} = 0$$

$$\Rightarrow p = 7/10$$

Interested readers can compute the second derivative $\dfrac{d^2 Log(L)}{dp^2}$ at $p = \dfrac{7}{10}$; you will for sure get a negative value, confirming that $p = \dfrac{7}{10}$ is indeed the point of maxima.

Some of you would have already had $\dfrac{7}{10}$ in mind without even going through maximum likelihood, just by the basic definition of probability. As you will see later, with this simple method, a lot of complex model parameters are estimated in the machine-learning and deep-learning world.

Let us look at another little trick that might come in handy while working on optimization. Computing the maxima of a function $f(x)$ is the same as computing the minima for the function $-f(x)$. The maxima for $f(x)$ and the minima for $-f(x)$ would take place at the same value of x. Similarly, the maxima for $f(x)$ and the minima for $1/f(x)$ would happen at the same value of x.

Often in machine-learning and deep-learning applications, we use advanced optimization packages, which only know to minimize a cost function to compute the model parameters. In such cases, we conveniently convert the maximization problem to a minimization problem by either changing the sign or taking the reciprocal of the function, whichever makes more sense. For example, in the preceding problem, we could have taken the negative of the log likelihood function—i.e., $LogL(p)$—and minimized it; we would have gotten the same probability estimate of 0.7.

Hypothesis Testing and p Value

Often, we need to do some hypothesis testing based on samples collected from a population. We start with a null hypothesis and, based on the statistical test performed, accept the null hypothesis or reject it.

Before we start with hypothesis testing, let us first consider one of the core fundamentals in statistics, which is the Central Limit theorem.

Let $x_1, x_2, x_3, \ldots, x_n$ be the n independent and identically distributed observation of a sample from a population with mean μ and finite variance σ^2.

The sample mean denoted by $\bar{x}$ follows normal distribution, with mean μ and variance $\dfrac{\sigma^2}{n}$; i.e.,

$$\bar{x} \sim Normal\left(\mu, \frac{\sigma^2}{n}\right) \text{ where } \bar{x} = \frac{x_1 + x_2 + x_3 + \ldots + x_n}{n}$$

This is called the Central Limit theorem. As the sample size n increases, the variance of $\bar{x}$ reduces and tends toward zero as $n \to \infty$.

Figure 1-32 illustrates a population distribution and a distribution of the mean of samples of fixed size n drawn from the population.

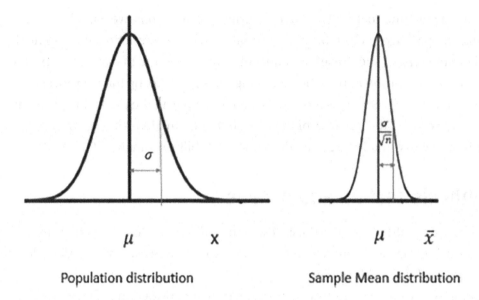

Figure 1-32. *Distribution for population and distribution for sample mean*

Please note that the sample mean follows normal distribution irrespective of whether the population variable is normally distributed or not. Now, let us consider a simple hypothesis-testing problem.

Boys who are 10 years old are known to have a mean weight of 85 pounds, with a standard deviation of 11.6. Boys in one county are checked as to whether they are obese. To test this, the mean weight of a group of 25 random boys from the county is collected. The mean weight is found to be 89.16 pounds.

We would have to form a null hypothesis and would reject it through the test if the evidence against the null hypothesis were strong enough.

Let us consider the null hypothesis: H_0. The children in the county are not obese, i.e., they come from the same population with a mean of $\mu = 85$.

Under the null hypothesis H_0, the sample mean is as follows:

$$\bar{x} \sim Normal\left((85, \frac{11.6^2}{25} \right)$$

The closer the sample mean observed is to the population mean, the better it is for the null hypothesis to be true. On the other hand, the further the sample mean observed is away from the population mean, the stronger the evidence is against the null hypothesis.

The standard normal variate $z = (\bar{x} - \mu)/(\sigma^2/n = (89.16 - 85)/(11.6/\sqrt{25}) = +1.75$

For every hypothesis test, we determine a p value. The p value of this hypothesis test is the probability of observing a sample mean that is further away from what is observed, i.e., $P(\bar{x} \geq 89.16)$ or $P(z \geq 1.75)$. So, the smaller the p value is, the stronger the evidence is against the null hypothesis.

When the p value is less than a specified threshold percentage α, which is called the type-1 error, the null hypothesis is rejected.

Please note that the deviation of the sample mean from the population can be purely the result of randomness since the sample mean has finite variance σ^2/n. The α gives us a threshold beyond which we should reject the null hypothesis even when the null hypothesis is true. We might be wrong, and the huge deviation might just be because of randomness. But the probability of that happening is very small, especially if we have a large sample size, since the sample mean standard deviation reduces significantly. When we do reject the null hypothesis even if the null hypothesis is true, we commit a type-1 error, and hence α gives us the probability of a type-1 error.

The p value for this test is $P(Z \geq 1.75) = 0.04$

The type-1 error α that one should choose depends on one's knowledge of the specific domain in which the test is performed. The type-1 error denotes the probability of rejecting the null hypothesis given that the null hypothesis is true. In practice, $\alpha = 0.05$ is a good enough type-1 error setting. Basically, we are taking a chance of only 5% of rejecting the null hypothesis by mistake. Since the p value computed is less than the type-1 error specified for the test, we cannot accept the null hypothesis. We say the test is statistically significant. The p value has been illustrated in Figure 1-33.

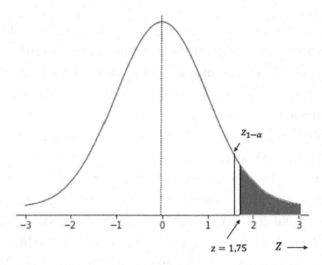

Figure 1-33. *Z test showing p value*

The dark-colored area corresponds to the p value, i.e., $P(z \geq 1.75)$. $Z_{1-\alpha}$ corresponds to the z value beyond which we are likely to commit a type-1 error given the null hypothesis is true. The area beyond $z_{1-\alpha}$—i.e., $P(z \geq Z_{1-\alpha})$—stands for the type-1 error probability. Since the p value is less than the type-1 error probability for the test, the null hypothesis cannot be taken as true. A Z test such as this is generally followed up by another good practice—the confidence interval test.

Also, the preceding test, popularly known as the Z test, is not always possible unless we have the population variance provided to us. For certain problems, we might not have the population variance. In such cases, the student T-test is more convenient since it uses sample variances instead of population variances.

The reader is encouraged to explore further regarding these statistical tests.

Formulation of Machine-Learning Algorithm and Optimization Techniques

The aim of modeling is to minimize the cost function of the model parameters given the data by using different optimization techniques. One may ask that if we set the derivative or gradient of the cost function to zero, we would have the model parameters. This is not always possible, since all solutions might not have a closed-form solution, or the closed-form solution might be computationally expensive or intractable. Further, when the data size is huge, there would be memory constraints when going for a closed-form solution. Hence, iterative methods are generally used for complex optimization problems.

Machine learning can be broadly classified into two types:

- Supervised machine learning

- Unsupervised machine learning

There is another paradigm of learning called reinforcement learning which has got lot of traction in the last few years. We will discuss on its generalized formulation later in this chapter.

Supervised Learning

In supervised learning, each training data point is associated with several input features—typically an input feature vector and its corresponding label. A model is constructed with several parameters that try to predict the output label given the input feature vector. The model parameters are derived by optimizing some form of cost function that is based on the error of prediction, i.e., the discrepancy between the actual labels and the predicted labels for the training data points. Alternatively, maximizing the likelihood of the training data would also provide us with the model parameters.

Linear Regression as a Supervised Learning Method

We might have a dataset that has the prices for houses as the target variable or output label, whereas features like area of the house, number of bedrooms, number of bathrooms, and so forth are its input feature vector. We can define a function that would predict the price of the house based on the input feature vector.

Let the input feature vector be represented by x' and the predicted value be y_p. Let the actual value of the housing price—i.e., the output label—be denoted by y. We can define a model where the output label is expressed as a function of the input feature vector, as shown in the following equation. The model is parameterized by several constants that we wish to learn via the training process.

$$y|x' = \theta'^T x' + b + \epsilon$$

Here ϵ is the random variation in prediction that can be modeled to be generated from zero mean finite variance Gaussian distribution, and hence we can write the following:

$$\epsilon \sim N\left(0, \sigma^2\right)$$

So, the housing price given an input y given x, i.e., $y \mid x$, is a linear combination of the input vector x' plus a bias term b and a random component ϵ, which follows a normal distribution with a 0 mean and a finite variance of σ^2.

As ϵ is a random component, it cannot be predicted, and the best we can predict is the mean of housing prices given a feature value, i.e.,

The predicted value $y_p \mid x = E[y \mid x] = \theta'^T x' + b$

Here, θ' is the linear combiner and b is the bias or the intercept. Both θ' and b are the model parameters that we wish to learn. We can express $y_p = \theta^T x$, where the bias has been added to the model parameter corresponding to the constant feature 1. This small trick makes the representation simpler.

Let us say we have m samples $(x^{(1)}, y^{(1)})$, $(x^{(2)}, y^{(2)})$.... $(x^{(m)}, y^{(m)})$. We can compute a cost function that takes the sum of the squares of the difference between the predicted and the actual values of the housing prices and try to minimize it to derive the model parameters.

The cost function can be defined as follows:

$$C(\theta) = \sum_{i=1}^{m} \left(\theta^T x_i - y^{(i)}\right)^2$$

We can minimize the cost function with respect to θ to determine the model parameter. This is a linear regression problem where the output label or target is continuous. Regression falls under the supervised class of learning. Figure 1-34 illustrates the relationship between housing prices and number of bedrooms.

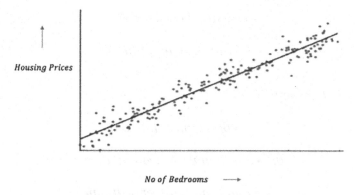

Housing Prices

No of Bedrooms ⟶

Figure 1-34. *Regression fit to the housing prices vs. number of bedrooms data. The red points denote the data points, and the blue line indicates the fitted regression line*

Let the input vector be $x' = [x_1 \ x_2 \ x_3]^T$, where

$$x_1 \rightarrow \text{the area of the house}$$

$$x_2 \rightarrow \text{the number of bedrooms}$$

$$x_3 \rightarrow \text{the number of bathrooms}$$

Let the parameter vector corresponding to the input feature vector be, where

$$\theta_1 \rightarrow \text{additional Cost per unit area}$$

$$\theta_2 \rightarrow \text{additional cost per bedroom}$$

$$\theta_3 \rightarrow \text{additional cost per bathroom}$$

After taking into consideration the bias term, the input feature vector becomes $x = [x_0 \ x_1 \ x_2 \ x_3]^T$, where

$$x_0 \rightarrow \text{constant value of } 1, i.e., \text{feature corresponding to the bias term}$$

$$x_1 \rightarrow \text{the area of the house}$$

$$x_2 \rightarrow \textit{the number of bedrooms}$$

$$x_3 \rightarrow \textit{the number of bathrooms}$$

and $\theta = [\theta_0\ \theta_1\ \theta_2\ \theta_3]$, where

$$\theta_0 \rightarrow \textit{bias term or intercept}$$

$$\theta_1 \rightarrow \textit{additional cost per unit area}$$

$$\theta_2 \rightarrow \textit{additional costper bedroom}$$

$$\theta_3 \rightarrow \textit{additional cost per bathroom}$$

Now that we have some understanding of how to construct a regression problem and its associated cost function, let's simplify the problem and proceed toward deriving the model parameters.

Model parameter $\theta^* = \underbrace{Arg\ Min}_{\theta} C(\theta) = \underbrace{Arg\ Min}_{\theta} \sum_{i=1}^{m} \left(\theta^T x_i - y^{(i)}\right)^2$

The input vectors for all the samples can be combined to a matrix X, and the corresponding target output can be represented as a vector Y.

$$X = \begin{bmatrix} x_0^{(1)} & x_1^{(1)} x_2^{(1)} x_3^{(1)} \\ x_0^{(2)} & x_1^{(2)} x_2^{(2)} x_3^{(2)} \\ x_0^{(3)} & x_1^{(3)} x_2^{(3)} x_3^{(3)} \\ & \cdot \\ & \cdot \\ x_0^{(m)} & x_1^{(m)} x_2^{(m)} x_3^{(m)} \end{bmatrix} \quad Y = \begin{bmatrix} y^{(1)} \\ y^{(2)} \\ y^{(3)} \\ \cdot \\ \cdot \\ y^{(m)} \end{bmatrix}$$

If we represent the vector of predictions as Yp, then $Yp = X\theta$. So, the error in prediction vector e can be represented as follows:

$$e = X\theta - Y$$

Hence, $C(\theta)$ can be expressed as the square of the l^2 norm of the error vector e, i.e.,

$$C(\theta) = \|e\|_2^2$$
$$= \left\| X\theta - Y_2^2 \right\|$$
$$= (X\theta - Y)^T (X\theta - Y)$$

Now that we have a simplified cost function in matrix form, it will be easy to apply different optimization techniques. These techniques would work for most cost functions, be it a convex cost function or a non-convex one. For non-convex ones, there are some additional things that need to be considered, which we will discuss in detail while considering neural networks.

We can directly derive the model parameters by computing the gradient and setting it to zero vector. You can apply the rules that we learned earlier to check if the conditions of minima are satisfied.

The gradient of the cost function with respect to the parameter vector θ is as seen here:

$$\nabla C(\theta) = 2X^T (X\theta - Y)$$

Setting $\nabla C(\theta) = 0$, we get $X^T X\theta = X^T Y => \hat{\theta} = (X^T X)^{-1} X^T\, Y$.

If one looks at this solution closely, one can observe that the pseudo-inverse of X— i.e., $(X^TX)^{-1}X^T$—comes into the solution of the linear regression problem. This is because a linear regression parameter vector can be looked at as a solution to the equation $X\theta = Y$, where X is an $m \times n$ rectangular matrix with $m > n$.

The preceding expression for $\hat{\theta}$ is the closed-form solution for the model parameter. Using this derived $\hat{\theta}$ for new data point x_{new}, we can predict the price of housing as $\hat{\theta}^T x_{new}$.

The computation of the inverse of (X^TX) is both cost- and memory-intensive for large datasets. Also, there are situations when the matrix X^TX is singular and hence its inverse is not defined. Therefore, we need to look at alternative methods to get to the minima point.

One thing to validate after building a linear regression model is the distribution of residual errors for the training data points. The errors should be approximately normally distributed with a 0 mean and some finite variance. The QQ plot that plots the actual

Since $X\theta$ is nothing but the linear combination of the column vectors of X, $X\theta$ stays in the same subspace as the ones spanned by the column vectors $c_i \; \forall \; i = \{1, 2, 3...n\}$.

Now the actual target-value vector Y lies outside the subspace spanned by the column vectors of X; thus, no matter what θ we combine X with, $X\theta$ can never equal or align itself in the direction of Y. There is going to be a nonzero error vector given by $e = Y - X\theta$.

Now that we know that we have an error, we need to investigate how to reduce the l^2 norm of the error. For the l^2 norm of the error vector to be at a minimum, it should be perpendicular to the prediction vector $X\theta$. Since $e = Y - X\theta$ is perpendicular to $X\theta$, it should be perpendicular to all vectors in that subspace.

So, the dot product of all the column vectors of X with the error vector $Y - X\theta$ should be zero, which gives us the following:

$$c_1^{T}\left[Y - X\theta\right] = 0, \; c_2^{T}\left[Y - X\theta\right] = 0,.....c_n^{T}\left[Y - X\theta\right] = 0$$

This can be rearranged in a matrix form as follows:

$$\left[c_1 \; c_2 ..... \; c_n\right]^{T}\left[Y - X\theta\right] = 0$$

$$=> X^{T}\left[Y - X\theta\right] = 0 => \hat{\theta} = \left(X^{T}X\right)^{-1}X^{T}\;Y$$

Also, please note that the error vector will be perpendicular to the prediction vector only if $X\theta$ is the projection of Y in the subspace spanned by the column vectors of X. The sole purpose of this illustration is to emphasize the importance of vector spaces in solving machine-learning problems.

Classification

Similarly, we may look at classification problems where instead of predicting the value of a continuous variable we predict the class label associated with an input feature vector. For example, we can try to predict whether a customer is likely to default based on his recent payment history and transaction details as well as his demographic and employment information. In such a problem, we would have data with the features just mentioned as input and a target indicating whether the customer has defaulted as the class label. Based on this labeled data, we can build a classifier that can predict a class label indicating whether the customer will default, or we can provide a probability score that the customer will default. In this scenario, the problem is a binary classification problem with two classes—the defaulter class and the non-defaulter class. When building such a classifier, the least square method might not give a good cost function since we are trying to guess the label and not predict a continuous variable. The popular cost functions for classification problems are generally log-loss cost functions that are based on maximum likelihood and entropy-based cost functions, such as Gini entropy and Shannon entropy.

The classifiers that have linear decision boundaries are called linear classifiers. Decision boundaries are hyperplanes or surfaces that separate the different classes. In the case of linear decision boundaries, the separating plane is a hyperplane.

Figures 1-36 and 1-37 illustrate linear and nonlinear decision boundaries, respectively, for the separation of two classes.

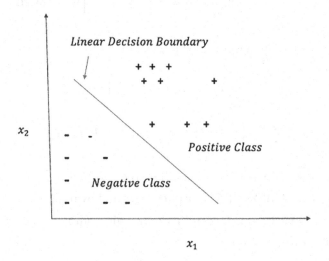

Figure 1-36. *Classification by a linear decision boundary*

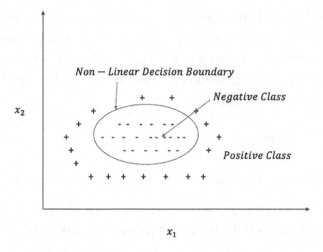

Figure 1-37. *Classification by a nonlinear decision boundary*

I would like to briefly discuss one of the most popular and simple classifiers, logistic regression, so that it becomes clear how the log-loss cost function comes into the picture via maximum likelihood methods in case of classification problems.

Suppose $(x^{(i)}, y^{(i)})$ are the labeled data points, where $x^{(i)} \in R^{n \times 1}$ $\forall i \in \{1, 2, .. m\}$ is the input vector, which includes the constant feature value of 1 as a component, and $y^{(i)}$ determines the class. The value of $y^{(i)}$ is set to 1 when the data point corresponds to a customer who has defaulted on his loan and 0 if the customer has not defaulted. The input vector $x^{(i)}$ can be represented in terms of its components as follows:

$$x^{(i)} = \begin{bmatrix} 1 & x_0^{(1)} & x_1^{(1)} & x_2^{(1)} x_3^{(1)} \end{bmatrix}$$

Logistic regression in general is a linear classifier and uses a squashing function to convert the linear score into probability. Let $\theta = [\theta_0 \ \theta_1 \ \theta_2 ... \theta_n]$ be the model parameter with each θ_j $\forall j \in \{0, 1, 2, ..., n\}$ representing the model parameter component with respect to the *jth* feature x_j of the input vector x.

The term θ_0 is the bias term. In linear regression, the bias is nothing but the intercept on the output y axis. We will look at what this bias term means for logistic regression (and for other linear classifiers) shortly.

The dot product $\theta^T x$ determines whether the given data point is likely to be a positive class or a negative class. For the problem at hand, the positive class is the event the customer defaults on his loan repayment, and the negative class is the event that the customer doesn't default. The probability that the customer will default given the input and the model is given by the following:

$$P\left(y=1/x,\theta\right)=1/\left(1+exp\left(-\theta^T x\right)\right)=p$$

$$P\left(y=0/x,\theta\right)=1-1/\left(1+exp\left(-\theta^T x\right)\right)=exp\left(-\theta^T x\right)/\left(1+exp\left(-\theta^T x\right)\right)=q$$

Now, let us look at the probability values for different values of $\theta^T x$:

- When $\theta^T x = 0$, then the probability of positive class is 1/2.

- When $\theta^T x > 0$, the probability of a positive class is greater than 1/2 and less than 1.

- When $\theta^T x < 0$, the probability of a positive class is less than 1/2 and greater than 0.

- When $\theta^T x$ is sufficiently large and positive, i.e., $\theta^T x \rightarrow \infty$, the probability $\rightarrow 1$.

- When $\theta^T x$ is sufficiently large and negative, i.e., $\theta^T x \rightarrow -\infty$, the probability $\rightarrow 0$.

The good thing about this probability formulation is that it keeps the values between 0 and 1, which would not have been possible with linear regression. Also, instead of the actual class, it gives continuous probability. Thus, depending on the problem at hand, the cutoff probability thresholds can be defined to determine the class.

This probability model function is called a logistic or sigmoid function. It has smooth, continuous gradients that make the model training mathematically convenient.

If we look carefully, we will see that the customer class y for each training sample follows a Bernoulli distribution, which we discussed earlier. For every data point, the class $y^{(i)}/x^{(i)} \sim Bernoulli(1, p_i)$. Based on the probability mass function of Bernoulli distribution, we can say the following:

$$P\left(y^{(i)}/x^{(i)},\theta\right)=\left(1-p_i\right)^{1-y^{(i)}} \text{ where } p_i = 1/(1+ exp\left(-\theta^T x\right))$$

Now, how do we define the cost function? We compute the likelihood of the data given the model parameters and then determine the model parameters that maximize the computed likelihood. We define the likelihood by L, and it can be represented as follows:

$$L = P(Data \,/\, model) = P\left(D^{(1)}D^{(2)}...D^{(m)} \,/\, \theta\right)$$

where $D^{(i)}$ represents the *ith* training sample $(x^{(i)}, y^{(i)})$.

Assuming the training samples are independent, given the model, L can be factorized as follows:

$$L = P\left(D^{(1)}D^{(2)}...D^{(m)} \,/\, \theta\right)$$

$$= P\left(D^{(1)} \,/\, \theta\right) P\left(D^{(2)} \,/\, \theta\right) .... P\left(D^{(m)} \,/\, \theta\right)$$

$$= \prod_{i=1}^{m} P\left(D^{(i)} \,/\, \theta\right)$$

We can take the log on both sides to convert the product of probabilities into a sum of the log of the probabilities. Also, the optimization remains the same since the maxima point for both L and $logL$ would be the same, as log is a monotonically increasing function.

Taking log on both sides, we get the following:

$$logL = \sum_{i=1}^{m} log\ P\left(D^{(i)} \,/\, \theta\right)$$

Now, $P(D^{(i)}/\theta) = P((x^{(i)}, y^{(i)})/\theta) = P(x^{(i)}/\ \theta)\ P(y^{(i)}/x^{(i)},\ \theta)$.

We are not concerned about the probability of the data point—i.e., $P(x^{(i)}/\ \theta)$—and assume that all the data points in the training are equally likely for the given model. So, $P(D^{(i)}/\theta) = k\ P(y^{(i)}/x^{(i)},\ \theta)$, where k is a constant.

Taking the log on both sides, we get the following:

$$log\ P\left(D^{(i)} \,/\, \theta\right) = logk + y^{(i)}logp_i + \left(1-y^{(i)}\right)log(1-p_i)$$

Summing over all data points, we get the following:

$$logL = \sum_{i=1}^{m} logk + y^{(i)}logp_i + \left(1-y^{(i)}\right)\log\left(1-p_i\right)$$

We need to maximize the *logL* to get the model parameter θ. Maximizing *logL* is the same as minimizing *−logL*, and so we can take the *−logL* as the cost function for logistic regression and minimize it. Also, we can drop the *logk* sum since it's a constant and the model parameter at the minima would be same irrespective of whether we have the *logk* sum. If we represent the cost function as $C(\theta)$, it can be expressed as seen here:

$$C(\theta) = \sum_{i=1}^{m} -y^{(i)}logp_i - \left(1-y^{(i)}\right)\log\left(1-p_i\right)$$

where $p_i = 1/(1 + exp\,(-\theta^T x))$

$C(\theta)$ is a convex function in θ, and the reader is encouraged to verify it with the rules learned earlier in the "Calculus" section. $C(\theta)$ can be minimized by common optimization techniques.

Hyperplanes and Linear Classifiers

Linear classifiers in some way or another are related to a hyperplane, so it makes sense to look at that relationship. In a way, learning a linear classifier is about learning about the hyperplane that separates the positive class from the negative class.

A hyperplane in an *n*-dimensional vector space is a plane of dimension $(n-1)$ that divides the *n*-dimensional vector space into two regions. One region consists of vectors lying above the hyperplane, and the other region consists of vectors lying below the hyperplane. For a two-dimensional vector space, straight lines act as a hyperplane. Similarly, for a three-dimensional vector space, a two-dimensional plane acts as a hyperplane.

A hyperplane is defined by two major parameters: its perpendicular distance from the origin represented by a bias term b' and the orientation of the hyperplane determined by a unit vector w perpendicular to the hyperplane surface as shown in Figure 1-38.

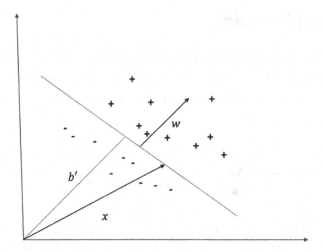

Figure 1-38. *Hyperplanes separating the two classes*

For a vector $x \in R^{n \times 1}$ to lie on the hyperplane, the projection of the vector in the direction of w should equal the distance of the hyperplane from origin—i.e., $w^T x = b'$. Thus, any point lying on the hyperplane must satisfy $w^T x - b' = 0$. Similarly, $w^T x - b' > 0$ must be satisfied for points lying above the hyperplane and $w^T x - b' < 0$ for points lying below the hyperplane.

In linear classifiers, we learn to model the hyperplane or learn the model parameters w and b. The vector w is generally aligned toward the positive class. The Perceptron and linear SVM are linear classifiers. Of course, the ways SVM and Perceptron learn the hyperplane are totally different, and hence they would come up with different hyperplanes, even for the same training data.

Even if we look at logistic regression, we see that it is based on a linear decision boundary. The linear decision boundary is nothing but a hyperplane, and points lying on the hyperplane (i.e., $w^T x - b' = 0$) are assigned a probability of 0.5 for either of the classes. And again, the way the logistic regression learns the decision boundary is totally different from the way SVM and Perceptron models do.

Unsupervised Learning

Unsupervised machine-learning algorithms aim at finding patterns or internal structures within datasets that contain input data points without labels or targets. K-means clustering, the mixture of Gaussians, and so on are methods of unsupervised learning. Even data-reduction techniques like principal component analysis (PCA), singular value decomposition (SVD), auto-encoders, and so forth are unsupervised learning methods.

Reinforcement Learning

Reinforcement learning (RL) deals with training an "Agent" that can take suitable "actions" given a specific "state" of the "environment" using a "policy" to maximize some long-term "reward." Each of the term in the definition is important to understand reinforcement learning.

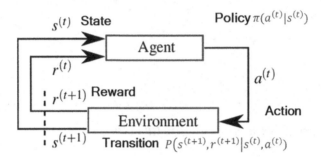

Figure 1-39. *High-level representation of a reinforcement learning paradigm*

In an RL environment, the Agent learns a policy by interacting with the environment in such a way to maximize some long-term reward.

A state in RL can be any condition of the environment. For instance, when one drives a car, the state can be defined by the position and speed of the car as well as the position and speed of the neighboring cars. In the diagram above, the state of the system at time t is defined as $s^{(t)}$.

Presented with a state $s^{(t)}$, the RL agent tries to determine an optimal action $a^{(t)}$ using the leant policy $\pi(.)$. A policy is a mapping from states to actions and can be probabilistic or deterministic. The RL agent learns the policy by interacting with the environment.

The reward is the environment's feedback of the action in each state. The immediate reward at time t on taking action $a^{(t)}$ at state $s^{(t)}$ is given by $r^{(t)}$.

When the agent takes an action $a^{(t)}$ at a given state $s^{(t)}$ along with the immediate reward $r^{(t)}$, it is presented with a new state $s^{(t+1)}$ based on the transition function of the environment $P(s^{(t+1)}, r^{(t+1)}| s^{(t)}, a^{(t)})$. The agent should learn the policy in such a way that it not only maximizes the immediate reward $r^{(t+)}$ but also moves to a favorable next step $s^{(t+1)}$ that favors high rewards. In essence, the action in each state should be taken in such a way so that the long-term reward is maximized.

To understand long-term reward, let us suppose an agent realizes the sequence of states, actions, and rewards following a policy π as the following:

Sequence of state, action, and rewards following a policy π

$$s^{(0)}, a^{(0)}, r^{(1)}, s^{(1)}, a^{(1)}, r^{(2)} \dots \dots \dots \dots \dots s^{(t)}, a^{(t)}, r^{(t+1)}, s^{(t+1)}, a^{(t+1)}, r^{(t+2)}, s^{(t+2)} \dots$$

$$s^{(t+N-1)}, a^{(t+N-1)}, r^{(t+N)}, s^{(t+N)}$$

Long-term reward at time t choosing action $a^{(t)}$ when presented with state $s^{(t)}$

$$G(t) = \sum_{i=1}^{N} r^{(t+i)}$$

Generally, the near-term rewards are given more precedence over the future rewards, and hence the long-term reward using a discount factor γ is expressed as follows:

$$G(t) = \sum_{i=1}^{N} \gamma^{i-1} r^{(t+i)}$$

The discount factor $\gamma < 1$, and hence it helps in ensuring the long-term reward does not diverge in case of very long or infinite horizon problems.

Optimization Techniques for Machine-Learning Gradient Descent

Gradient descent, along with its several variants, is probably the most widely used optimization technique in machine learning and deep learning. It's an iterative method that starts with a random model parameter and uses the gradient of the cost function with respect to the model parameter to determine the direction in which the model parameter should be updated.

Suppose we have a cost function $C(\theta)$, where θ represents the model parameters. We know the gradient of the cost function with respect to θ gives us the direction of maximum increase of $C(\theta)$ in a linear sense at the value of θ at which the gradient is evaluated. So, to get the direction of maximum decrease of $C(\theta)$ in a linear sense, one should use the negative of the gradient.

The update rule of the model parameter θ at iteration $(t + 1)$ is given by the following:

$$\theta^{(t+1)} = \theta^{(t)} - \eta \nabla C\left(\theta^{(t)}\right)$$

where η represents the learning rate and $\theta^{(t+1)}$ and $\theta^{(t)}$ represent the parameter vector at iteration $(t+1)$ and t, respectively.

Once minima is reached through this iterative process, the gradient of the cost function at minima would technically be zero, and hence there would be no further updates to θ. However, because of rounding errors and other limitations of computers, converging to true minima might not be achievable. Thus, we would have to come up with some other logic to stop the iterative process when we believe we have reached minima—or at least close enough to the minima—to stop the iterative process of model training. One of the ways generally used is to check the magnitude of the gradient vector, and if it is less than some predefined minute quantity—say, ε—stop the iterative process. The other crude way that can be used is to stop the iterative process of the parameter update after a fixed number of iterations, like 1000.

Learning rate plays a very vital role in the convergence of the gradient descent to the minima point. If the learning rate is large, the convergence might be faster but might lead to severe oscillations around the point of minima. A small learning rate might take a longer time to reach the minima, but the convergence is generally oscillation-free.

To see why gradient descent works, let's take a model that has one parameter and a cost function $C(\theta) = (\theta - a)^2 + b$ and see why the method works.

As we can see from Figure 1-37, the gradient (in this case it's just the derivative) at point θ_1 is positive, and thus if we move along the gradient direction, the cost function would increase. Instead, at θ_1 if we move along the negative of the gradient, the cost reduces. Again, let's take the point θ_2 where the gradient is negative. If we take the gradient direction here for updating θ, the cost function would increase. Taking the negative direction of the gradient would ensure that we move toward the minima. Once we reach the minima at $\theta = a$, the gradient is 0, and hence there would be no further update to θ.

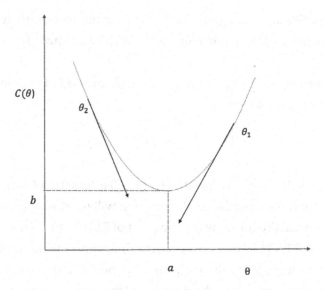

Figure 1-40. *Gradient-descent intuition with simple cost function in one variable*

Gradient Descent for a Multivariate Cost Function

Now that we have some intuition about gradient descent from looking at a cost function based on one parameter, let us look at gradient descent when the cost function is based on multiple parameters.

Let us look at a Taylor series expansion of a function of multiple variables. Multiple variables can be represented by the vector θ. Let us consider a cost function $C(\theta)$ where $\theta \in R^{n \times 1}$.

As discussed earlier, the Taylor series expansion around a point θ can be expressed in matrix notation, as shown here:

$$C(\theta + \Delta\theta) = C(\theta) + \Delta\theta^T \nabla C(\theta) + \frac{1}{2} \Delta\theta^T H(\theta) \Delta\theta + higher\ order\ terms$$

$$\Delta\theta \rightarrow Change\ in\ \theta\ vector$$

$$\nabla C(\theta) \rightarrow Gradient\ vector\ of\ C(\theta)$$

$$H(\theta) \rightarrow Hessian\ matrix\ of\ C(\theta)$$

Let us suppose at iteration t of gradient descent the model parameter is θ and we want update θ to $(\theta + \Delta\theta)$ by making an update of $\Delta\theta$ such that $C(\theta + \Delta\theta)$ is lesser than $C(\theta)$.

If we assume linearity of the function in the neighborhood of θ, then from Taylor series expansion, we get the following:

$$C(\theta + \Delta\theta) = C(\theta) + \Delta\theta^T \nabla C(\theta)$$

We want to choose $\Delta\theta$ in such a way that $C(\theta + \Delta\theta)$ is less than $C(\theta)$.

For all $\Delta\theta$ with the same magnitude, the one that will maximize the dot product $\Delta\theta^T \nabla C(\theta)$ should have a direction the same as that of $\nabla C(\theta)$. But that would give us the maximum possible $\Delta\theta^T \nabla C(\theta)$. Hence, to get the minimum value of the dot product $\Delta\theta^T \nabla C(\theta)$, the direction of $\Delta\theta$ should be the exact opposite of that of $\nabla C(\theta)$. In other words, $\Delta\theta$ should be proportional to the negative of the gradient vector :

$$\Delta\theta \propto -\nabla C(\theta)$$

=> $\Delta\theta = -\eta \nabla C(\theta)$, where η is the learning rate

=>$\theta + \Delta\theta = \theta - \eta\nabla C(\theta)$

=>$\theta^{(t+1)} = \theta^{(t)} - \eta \nabla C(\theta^{(t)})$

which is the famous equation for gradient descent. To visualize how the gradient descent proceeds to the minima, we need to have some basic understanding of contour plots and contour lines.

Contour Plot and Contour Lines

Let us consider a function $C(\theta)$ where $\theta \in R^{n \times 1}$. A contour line is a line/curve in the vector space of θ that connects points that have the same value of the function $C(\theta)$. For each unique value of $C(\theta)$, we would have separate contour lines.

Let us plot the contour lines for a function $C(\theta) = \theta^T A \theta$, where $\theta = [\theta_1 \theta_2]^T \in R^{2 \times 1}$ and

$$A = \begin{bmatrix} 7 & 2 \\ 2 & 5 \end{bmatrix}$$

Expanding the expression for $C(\theta)$, we get the following:

$$C(\theta_1, \theta_2) = 7\theta_1^2 + 5\theta_2^2 + 4\theta_1\theta_2$$

The contour plot for the cost function is depicted in Figure 1-38.

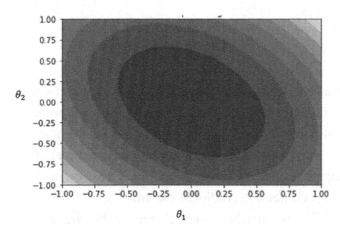

Figure 1-41. *Contour plots*

Each of the ellipses are contour lines specific to a fixed value of the function $C(\theta_1, \theta_2)$. If $C(\theta_1, \theta_2) = a$, where a is a constant, then the equation

$$a = 7\theta_1^2 + 5\theta_2^2 + 4\theta_1\theta_2 \text{ represents an ellipse.}$$

For different values of constant, *we* get different ellipses, as depicted in Figure 1-38. All points on a specific contour line have the same value of the function.

Now that we know what a contour plot is, let us look at gradient-descent progression in a contour plot for the cost function $C(\theta)$, where $\theta \in R^{2 \times 1}$. The gradient-descent steps have been illustrated in Figure 1-39.

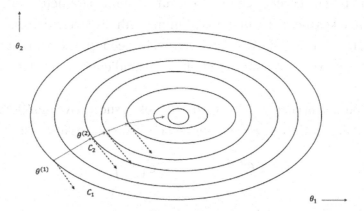

Figure 1-42. *Gradient descent for a two-variable cost function*

Let us take the largest ellipse corresponding to cost C_1 and assume our current θ is at the point $\theta^{(1)}$ in C_1.

Assuming the linearity of the $C(\theta)$ about θ, the change in cost function $C(\theta)$ can be presented as seen here:

$$\Delta C(\theta) = \Delta \theta^T \nabla C(\theta)$$

If we take a small change in cost between two points very close to each other in the same contour line, then $\Delta C(\theta) = 0$, since all points on the same contour line have the same fixed value. Also, it should be noted that when we take two points very close to each other on the same contour line, the $\Delta \theta$ represents the tangent to the contour line represented by tangential arrows to the contour line. Please do not confuse this $\Delta \theta$ to be the $\Delta \theta$ update to the parameter in gradient descent.

$\Delta C(\theta) = 0 \rightarrow \Delta \theta^T \nabla C(\theta) = 0$, which basically means that the gradient is perpendicular to the tangent at the point $\theta^{(1)}$ in the contour line C_1. The gradient would have pointed outward, whereas the negative of the gradient points inward, as depicted by the arrow perpendicular to the tangent. Based on the learning rate, it will reach a point $\theta^{(2)}$ in a different contour line represented by C_2, whose cost-function value would be less than that of C_1. Again, the gradient would be evaluated at $\theta^{(2)}$, and the same process would be repeated for several iterations until it reached the minima point, where the gradient would drop to 0 technically, after which there would be no more updates to θ.

Steepest Descent

Steepest descent is a form of gradient descent where the learning rate is not constant but rather is computed at every iteration to ensure that the parameter update through gradient descent takes the cost function to minima with respect to the learning rate. In other words, the learning rate at every iteration in steepest descent is optimized to ensure that the movement in the direction of the gradient is utilized to the maximum extent.

Let us take our usual cost function $C(\theta)$ and look at successive iterations t and $(t + 1)$. As with gradient descent, we have the parameter update rule as follows:

$$\theta^{(t+1)} = \theta^{(t)} - \eta \nabla C\left(\theta^{(t)}\right)$$

So, the cost function at iteration $(t + 1)$ can be expressed as follows:

$$C\left(\theta^{(t+1)}\right) = C\left(\theta^{(t)} - \eta \nabla C\left(\theta^{(t)}\right)\right)$$

To minimize the cost function at iteration $(t+1)$ with respect to the learning rate, see the following:

$$\frac{\partial C\left(\theta^{(t+1)}\right)}{\partial \eta}=0$$

$$\Rightarrow \nabla C\left(\theta^{(t+1)}\right)\frac{\partial\left[C(\theta^{(t)}-\eta\nabla C\left(\theta^{(t)}\right)\right]}{\partial \eta}=0$$

$$\Rightarrow -\nabla C\left(\theta^{(t+1)}\right)^{T}\nabla C\left(\theta^{(t)}\right)=0$$

$$\Rightarrow \nabla C\left(\theta^{(t+1)}\right)^{T}\nabla C\left(\theta^{(t)}\right)=0$$

So, for steepest descent, the dot product of the gradients at $(t+1)$ and t is 0, which implies that the gradient vector at every iteration should be perpendicular to the gradient vector at its previous iteration.

Stochastic Gradient Descent

Both steepest descent and gradient descent are full-batch models; i.e., the gradients are computed based on the whole training dataset. So, if the dataset is huge, the gradient computation becomes expensive, and the memory requirement increases. Also, if the dataset has huge redundancy, then computing the gradient on the full dataset is not useful since similar gradients can be computed by using much smaller batches called mini-batches. The most popular method to overcome the preceding problems is to use an optimization technique called stochastic gradient descent.

Stochastic gradient descent is a technique for minimizing a cost function based on the gradient-descent method where the gradient at each step is not based on the entire dataset but rather on single data points.

Let $C(\theta)$ be the cost function based on m training samples. The gradient descent at each step is not based on $C(\theta)$ but rather on $C^{(i)}(\theta)$, which is the cost function based on the *ith* training sample. So, if we must plot the gradient vectors at each iteration against the overall cost function, the contour lines would not be perpendicular to the tangents since they are based on gradients of $C^{(i)}(\theta)$ and not on the overall C(θ).

The cost functions $C^{(i)}(\theta)$ are used for gradient computation at each iteration, and the model parameter vector θ is updated by the standard gradient descent in each iteration until we have made a pass over the entire training dataset. We can perform several such passes over the entire dataset until a reasonable convergence is obtained.

Since the gradients at each iteration are not based on the entire training dataset but rather on single training samples, they are generally very noisy and may change direction rapidly. This may lead to oscillations near the minima of the cost function, and hence the learning rate should be less while converging to the minima so that the update to the parameter vector is as small as possible. The gradients are cheaper and faster to compute, and so the gradient descent tends to converge faster.

One thing that is important in stochastic gradient descent is that the training samples should be as random as possible. This will ensure that a stochastic gradient descent over a period of a few training samples provides a similar update to the model parameter as that resulting from an actual gradient descent, since the random samples are more likely to represent the total training dataset. If the samples at each iteration of stochastic gradient descent are biased, they don't represent the actual dataset, and hence the update to the model parameter might be in a direction that would result in it taking a long time for the stochastic gradient descent to converge.

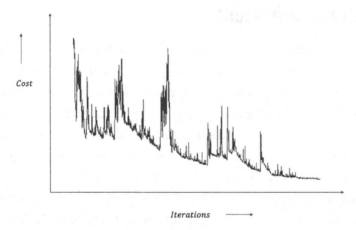

Figure 1-43. *Fluctuation in the total cost-function value over iterations in stochastic gradient descent*

As illustrated in Figure 1-44, the gradients at each step for stochastic gradient descent are not perpendicular to the tangents at the contour lines. However, they would be perpendicular to the tangents to the contour lines for individual training samples had we plotted them. Also, the associated cost reduction over iterations is noisy because of the fluctuating gradients.

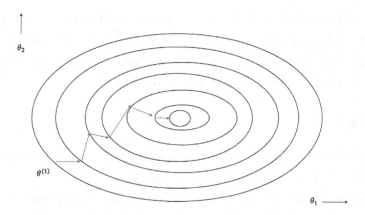

Figure 1-44. *Stochastic gradient-descent parameter update*

The gradient computations become very cheap when we use single training data points. Also, convergence is quite fast, but it does come with its own set of disadvantages, as follows:

- Since the estimated gradients at each iteration are based not on the total cost function but rather on the cost function associated with single data points, the gradients are very noisy. This leads to convergence problems at the minima and may lead to oscillations.

- The tuning of the learning rate becomes important since a high learning rate might lead to oscillations while converging to the minima. This is because the gradients are very noisy, and hence if the gradient estimates at the convergence are not near to zero, a high learning rate will take the update well past the minima point, and the process can repeat on either side of the minima.

- Since the gradients are noisy, the model parameter values after each iteration are also very noisy, and thus heuristics need to be added to the stochastic gradient descent to determine which value of model parameter to take. This also brings about another question: when to stop the training.

A compromise between full-batch model and stochastic gradient descent is a mini-batch approach wherein the gradient is based neither on the full training dataset nor on the single data points. Rather, it uses a mini-batch of training data points to compute the cost function. Most of the deep-learning algorithms use a mini-batch approach for stochastic gradient descent. The gradients are less noisy and at the same time don't cause many memory constraints because the mini-batch sizes are moderate.

We will discuss mini-batches in more detail in Chapter 2.

Newton's Method

Before we start Newton's method for optimizing a cost function for its minima, let us look at the limitations of gradient-descent techniques.

Gradient-descent methods rely on the linearity of the cost function between successive iterations; i.e., the parameter value at iteration $(t + 1)$ can be reached from the parameter value at time t by following the gradient, since the path of the cost function $C(\theta)$ from t to $(t + 1)$ is linear or can be joined by a straight line. This is a very simplified assumption and would not yield good directions for gradient descent if the cost function is highly nonlinear or has curvature. To get a better idea of this, let us look at the plot of a cost function with a single variable for three different cases.

Linear Curve

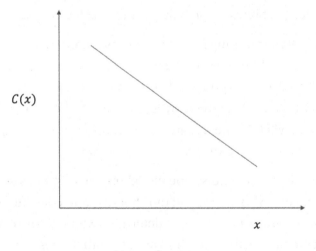

Linear function

Figure 1-45. *Linear cost function*

Negative Curvature

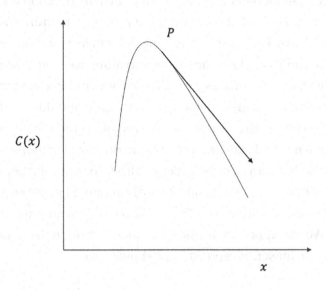

Negative Curvature

Figure 1-46. *Cost function with negative curvature at point P*

Positive Curvature

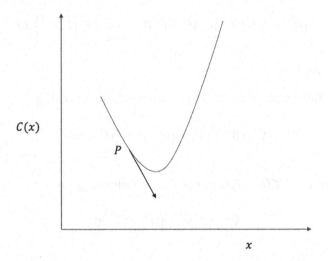

Figure 1-47. *Cost function with positive curvature at point P*

For the linear cost function, as shown in Figure 1-45, the negative of the gradient would give us the best direction for reaching the minima since the function is linear and does not have any curvature. For both the negative and positive curvature cost functions, shown in Figures 1-46 and 1-47, respectively, the derivative would not give us a good direction for minima, and so to take care of the curvature, we would need the Hessian along with the derivative. Hessians, as we have seen, are nothing but a matrix of second derivatives. They contain information about the curvature and thus would give a better direction for a parameter update as compared to a simple gradient descent.

Gradient-descent methods are first-order approximation methods for optimization, whereas Newton's methods are second-order methods for optimization since they use the Hessian along with the gradient to take care of curvatures in the cost function.

Let us take our usual cost function $C(\theta)$, where $\theta \in R^{n \times 1}$ is an n-dimensional model parameter vector. We can approximate the cost function $C(\theta)$ in the neighborhood of θ by its second-order Taylor series expansion, as shown here:

$$C(\theta + \Delta\theta) = C(\theta) + \Delta\theta^T \nabla C(\theta) + \frac{1}{2}\Delta\theta^T H(\theta)\Delta\theta$$

$\nabla C(\theta)$ is the gradient and $H(\theta)$ is the Hessian of the cost function $C(\theta)$.

Now, if θ is the value of the model parameter vector at iteration t, and $(\theta + \Delta\theta)$ is the value of the model parameter at iteration, then

$$C\left(\theta^{(t+1)}\right) = C\left(\theta^{(t)}\right) + \Delta\theta^T \nabla C\left(\theta^{(t)}\right) + \frac{1}{2}\Delta\theta^T H\left(\theta^{(t)}\right)\Delta\theta$$

where $\Delta\theta = \theta^{(t+1)} - \theta^{(t)}$.

Taking the gradient with respect to $\theta^{(t+1)}$, we have the following:

$$\nabla C\left(\theta^{(t+1)}\right) = \nabla C\left(\theta^{(t)}\right) + H\left(\theta^{(t)}\right)\Delta\theta$$

Setting the gradient $\nabla C(\theta^{(t+1)})$ to 0, we get the following:

$$\nabla C\left(\theta^{(t)}\right) + H\left(\theta^{(t)}\right)\Delta\theta = 0$$

$$=> \Delta\theta = -H\left(\theta^{(t)}\right)^{-1}\nabla C\left(\theta^{(t)}\right)$$

So, the parameter update for Newton's method is as follows:

$$=> \theta^{(t+1)} = \theta^{(t)} - H\left(\theta^{(t)}\right)^{-1} \nabla C\left(\theta^{(t)}\right)$$

We do not have a learning rate for Newton's method, but one may choose to use a learning rate, much like with gradient descent. Since the directions for nonlinear cost functions are better with Newton's method, the iterations to converge to the minima would be fewer as compared to gradient descent. One thing to note is that if the cost function that we are trying to optimize is a quadratic cost function, such as the one in linear regression, then Newton's method would technically converge to the minima in one step.

However, computing the Hessian matrix and its inverse is computationally expensive or intractable at times, especially when the number of input features is large. Also, at times there might be functions for which the Hessian is not even properly defined. So, for large machine-learning and deep-learning applications, gradient-descent—especially stochastic gradient descent—techniques with mini-batches are used since they are relatively less computationally intensive and scale up well when the data size is large.

Constrained Optimization Problem

In a constrained optimization problem, along with the cost function that we need to optimize, we have a set of constraints that we need to adhere to. The constraints might be equations or inequalities.

Whenever we want to minimize a function that is subject to an equality constraint, we use the Lagrange formulation. Let's say we must minimize $f(\theta)$ subject to $g(\theta) = 0$ where $\theta \in R^{n \times 1}$. For such a constrained optimization problem, we need to minimize a function $L(\theta, \ \lambda) = f(\theta) + \lambda g(\theta)$. Taking the gradient of L, which is called the Lagrangian, with respect to the combined vector θ, λ, and setting it to 0 would give us the required θ that minimizes $f(\theta)$ and adheres to the constraint. λ is called the Lagrange multiplier. When there are several constraints, we need to add all such constraints, using a separate Lagrange multiplier for each constraint. Let's say we want to minimize $f(\theta)$ subject to m constraints $g_i(\theta) = 0 \ \forall \ i \ \in \{1, 2, 3, \ldots m\}$; the Lagrangian can be expressed as follows:

$$L\left(\theta, \ \lambda\right) = f\left(\theta\right) + \sum_{i=1}^{m} \lambda_i g_i\left(\theta\right)$$

where $\lambda = [\lambda_1 \lambda_2 \ .. \ \lambda_m]^T$

   To minimize the function, the gradient of $L(\theta, \lambda)$ with respect to both θ and λ vectors should be a zero vector, i.e.,

$$\nabla_\theta (\theta, \ \lambda) = 0$$

$$\nabla_\lambda (\theta, \ \lambda) = 0$$

   The preceding method cannot be directly used for constraints with inequality. In such cases, a more generalized approach called the Karush-Kuhn-Tucker method can be used.

   Let $C(\theta)$ be the cost function that we wish to minimize, where . Also, let there be k number of constraint on θ such that the following:

$$f_1(\theta) = a_1$$

$$f_2(\theta) = a_2$$

$$f_3(\theta) \le a_3$$

$$f_4(\theta) \ge a_4$$

...
...

$$f_k(\theta) = a_k$$

   This becomes a constrained optimization problem since there are constraints that θ should adhere to. Every inequality can be transformed into a standard form where a certain function is less than or less than equal to zero. For example,

$$f_4(\theta) \ge a_4 => -f_4(\theta) \le -a_4 => -f_4(\theta) + a_4 \le 0$$

   Let each such constraint strictly less than, or less than equal to, zero be represented by $g_i(\theta)$. Also, let there be some strict equality equations $e_j(\theta)$. Such minimization problems are solved through the Karush-Kuhn-Tucker version of Lagrangian formulation.

Instead of minimizing $C(\theta)$, we need to minimize a cost function $L(\theta, \alpha, \beta)$ as follows:

$$L(\theta, \alpha, \beta) = C(\theta) + \sum_{i=1}^{k_1} \alpha_i g_i(\theta) + \sum_{j=1}^{k_2} \beta_j e_j(\theta)$$

The scalers $\alpha_i \forall \ i \in \{1, 2, 3, ..k_1\}$ and $\beta_j \forall j \in \{1, 2, 3, ..k_2\}$ are called the Lagrangian multipliers, and there would be k of them corresponding to k constraints. So, we have converted a constrained minimization problem into an unconstrained minimization problem.

To solve the problem, the Karush-Kuhn-Tucker conditions should be met at the minima point as follows:

- The gradient of $L(\theta, \alpha, \beta)$ with respect to θ should be the zero vector, i.e.,

$$\nabla_\theta(\theta, \alpha, \beta) = 0$$

$$\Rightarrow \nabla_\theta C(\theta) + \sum_{i=1}^{k_1} \alpha_i \nabla_\theta g_i(\theta) + \sum_{j=1}^{k_2} \beta_j \nabla_\theta e_j(\theta) = 0$$

- The gradient of the $L(\theta, \alpha, \beta)$ with respect to β, which is the Lagrange multiplier vector corresponding to the equality conditions, should be zero:

$$\nabla_\beta(\theta, \alpha, \beta) = 0$$

$$\Rightarrow \nabla_\beta C(\theta) + \sum_{i=1}^{k_1} \alpha_i \nabla_\beta g_i(\theta) + \sum_{j=1}^{k_2} \beta_j \nabla_\beta e_j(\theta) = 0$$

- The inequality conditions should become equality conditions at the minima point. Also, the inequality Lagrange multipliers should be nonnegative:

$$\alpha_i g_i(\theta) = 0 \ and \ \alpha_i \geq 0 \quad \forall i \in \{1, 2, ..., k_1\}$$

Solving for the preceding conditions would provide the minima to the constrained optimization problem.

A Few Important Topics in Machine Learning

In this section, we will discuss a few important topics that are very much relevant to machine learning. Their underlying mathematics is very rich.

Dimensionality-Reduction Methods

Principal component analysis and singular value decomposition are the most used dimensionality-reduction techniques in the machine-learning domain. We will discuss these techniques to some extent here. Please note that these data-reduction techniques are based on linear correlation and do not capture nonlinear correlation such as co-skewness, co-Kurtosis, and so on. We will talk about a few dimensionality-reduction techniques that are based on artificial neural networks, such as auto-encoders, in the latter part of the book.

Principal Component Analysis

Principal component analysis is a dimensionality-reduction technique that ideally should have been discussed in the "Linear Algebra" section. However, to make its mathematics much easier to grasp, I intentionally kept it for after the constrained optimization problem. Let us look at the two-dimensional data plot in Figure 1-45. As we can see, the maximum variance of the data is neither along the x direction nor along the y direction, but somewhat in a direction in between. So, had we projected the data in a direction where the variance is at maximum, it would have covered most of the variability in the data. Also, the rest of the variance could have been ignored as noise.

The data along the x direction and the y direction are highly correlated (see Figure 1-48). The covariance matrix would provide the required information about the data to reduce the redundancy. Instead of looking at the x and y directions, we can look at the a_1 direction, which has the maximum variance.

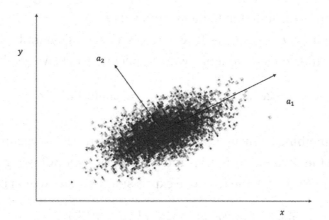

Figure 1-48. *Correlated 2-D data. a_1 and a_2 are the directions along which the data is uncorrelated and are the principal components*

Now, let us get into some math and assume that we do not have the plot and we only have the data for m number of samples $x^{(i)} \in R^{n \times 1}$.

We want to find out independent directions in the n-dimensional plane in the order of decreasing variances. By independent directions, I mean the covariance between those directions should be 0. Let a_1 be the unit vector along which the variance of the data is maximum. We first subtract the mean of the data vector to center the data on the origin. Let μ be the mean vector of the data vector x; i.e., $E[x] = \mu$.

A component of the $(x - \mu)$ vector in the direction of a_1 is the projection of $(x - \mu)$ on a_1; let it be denoted by z_1:

$$z_1 = a_1^T \left(x - \mu \right)$$

$var(z_1) = var[a_1^T(x - \mu)] = a_1^T cov(x) a_1$, where var denotes variance and $cov(x)$ denotes the covariance matrix.

For given data points, the variance is a function of a_1. So, we would have to maximize the variance with respect to a_1 given that a_1 is a unit vector:

$$a_1^T a_1 = 1 \quad => a_1^T a_1 - 1 = 0$$

So, we can express the function to be maximized as

$L(a_1, \lambda) = a_1^T cov(x)a_1 - \lambda(a_1^T a_1 - 1)$, where λ is a Lagrangian multiplier.

For maxima, setting the gradient of L with respect to a_1 to 0, we get the following:

$$\nabla L = 2cov(x)a_1 - 2\lambda a_1 = 0 \quad => cov(x)a_1 = \lambda a_1$$

We can see something come up that we studied earlier. The a_1 vector is nothing but an Eigen vector of the covariance matrix, and λ is the corresponding Eigen value.

Now, substituting this into the variance expression along a_1, we get the following:

$$var(z_1) = a_1^T cov(x)a_1 = a_1^T \lambda a_1 = \lambda a_1^T a_1 = \lambda$$

Since the expression for variance along a_1 is the Eigen value itself, the Eigen vector corresponding to the highest Eigen value gives us the first principal component or the direction along which the variance in data is maximum.

Now, let us get the second principal component, or the direction along which the variance is maximum right after a_1.

Let the direction of the second principal component be given by the unit vector a_2. Since we are looking for orthogonal components, the direction of a_2 should be perpendicular to a_1.

A projection of the data along a_2 can be expressed by the variable $z_2 = a_2^T(x - \mu)$.

Hence, the variance of the data along a_2 is Var (z_2) = Var $[a_2^T(x - \mu)] = a_2^T cov(x)a_2$.

We would have to maximize $Var(z_2)$ subject to the constraints $a_2^T a_2 = 1$ since a_2 is a unit vector and $a_2^T a_1 = 0$ since a_2 should be orthogonal to a_1.

We need to maximize the following function $L(a_2, \alpha, \beta)$ with respect to the parameters a_2, α, β:

$$L(a_2, \alpha, \beta) = a_2^T cov(x)a_2 - \alpha\left(a_2^T a_2 - 1\right) - \beta\left(a_2^T a_1\right)$$

By taking a gradient with respect to a_2 and setting it to zero vector, we get the following:

$$\nabla L = 2cov(x)a_2 - 2\alpha a_2 - \beta a_1 = 0$$

By taking the dot product of the gradient ∇L with vector a_1, we get the following:

$$2a_1^T cov(x)a_2 - 2\alpha a_1^T a_2 - \beta a_1^T a_1 = 0$$

$a_1^T cov(x)a_2$ is a scalar and can be written as $a_2^T cov(x)a_1$.

On simplification, $a_2^T cov(x)a_1 = a_2^T \lambda a_1 = \lambda a_2^T a_1 = 0$. Also, the term $2\alpha a_1^T a_2$ is equal to 0, which leaves $\beta a_1^T a_1 = 0$. Since $a_1^T a_1 = 1$, β must be equal to 0.

Substituting $\beta = 0$ in the expression for $\nabla L = 2\ cov\ (x)a_2 - 2\alpha a_2 - \beta a_1 = 0$, we have the following:

$$2cov(x)a_2 - 2\alpha a_2 = 0 \ \ i.e. \ \ cov(x)a_2 = \alpha a_2$$

Hence, the second principal component is also an Eigen vector of the covariance matrix, and the Eigen value α must be the second largest Eigen value just after λ. In this way, we would get n Eigen vectors from the covariance matrix $cov(x) \in R^{n \times n}$, and the variance of the data along each of those Eigen vector directions (or principal components) would be represented by the Eigen values. One thing to note is that the covariance matrix is always symmetrical and thus the Eigen vectors would always be orthogonal to each other and so would give independent directions.

The covariance matrix is always positive semi-definite.

This is true because the Eigen values of the covariance matrix represent variances, and variance cannot be negative. If $cov(x)$ is positive definite, i.e., $a^T cov\ (x)a > 0$, then all the Eigen values of covariance matrices are positive.

Figure 1-49 illustrates a principal component analysis transformation of the data. As we can see, PCA has centered the data and got rid of correlation among the PCA transformed variables.

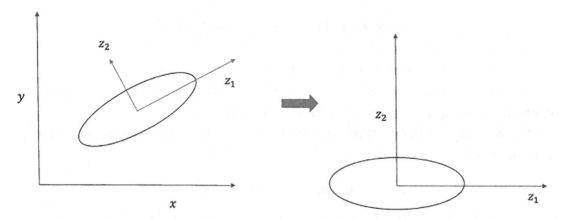

Figure 1-49. *Principal component analysis centers the data and then projects the data into axes along which variance is maximum. The data along z_2 can be ignored if the variance along it is negligible*

When Will PCA Be Useful in Data Reduction?

When there is high correlation between the different dimensions of the input, there would only be a few independent directions along which the variance of data would be high, and along other directions the variance would be insignificant. With PCA, one can keep the data components in the few directions in which variance is high and make a significant contribution to the overall variance, ignoring the rest of the data.

How Do You Know How Much Variance Is Retained by the Selected Principal Components?

If the z vector presents the transformed components for input vector x, then the $cov(z)$ would be a diagonal matrix containing the Eigen values of the $cov(x)$ matrix as its diagonal entries.

$$cov(z) = \begin{bmatrix} \lambda_1 & \cdots & 0 \\ \vdots & \lambda_2 & \vdots \\ 0 & \cdots & \lambda_n \end{bmatrix}$$

Also, let us suppose the Eigen values are ordered like $\lambda_1 > \lambda_2 > \lambda_3 \, .. > \lambda_n$.

Let us suppose we choose to keep only the first k principal components; the proportion of variance in the data captured is as follows:

$$\frac{\lambda_1 + \lambda_2 + \lambda_3 + \ldots + \lambda_k}{\lambda_1 + \lambda_2 + \lambda_3 + \ldots + \lambda_k + \ldots + \lambda_n}$$

Singular Value Decomposition

Singular value decomposition is a dimensionality-reduction technique that factorizes a matrix $A \in \mathrm{R}^{m \times n}$ $A \in \mathrm{R}^{m \times n}$ into a product of three matrices, as $A = USV^T$ where

$U \in R^{m \times m}$ and is composed of all the Eigen vectors of the matrix AA^T.

$V \in R^{n \times n}$ and is composed of all the Eigen vectors of the matrix A^TA.

$S \in R^{m \times n}$ and is composed of k square root of the Eigen vectors of both A^TA and AA^T, where k is the rank of matrix A.

The column vectors of U are all orthogonal to each other and hence form an orthogonal basis. Similarly, the column vectors of V also form an orthogonal basis:

$$U = \begin{bmatrix} u_1 u_2 \ldots u_m \end{bmatrix}$$

where $u_i \in R^{m \times 1}$ are the column vectors of U.

$$V = \begin{bmatrix} v_1 v_2 \ldots v_m \end{bmatrix}$$

where $v_i \in R^{n \times 1}$ are the column vectors of V.

$$S = \begin{bmatrix} \sigma_1 & \cdots & 0 \\ \vdots & \sigma_2 & \vdots \\ 0 & \cdots & 0 \end{bmatrix}$$

Depending on the rank of A, there would be $\sigma_1, \sigma_2, \ldots \ldots \ldots \sigma_k$ diagonal entries corresponding to the rank k of matrix A:

$$A = \sigma_1 u_1 v_1^T + \sigma_2 u_2 v_2^T + \ldots \ldots + \sigma_k u_k v_k^T$$

The $\sigma_i\ \forall i\ \in\{1,2,3,..k\}$, also called singular values, are the square root of the Eigen values of both A^TA and AA^T, and hence they are measures of variance in the data. Each of the $\sigma_i u_i v_i^T \forall i\ \in\{1,2,3,..k\}$ is a rank-one matrix. We can only keep the rank-one matrices for which the singular values are significant and explain a considerable proportion of the variance in the data.

If one takes only the first p rank-one matrices corresponding to the first p singular values of largest magnitude, then the variance retained in the data is given by the following:

$$\frac{\sigma_1^{\ 2}+\sigma_2^{\ 2}+\sigma_3^{\ 2}+...+\sigma_p^{\ 2}}{\sigma_1^{\ 2}+\sigma_2^{\ 2}+\sigma_3^{\ 2}+...+\sigma_p^{\ 2}+...+\sigma_k^{\ 2}}$$

Images can be compressed using singular value decomposition. Similarly, singular value decomposition is used in collaborative filtering to decompose a user-rating matrix to two matrices containing user vectors and item vectors. The singular value decomposition of a matrix is given by USV^T. The user-rating matrix R can be decomposed as follows:

$$R=USV^T=US^{\frac{1}{2}}S^{\frac{1}{2}}V^T=U'V'^T$$

where U' is the user-vector matrix and is equal to $US^{\frac{1}{2}}$ and V' is the items-vector matrix where $V'=S^{\frac{1}{2}}V^T$.

Regularization

The process of building a machine-learning model involves deriving parameters that fit the training data. If the model is simple, then the model lacks sensitivity to the variation in data and suffers from high bias. However, if the model is too complex, it tries to model for as much variation as possible and in the process models for random noise in the training data. This removes the bias produced by simple models but introduces high variance, i.e., the model is sensitive to very small changes in the input. High variance for a model is not a good thing, especially if the noise in the data is considerable. In such cases, the model in the pursuit of performing too well on the training data performs poorly on the test dataset since the model loses its capability to generalize well with the new data. This problem of models' suffering from high variance is called *overfitting*.

As we can see in Figure 1-50, we have three models fit to the data. The one parallel to the horizontal is suffering from high bias, while the curvy one is suffering from high variance. The straight line in between at around 45 degrees to the horizontal has neither high variance nor high bias.

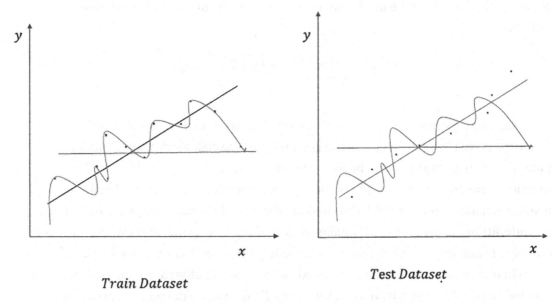

Train Dataset   **Test** *Dataset*

Figure 1-50. *Illustration of models with high variance and high bias*

The model with high variance does well on the train data but fails to do well on the test dataset, even when the dataset's nature has not changed much. The model represented in blue color may not fit the training perfectly, but it does better on the test data since the model does not suffer from high variance. The trick is to have a model that does not suffer from high bias and at the same time is not so complex that it models for random noise as well.

Models with high variance typically have model parameters with a large magnitude since the sensitivity of the models to small changes in data is high. To overcome the problem of overfitting resulting from high model variance, a popular technique called regularization is widely used.

To put things into perspective, let us look at the linear regression cost function that we looked at earlier:

$$C(\theta) = \|X\theta - Y\|_2^2$$

$$= (X\theta - Y)^T (X\theta - Y)$$

103

As discussed earlier, models with high variance have model parameters with a large magnitude. We can put an extra component into the cost function $C(\theta)$ that penalizes the overall cost function in case the magnitude of the model parameter vector is high.

So, we can have a new cost function, $L(\theta) = \|X\theta - Y\|_2^2 + \lambda \|\theta\|_2^2$, where $\|\theta\|_2^2$ is the square of the l^2 norm of the model parameter vector. The optimization problem becomes as follows:

$$\theta^* = \underbrace{Arg\ Min}_{\theta} L(\theta) = \|X\theta - Y\|_2^2 + \lambda \|\theta\|_2^2$$

Taking the gradient ∇L with respect to θ and setting it to 0 give us $\theta^* = (X^TX + \lambda I)^{-1}X^TY$.

Now, as we can see because of the $\|\theta\|_2^2$ term in the cost function, the model parameter's magnitude cannot be too large since it would penalize the overall cost function. λ determines the weight of the regularization term. A higher value for λ would result in smaller values of $\|\theta\|_2^2$, thus making the model simpler and prone to high bias or underfitting. In general, even smaller values of λ go a long way in reducing the model complexity and the model variance. λ is generally optimized using cross-validation.

When the square of the l^2 norm is used as the regularization term, the optimization method is called l^2 regularization. At times, the l^1 norm of model parameter vectors is used as the regularization term, and the optimization method is termed l^1 regularization. l^2 regularization applied to regression problems is called ridge regression, whereas l^1 regularization applied to such regression problems is termed lasso regression.

For l^1 regularization, the preceding regression problem becomes the following:

$$\theta^* = \underbrace{Arg\ Min}_{\theta} L(\theta) = \|X\theta - Y\|_2^2 + \lambda \|\theta\|_1$$

Ridge regression is mathematically more convenient since it has a closed-form solution, whereas lasso regression does not have a closed-form solution. However, lasso regression is much more robust to outliers in comparison with ridge regression. Lasso problems give sparse solutions, so it is good for feature selection, especially when there is moderate to high correlation among the input features.

Regularization Viewed as a Constraint Optimization Problem

Instead of adding the penalty term, we can add a constraint on the magnitude of the model parameter vector to be less than or equal to some constant value. We can then have an optimization problem as follows:

$$\theta^* = argmin_\theta \ C(\theta) = \|X\theta - Y\|_2^2$$

such that $\|\theta\|_2^2 \leq b$

where b is a constant.

We can convert this constrained minimization problem to an unconstrained minimization problem by creating a new Lagrangian formulation, as seen here:

$$L(\theta, \lambda) = \|X\theta - Y\|_2^2 + \lambda\left(\|\theta\|_2^2 - b\right)$$

To minimize the Lagrangian cost function per the Karush-Kuhn-Tucker conditions, the following are important:

- The gradient of L with respect to θ; i.e., $\nabla_\theta L(\theta, \lambda)$ should be the zero vector, which on simplification gives the following:

$$\theta = \left(X^T X + \lambda I\right)^{-1} X^T Y \qquad (1)$$

- Also, at the optimal point $\lambda\left(\|\theta\|_2^2 - b\right) = 0$ and $\lambda \geq 0$

    If we consider regularization, i.e., $\lambda > 0$, then $\|\theta\|_2^2 - b = 0$ (2)

As we can see from (1), θ obtained is a function of λ. λ should be adjusted such that the constraint from (2) is satisfied.

The solution $\theta = (X^T X + \lambda I)^{-1} X^T Y$ from (1) is the same as what we get from l^2 regularization. In machine-learning applications, the Lagrange multiplier is generally optimized through hyperparameter tuning or cross-validation since we have no knowledge of what a good value for b would be. When we take small values of λ, the value of b increases and so does the norm of θ, whereas larger values of λ provide smaller b and hence a smaller norm for θ.

Coming back to regularization, any component in the cost function that penalizes the complexity of the model provides regularization. In tree-based models, as we increase the number of leaf nodes, the complexity of the tree grows. We can add a term to the cost function that is based on the number of leaf nodes in the tree, and it will provide regularization. A similar thing can be done for the depth of the tree.

Even stopping the model-training process early provides regularization. For instance, in gradient-descent method, the more iterations we run, the more complex the model gets since with each iteration, the gradient descent tries to reduce the cost-function value further. We can stop the model-learning process early based on some criteria, such as an increase in the cost-function value for the test dataset in the iterative process. Whenever in the iterative process of training the training cost-function value decreases while the test cost-function value increases, it might be an indication of the onset of overfitting, and thus it makes sense to stop the iterative learning.

Whenever training data is less in comparison with the number of parameters the model must learn, there is a high chance of overfitting, because the model will learn too many rules for a small dataset and might fail to generalize well to the unseen data. If the dataset is adequate in comparison with the number of parameters, then the rules learned are over a good proportion of the population data, and hence the chances of model overfitting go down.

Bias and Variance Trade-Off

As we have discussed earlier, a model which is too simple in comparison with the system or relationship that we are trying to learn suffers from high bias. Whereas when the model is too complex in comparison with the desired relationship it must learn, it suffers from high variance.

Mathematically to define bias precisely, let us assume the world model represented as $y = f(X; \theta)$ where function f parameterized by $\theta \in R^n$ maps input $X \in R^n$ to output $y \in R$.

With a ML model we would estimate θ as $\hat{\theta}$. If we use different datasets, we will get different estimate of $\hat{\theta}$ in general. The model is said to suffer from high bias if the expected value of $\hat{\theta}$ differs from the world value or oracle value θ by a large amount. Bias can hence be expressed as follows:

$$Bias = E\left[\hat{\theta}\right] - \theta$$

Variance on the model on the other hand is the variability in the estimates $\hat{\theta}$ of the parameter θ and can be expressed as $E\left[\hat{\theta}-E\left[\hat{\theta}\right]\right]^2$ when θ is a scalar. When θ is a vector, we can investigate the covariance matrix instead for variance of the model as shown in the following:

$$E\left[\left(\hat{\theta}-E\left[\hat{\theta}\right]\right)\left(\hat{\theta}-E\left[\hat{\theta}\right]\right)^T\right]$$

The model error in general can be broken down into variance of the model and square of the bias of the model. To prove the same, let us assume we have a model with one parameter the oracle value of which is θ. We estimate the parameter of the model θ as $\hat{\theta}$. Since training on different sample of the data would yield different $\hat{\theta}$, $\hat{\theta}$ here is a random variable with some mean $E\left[\hat{\theta}\right]$ and finite variance. The expectation on the square of the difference between $\hat{\theta}$ and θ as shown in the following would give us our model error.

We can expand the above expression as shown in the following:

$$E\left[\left(\hat{\theta}-\theta\right)^2\right]=E\left[\left(\hat{\theta}-E\left[\hat{\theta}\right]+E\left[\hat{\theta}\right]-\theta\right)^2\right]$$

$$=E\left[\left(\hat{\theta}-E\left[\hat{\theta}\right]\right)^2\right]+E\left[\left(E\left[\hat{\theta}\right]-\theta\right)^2\right]+2\,E\left[\left(\hat{\theta}-E\left[\hat{\theta}\right]\right)\right]\left[E\left[\hat{\theta}\right]-\theta\right]$$

The first term $E\left[\left(\hat{\theta}-E\left[\hat{\theta}\right]\right)^2\right]$ is the variance of the model. Everything in the second term is constant and hence $E\left[\left(E\left[\hat{\theta}\right]-\theta\right)^2\right]=\left(E\left[\hat{\theta}\right]-\theta\right)^2$. A closer look at $\left(E\left[\hat{\theta}\right]-\theta\right)^2$ suggests that $\left(E\left[\hat{\theta}\right]-\theta\right)^2$ is the bias squared. In the third term $E\left[\left(\hat{\theta}-E\left[\hat{\theta}\right]\right)\right]\left[E\left[\hat{\theta}\right]-\theta\right]$ since $E\left[\hat{\theta}\right]$ and θ are constants, the expression simplifies to as the following:

$$2\,E\left[\left(\hat{\theta}-E\left[\hat{\theta}\right]\right)\right]\left[E\left[\hat{\theta}\right]-\theta\right]=2*\left[E\left[\hat{\theta}\right]-\theta\right]*E\left[\left(\hat{\theta}-E\left[\hat{\theta}\right]\right)\right]$$

Since mean of $\hat{\theta}$ is $E\left[\hat{\theta}\right]$ hence $E\left[\left(\hat{\theta}-E\left[\hat{\theta}\right]\right)\right]=0$.

Hence, we can see that model error is a combination of variance and bias in the model estimate. So, for a given model error, sum of variance of the model and square of the bias remain constant—if we try to reduce one, the other would increase.

Summary

In this chapter, we have touched upon all the required mathematical concepts for proceeding with machine-learning and deep-learning concepts. The reader is still advised to go through proper textbooks pertaining to these subjects in his or her spare time for more clarity. However, this chapter is a good starting point. In the next chapter, we will start with artificial neural networks and the basics of TensorFlow.

CHAPTER 2

Introduction to Deep-Learning Concepts and TensorFlow

Deep Learning and Its Evolution

Deep learning evolved from artificial neural networks, which have existed since the 1940s. Neural networks are interconnected networks of processing units called artificial neurons that loosely mimic the axons found in a biological brain. In a biological neuron, dendrites receive input signals from various neighboring neurons, typically more than 1000 of them. These modified signals are then passed on to the cell body or soma of the neuron, where these signals are summed together and then passed on to the axon of the neuron. If the received input signal is more than a specified threshold, the axon will release a signal, which will be passed on to the neighboring dendrites of other neurons. Figure 2-1 depicts the structure of a biological neuron for reference.

© Santanu Pattanayak 2023
S. Pattanayak, *Pro Deep Learning with TensorFlow 2.0*, https://doi.org/10.1007/978-1-4842-8931-0_2

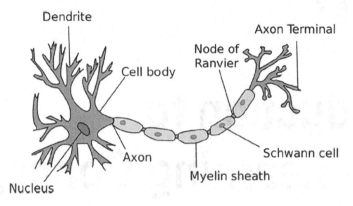

Figure 2-1. *Structure of a biological neuron*

Artificial neuron units are inspired by the biological neurons, with some modifications for convenience. Much like the dendrites, the input connections to the neuron carry the attenuated or amplified input signals from other neighboring neurons. The signals are passed on to the neuron, where the input signals are summed up and then a decision is made as to what to output based on the total input received. For instance, for a binary threshold neuron, an output value of 1 is provided when the total input exceeds a predefined threshold; otherwise, the output stays at 0. Several other types of neurons are used in artificial neural networks, and their implementation only differs with respect to the activation function on the total input to produce the neuron output. In Figure 2-2, the different biological equivalents are tagged in the artificial neuron for easy analogy and interpretation.

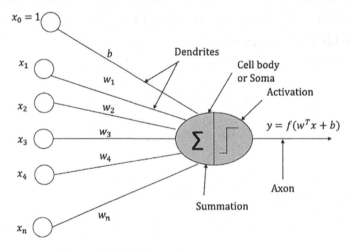

Figure 2-2. *Structure of an artificial neuron*

Artificial neural networks started with a lot of promise in the early 1940s. We will go through the chronology of major events in the artificial neural network community to get a feel of how this discipline has evolved over the years and what challenges were faced along the way.

- Warren McCulloch and Walter Pitts, two electrical engineers, published a paper titled "A Logical Calculus of the Ideas Immanent in Nervous Activity," related to neural networks, in 1943. The paper can be located at www.cs.cmu.edu/~epxing/Class/10715/reading/ McCulloch.and.Pitts.pdf. Their neurons have a binary output state, and there are two types of input to the neurons: excitatory inputs and inhibitory inputs. All excitatory inputs to the neuron have equal positive weights. If all the inputs to the neuron are excitatory and if the total input $\sum_i w_i x_i > 0$, the neuron would output 1. In cases where any of the inhibitory inputs are active or $\sum_i w_i x_i \leq 0$, the output would be 0. Using this logic, all the Boolean logic functions can be implemented by one or more such neurons. The drawback of these networks was that they had no way of learning the weights through training. One must figure out the weights manually and combine the neurons to achieve the required computation.

- The next big thing was the Perceptron, which was invented by Frank Rosenblatt in 1957. He, along with his collaborators, Alexander Stieber and Robert H. Shatz, documented their invention in a report titled "The Perceptron—A Perceiving and Recognizing Automaton," which can be located at https://blogs.umass.edu/brain-wars/ files/2016/03/rosenblatt-1957.pdf. The Perceptron was built with the motive of binary classification tasks. Both the weights and bias to the neuron can be trained through the Perceptron learning rule. The weights can be both positive and negative. There were strong claims made by Frank Rosenblatt about the capabilities of the Perceptron model. Unfortunately, not all of them were true.

- Marvin Minsky and Seymour A. Papert wrote a book titled
  Perceptrons: An Introduction to Computational Geometry in 1969
  (MIT Press), which showed the limitations of the Perceptron learning
  algorithm even on simple tasks such as developing the XOR Boolean
  function with a single Perceptron. A better part of the artificial neural
  network community perceived that these limitations showed by
  Minsky and Papert applied to all neural networks, and hence the
  research in artificial neural networks nearly halted for a decade, until
  the 1980s.

- In the 1980s, interest in artificial neural networks was revived by
  Geoffrey Hilton, David Rumelhart, Ronald Williams, and others,
  primarily because of the backpropagation method of learning multi-
  layered problems and because of the ability of neural networks to
  solve nonlinear classification problems.

- In the 1990s, support vector machines (SVM), invented by V. Vapnik
  and C. Cortes, became popular since neural networks were not
  scaling up to large problems.

- Artificial neural networks were renamed deep learning in 2006 when
  Geoffrey Hinton and others introduced the idea of unsupervised
  pretraining and deep-belief networks. Their work on deep-belief
  networks was published in a paper titled "A Fast Learning Algorithm
  for Deep Belief Nets." The paper can be located at `www.cs.toronto.`
  `edu/~hinton/absps/fastnc.pdf`.

- ImageNet, a large collection of labeled images, was created and
  released by a group in Stanford in 2010.

- In 2012, Alex Krizhevsky, Ilya Sutskever, and Geoffrey Hinton won the
  ImageNet Competition for achieving an error rate of 16%, whereas
  in the first two years, the best models had around 28% and 26% error
  rates. This was a huge margin of win. The implementation of the
  solution had several aspects of deep learning that are standard in any
  deep-learning implementation today.

  - Graphical processing units (GPUs) were used to train the model. GPUs are
    very good at doing matrix operations and are computationally very fast
    since they have thousands of cores to do parallel computing.

– Dropout was used as a regularization technique to reduce overfitting.

– Rectified linear units (ReLU) were used as activation functions for the hidden layers.

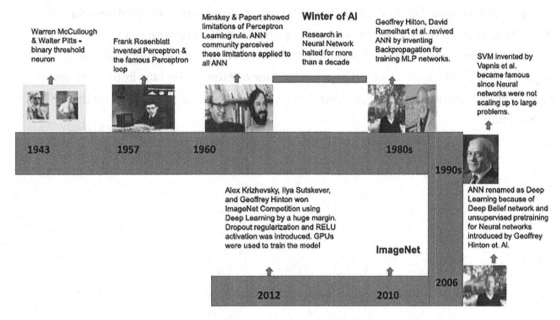

Figure 2-3. *Evolution of artificial neural networks*

Figure 2-3 shows the evolution of the artificial neural network into deep learning. ANN stands for artificial neural network, MLP stands for multi-layer Perceptron, and AI stands for artificial intelligence.

Perceptrons and Perceptron Learning Algorithm

Although there are limitations to what Perceptron learning algorithms can do, they are the precursor to the advanced techniques in deep learning we see today. Hence, a detailed study of Perceptrons and the Perceptron learning algorithm is worthwhile. Perceptrons are linear binary classifiers that use a hyperplane to separate the two classes. The Perceptron learning algorithm is guaranteed to fetch a set of weights and bias that classifies all the inputs correctly, provided such a feasible set of weights and bias exist.

Perceptron is a linear classifier, and as we saw in Chapter 1, linear classifiers generally perform binary classification by constructing a hyperplane that separates the positive class from the negative class.

The hyperplane is represented by a unit weight vector $w' \in R^{n \times 1}$ that is perpendicular to the hyperplane and a bias term b that determines the distance of the hyperplane from the origin. The vector $w' \in R^{n \times 1}$ is chosen to point toward the positive class.

As illustrated in Figure 2-4, for any input vector $x' \in R^{n \times 1}$, the dot product with the negative of the unit vector w' would give the distance b of the hyperplane from the origin since x' and w' are on the side opposite of the origin. Formally, for points lying on the hyperplane,

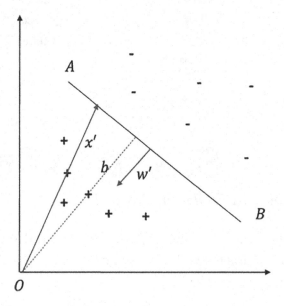

Figure 2-4. *Hyperplane separating two classes*

$$-w'^{T} x' = b => w'^{T} x' + b = 0$$

Similarly, for points lying below the hyperplane, i.e., the input vectors $x'_{+} \in R^{n \times 1}$ belonging to the positive class, the negative of the projection of x'_{+} on w' should be less than b. So, for the points belonging to the positive class,

$$-w'^{T} x' \langle b =) w'^{T} x' + b > 0$$

Similarly, for points lying above the hyperplane, i.e., the input vectors $x'_+ \in R^{n \times 1}$ belonging to the negative class, the negative of the projection of x'_+ on w' should be greater than b. So, for the points belonging to the negative class,

$$-w'^T x' > b => w'^T x' + b < 0$$

Summarizing the preceding deductions, we can conclude the following:

- $w'^T x' + b = 0$ corresponds to the hyperplane, and all $x' \in R^{n \times 1}$ that lie on the hyperplane will satisfy this condition. Generally, points on the hyperplane are taken to belong to the negative class.

- $w'^T x' + b > 0$ corresponds to all points in the positive class.

- $w'^T x' + b \leq 0$ corresponds to all points in the negative class.

However, for Perceptron we don't keep the weight vector w' as a unit vector but rather keep it as any general vector. In such cases, the bias b would not correspond to the distance of the hyperplane from the origin, but rather it would be a scaled version of the distance from the origin, the scaling factor being the magnitude or the l^2 norm of the vector w', i.e., $\|w'\|_2$. Just to summarize, if w' is any general vector perpendicular to the hyperplane and is pointing toward the positive class, then $w'^T x + b = 0$ still represents a hyperplane where b represents the distance of the hyperplane from the origin times the magnitude of w'.

In the machine-learning domain, the task is to learn the parameters of the hyperplane (i.e., w' and b). We generally tend to simplify the problem to get rid of the bias term and consume it as a parameter within w corresponding to a constant input feature of 1,1 as we discussed earlier in Chapter 1.

Let the new parameter vector after adding the bias be $w \in R^{(n+1) \times 1}$ and the new input features vector after adding the constant term 1 be $x \in R^{(n+1) \times 1}$, where

$$x' = \begin{bmatrix} x_1 & x_2 x_3 .. & x_n \end{bmatrix}^T$$

$$x = \begin{bmatrix} 1 & x_1 & x_2 x_3 .. & x_n \end{bmatrix}^T$$

$$w' = \begin{bmatrix} w_1 & w_2 w_3 .. & w_n \end{bmatrix}^T$$

$$w = \begin{bmatrix} b & w_1 & w_2 w_3 .. & w_n \end{bmatrix}^T$$

By doing the preceding manipulation, we've made the hyperplane in R^n at a distance from the origin pass through the origin in the $R^{(n+1)}$ vector space. The hyperplane is now only determined by its weight parameter vector $w \in R^{(n+1) \times 1}$, and the rules for classification get simplified as follows:

- $w^T x = 0$ corresponds to the hyperplane, and all $x \in R^{(n+1) \times 1}$ that lies on the hyperplane will satisfy this condition.

- $w^T x > 0$ corresponds to all points in the positive class. This means the classification is now solely determined by the angle between the vectors w and x. If input vector x makes an angle within -90 degrees to $+90$ degrees with the weight parameter vector w, then the output class is positive.

- $w^T x \leq 0$ corresponds to points in the negative class. The equality condition is treated differently in different classification algorithms. For Perceptrons, points on the hyperplanes are treated as belonging to the negative class.

Now we have everything we need to proceed with the Perceptron learning algorithm. Let $x^{(i)} \in R^{(n+1) \times 1} \ \forall \ i \in \{1, 2, \ldots m\}$ represent the m input feature vectors and $y^{(i)} \in \{0, 1\} \ \forall \ i = \{1, 2, \ldots, m\}$ the corresponding class label.

The Perceptron learning problem is as follows:

Step 1: Start with a random set of weights $w \in R^{(n+1) \times 1}$.

Step 2: Evaluate the predicted class for a data point. For an input data point $x^{(i)}$, if $w^T x^{(i)} > 0$, then the predicted class $y_p^{(i)} = 1$, else $y_p^{(i)} = 0$. For the Perceptron classifier, points on the hyperplane are generally considered to belong to the negative class.

Step 3: Update the weight vector w as follows:

If $y_p^{(i)} = 0$ and the actual class $y^{(i)} = 1$, update the weight vector as $w = w + x^{(i)}$.

If $y_p^{(i)} = 1$ and the actual class $y^{(i)} = 0$, update the weight vector as $w = w - x^{(i)}$.

If $y_p^{(i)} = y^{(i)}$, no updates are required to w.

Step 4: Go to Step 2 and process the next data point.

Step 5: Stop when all the data points have been correctly classified.

Perceptron will only be able to classify the two classes properly if there exists a feasible weight vector w that can linearly separate the two classes. In such cases, the Perceptron Convergence theorem guarantees convergence.

Geometrical Interpretation of Perceptron Learning

The geometrical interpretation of Perceptron learning sheds some light on the feasible weight vector w that represents a hyperplane separating the positive and the negative classes.

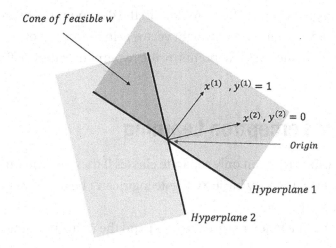

Vector Space of weights w

Figure 2-5. *Hyperplanes in weight space and feasible set of weight vectors*

Let us take two data points, $(x^{(1)}, y^{(1)})$ and $(x^{(2)}, y^{(2)})$, as illustrated in Figure 2-5. Further, let $x^{(i)} \in R^{3 \times 1}$ $x^{(i)} \in R^{3 \times 1}$ include the constant feature of 1 for the intercept term. Also, let's take $y^{(1)} = 1$ and $y^{(2)} = 0$ (i.e., data point 1 belongs to the positive class, while data point 2 belongs to the negative class).

In an input feature vector space, the weight vector determines the hyperplane. Along the same lines, we need to consider the individual input vectors as being representative of hyperplanes in the weight space to determine the feasible set of weight vectors for classifying the data points correctly.

In Figure 2-5, hyperplane 1 is determined by input vector $x^{(1)}$, which is perpendicular to hyperplane 1. Also, the hyperplane passes through the origin since the bias term has been consumed as a parameter within the weight vector w. For the first data point, $y^{(1)} = 1$. The prediction for the first data point would be correct if $w^T x^{(1)} > 0$. All the weight vectors w that are within an angle of -90 to +90 degrees from the input vector $x^{(1)}$ would satisfy the condition $w^T x^{(1)} > 0$. They form the feasible set of weight vectors for the first data point, as illustrated by the shaded region above hyperplane 1 in Figure 2-5.

Similarly, hyperplane 2 is determined by input vector $x^{(2)}$, which is perpendicular to hyperplane 2. For the second data point, $y^{(2)} = 0$. The prediction for the second data point would be correct if $w^T x^{(2)} \leq 0$. All the weight vectors w that are beyond an angle of -90 to +90 degrees from the input vector $x^{(2)}$ would satisfy the condition $w^T x^{(2)} \leq 0$. They form the feasible set of weight vectors for the second data point as illustrated by the shaded region below hyperplane 2 in Figure 2-5.

So, the set of weight vectors w that satisfy both the data points is the region of overlap between the two shaded regions. Any weight vector w in the region of overlap would be able to linearly separate the two data points by the hyperplanes they define in the input vector space.

Limitations of Perceptron Learning

The Perceptron learning rule can only separate classes if they are linearly separable in the input space. Even the very basic XOR gate logic can't be implemented by the Perceptron learning rule.

For the XOR logic, the following are the input and the corresponding output label or classes:

$$x_1 = 1, x_2 = 0 \quad y = 1$$

$$x_1 = 0, x_2 = 1 \quad y = 1$$

$$x_1 = 1, x_2 = 1 \quad y = 0$$

$$x_1 = 0, x_2 = 0 \quad y = 0$$

Let us initialize the weight vector $w \rightarrow [0\ 0\ 0]^T$, where the first component of the weight vector corresponds to the bias term. Similarly, all the input vectors would have their first components as 1.

- For $x_1 = 1$, $x_2 = 0$, $y = 1$, the prediction is $w^T x = \begin{bmatrix} 0\ 0\ 0 \end{bmatrix} \begin{bmatrix} 1 \\ 1 \\ 0 \end{bmatrix} = 0$. Since

  $w^T x = 0$, the data point would be classified as 0, which doesn't match the actual class of 1. Hence, the updated weight vector as per the

  Perceptron rule should be $w \rightarrow w + x = \begin{bmatrix} 0 \\ 0 \\ 0 \end{bmatrix} + \begin{bmatrix} 1 \\ 1 \\ 0 \end{bmatrix} = \begin{bmatrix} 1 \\ 1 \\ 0 \end{bmatrix}$.

- For $x_1 = 0$, $x_2 = 1$, $y = 1$, the prediction is $w^T x = \begin{bmatrix} 1\ 1\ 0 \end{bmatrix} \begin{bmatrix} 1 \\ 0 \\ 1 \end{bmatrix} = 1$. Since

  $w^T x = 1 > 0$, the data point would be correctly classified as 1. Hence,

  there would be no update to the weight vector, and it stays at $\begin{bmatrix} 1 \\ 1 \\ 0 \end{bmatrix}$.

- For $x_1 = 1$, $x_2 = 1$, $y = 0$, the prediction is $w^T x = \begin{bmatrix} 1\ 1\ 0 \end{bmatrix} \begin{bmatrix} 1 \\ 1 \\ 1 \end{bmatrix} = 2$. Since

  $w^T x = 2$, the data point would be classified as 1, which doesn't match the actual class of 0. Hence, the updated weight vector

  should be $w \rightarrow w - x = \begin{bmatrix} 1 \\ 1 \\ 0 \end{bmatrix} - \begin{bmatrix} 1 \\ 1 \\ 1 \end{bmatrix} = \begin{bmatrix} 0 \\ 0 \\ -1 \end{bmatrix}$.

- For $x_1 = 0$, $x_2 = 0$, $y = 0$, the prediction is $w^T x = \begin{bmatrix} 0\ 0\ -1 \end{bmatrix} \begin{bmatrix} 1 \\ 0 \\ 0 \end{bmatrix} = 0$. Since

  $w^T x = 0$, the data point would be correctly classified as 0. Hence, there would be no update to the weight vector w.

So, the weight vector after the first pass over the data points is $w = [0\ 0 - 1]^T$. Based on the updated weight vector w, let's evaluate how well the points have been classified.

- For data point 1, $w^T x = \begin{bmatrix} 0 & 0 & -1 \end{bmatrix} \begin{bmatrix} 1 \\ 1 \\ 0 \end{bmatrix} = 0$, so it is wrongly classified as class 0.

- For data point 2, $w^T x = \begin{bmatrix} 0 & 0 & -1 \end{bmatrix} \begin{bmatrix} 1 \\ 0 \\ 1 \end{bmatrix} = -1$, so it is wrongly classified as class 0.

- For data point 3, $w^T x = \begin{bmatrix} 0 & 0 & -1 \end{bmatrix} \begin{bmatrix} 1 \\ 1 \\ 1 \end{bmatrix} = -1$, so it is correctly classified as class 0.

- For data point 4, $w^T x = \begin{bmatrix} 0 & 0 & -1 \end{bmatrix} \begin{bmatrix} 1 \\ 0 \\ 0 \end{bmatrix} = 0$, so it is correctly classified as class 0.

Based on the preceding classifications, we see after the first iteration that the Perceptron algorithm managed to classify only the negative class correctly. If we were to apply the Perceptron learning rule over the data points again, the updates to the weight vector w in the second pass would be as follows:

- For data point 1, $w^T x = \begin{bmatrix} 0 & 0 & -1 \end{bmatrix} \begin{bmatrix} 1 \\ 1 \\ 0 \end{bmatrix} = 0$, so it is wrongly classified

  as class 0. Hence, the updated weight as per the Perceptron

  rule is $w \rightarrow w + x = \begin{bmatrix} 0 \\ 0 \\ -1 \end{bmatrix} + \begin{bmatrix} 1 \\ 1 \\ 0 \end{bmatrix} = \begin{bmatrix} 1 \\ 1 \\ -1 \end{bmatrix}$.

- For data point 2, $w^T x = \begin{bmatrix} 1 & 1 & -1 \end{bmatrix} \begin{bmatrix} 1 \\ 0 \\ 1 \end{bmatrix} = 0$, so it is wrongly classified

  as class 0. Hence, the updated weight as per the Perceptron

  rule is $w \rightarrow w + x = \begin{bmatrix} 1 \\ 1 \\ -1 \end{bmatrix} + \begin{bmatrix} 1 \\ 0 \\ 1 \end{bmatrix} = \begin{bmatrix} 2 \\ 1 \\ 0 \end{bmatrix}$.

- For data point 3, $w^T x = \begin{bmatrix} 2 & 1 & 0 \end{bmatrix} \begin{bmatrix} 1 \\ 1 \\ 1 \end{bmatrix} = 3$, so it is wrongly classified

  as class 1. Hence, the updated weight as per Perceptron

  rule is $w \rightarrow w - x = \begin{bmatrix} 2 \\ 1 \\ 0 \end{bmatrix} - \begin{bmatrix} 1 \\ 1 \\ 1 \end{bmatrix} = \begin{bmatrix} 1 \\ 0 \\ -1 \end{bmatrix}$.

- For data point 4, $w^T x = \begin{bmatrix} 1 & 0 & -1 \end{bmatrix} \begin{bmatrix} 1 \\ 0 \\ 0 \end{bmatrix} = 1$, so it is wrongly classified

  as class 1. Hence, the updated weight as per Perceptron

  rule is $w \rightarrow w - x = \begin{bmatrix} 1 \\ 0 \\ -1 \end{bmatrix} - \begin{bmatrix} 1 \\ 0 \\ 0 \end{bmatrix} = \begin{bmatrix} 0 \\ 0 \\ -1 \end{bmatrix}$.

The weight vector after the second pass is $[0\ 0\ -1]^T$, which is the same as the weight vector after the first pass. From the observations made during the first and second passes of Perceptron learning, it's clear that no matter how many passes we make over the data points, we will always end up with the weight vector $[0\ 0\ -1]^T$. As we saw earlier, this weight vector can only classify the negative class correctly, and so we can safely infer without loss of generality that the Perceptron algorithm will always fail to model the XOR logic.

Need for Nonlinearity

As we have seen, the Perceptron algorithm can only learn a linear decision boundary for classification and hence cannot solve problems where nonlinearity in the decision boundary is a need. Through the illustration of the XOR problem, we saw that the Perceptron is incapable of linearly separating the two classes properly.

We need to have two hyperplanes to separate the two classes, as illustrated in Figure 2-6, and the one hyperplane learned through the Perceptron algorithm does not suffice to provide the required classification. In Figure 2-6, the data points in between the two hyperplane lines belong to the positive class, while the other two data points belong to the negative class. Requiring two hyperplanes to separate two classes is the equivalent of having a nonlinear classifier.

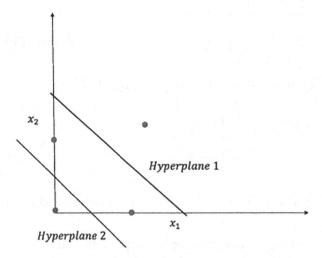

Figure 2-6. *XOR problem with two hyperplanes to separate the two classes*

A multi-layer Perceptron (MLP) can provide nonlinear separation between classes by introducing nonlinearity in the hidden layers. Do note that when a Perceptron outputs a 0 or 1 based on the total input received, the output is a nonlinear function of its input. Everything said and done while learning the weights of the multi-layer Perceptron is not possible to achieve via the Perceptron learning rule.

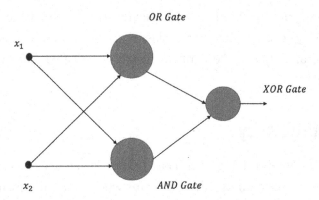

Figure 2-7. *XOR logic implementation with multi-layer Perceptron network*

In Figure 2-7, the XOR logic has been implemented through a multi-layer Perceptron network. If we have a hidden layer comprising two Perceptrons, with one capable of performing the OR logic while the other is capable of performing the AND logic, then the whole network would be able to implement the XOR logic. The Perceptrons for the OR and AND logic can be trained using the Perceptron learning rule. However, the network as a whole can't be trained through the Perceptron learning rule. If we look at the final input to the XOR gate, it will be a nonlinear function of its input to produce a nonlinear decision boundary.

Hidden-Layer Perceptrons' Activation Function for Nonlinearity

If we make the activation functions of the hidden layers linear, then the output of the final neuron would be linear, and hence we would not be able to learn any nonlinear decision boundaries. To illustrate this, let's try to implement the XOR function through hidden-layer units that have linear activation functions.

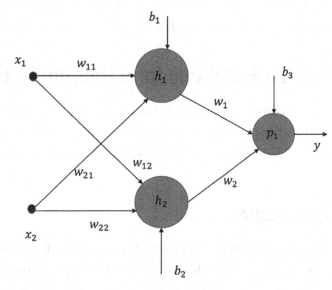

Figure 2-8. *Linear output hidden layers in a two-layer Perceptron network*

Figure 2-8 shows a two-layer Perceptron network with one hidden layer. The hidden layer consists of two neuron units. We look at the overall output of the network when the activations in the hidden units are linear:

Output of the hidden unit $h_1 = w_{11}x_1 + w_{21}x_2 + b_1$

Output of the hidden unit $h_2 = w_{12}x_1 + w_{22}x_2 + b_2$

Output of the output unit $p_1 = w_1(w_{11}x_1 + w_{21}x_2 + b_1) + w_2(w_{12}x_1 + w_{22}x_2 + b_2) + b_3$

$$= (w_1w_{11} + w_2w_{12})x_1 + (w_1w_{21} + w_2w_{22})x_2 + w_1b_1 + w_2b_2 + b_3$$

As deduced in the preceding lines, the final output of the network—i.e., the output of unit p_1— is a linear function of its inputs, and thus the network can't produce a nonlinear separation between classes.

If instead of a linear output being produced by the hidden layer we introduce an activation function represented as $f(x) = 1/(1 + e^{-x})$, then the output of the hidden unit $h_1 = 1/\left(1 + e^{-(w_{11}x_1 + w_{21}x_2 + b_1)}\right)$.

Similarly, the output of the hidden unit $h_2 = 1/\left(1 + e^{-(w_{12}x_1 + w_{22}x_2 + b_2)}\right)$.

The output of the output unit $p_1 = w_1/\left(1 + e^{-(w_{11}x_1 + w_{21}x_2 + b_1)}\right) + w_2/\left(1 + e^{-(w_{12}x_1 + w_{22}x_2 + b_2)}\right) + b_3$.

Clearly, the preceding output is nonlinear in its inputs and hence can learn more complex nonlinear decision boundaries rather than using a linear hyperplane for classification problems. The activation function for the hidden layers is called a sigmoid function, and we will discuss it in more detail in the subsequent sections.

Different Activation Functions for a Neuron/Perceptron

There are several activation functions for neural units, and their use varies with respect to the problem at hand and the topology of the neural network. In this section, we are going to discuss all the relevant activation functions that are used in today's artificial neural networks.

Linear Activation Function

In a linear neuron, the output is linearly dependent on its inputs. If the neuron receives three inputs x_1, x_2, and x_3, then the output y of the linear neuron is given by $y = w_1x_1 + w_2x_2 + w_3x_3 + b$, where w_1, w_2, and w_3 are the synaptic weights for the input x_1, x_2, and x_3, respectively, and b is the bias at the neuron unit.

In vector notation, we can express the output $y = w^T x + b$.

If we take $w^T x + b = z$, then the output with respect to the net input z will be as represented in Figure 2-9.

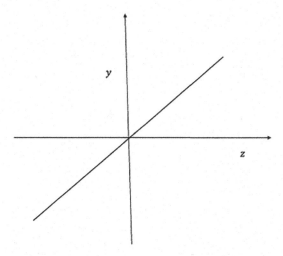

Figure 2-9. *Linear output hidden layers in a two-layer Perceptron network*

Binary Threshold Activation Function

In a binary threshold neuron (see Figure 2-10), if the net input to the neuron exceeds a specified threshold, then the neuron is activated; i.e., outputs 1 or else it outputs 0. If the net linear input to the neuron is $z = w^T x + b$ and k is the threshold beyond which the neuron activates, then

$$y = 1 \qquad if \ z > k$$

$$y = 0 \qquad if \ z \leq k$$

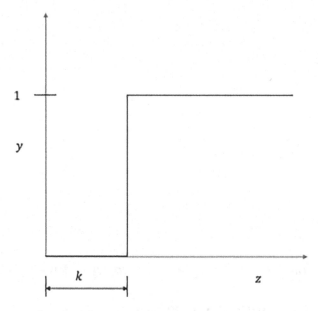

Figure 2-10. *Binary threshold neuron*

Generally, the binary threshold neuron is adjusted to activate at threshold 0 by adjusting the bias. The neuron is activated when $w^T x + b > k => w^T x + (b - k) > 0$.

Sigmoid Activation Function

The input–output relation for a sigmoid neuron is expressed as the following:

$$y = 1/\left(1 + e^{-z}\right)$$

where $z = w^T x + b$ is the net input to the *sigmoid* activation function.

- When the net input z to a sigmoid function is a positive large number $e^{-z} \blacktriangleright 0$ and so $y \blacktriangleright 1$.

- When the net input z to a sigmoid is a negative large number $e^{-z} \blacktriangleright \infty$ and so $y \blacktriangleright 0$.

- When the net input z to a sigmoid function is 0, then $e^{-z} = 1$ and so $y = \dfrac{1}{2}$.

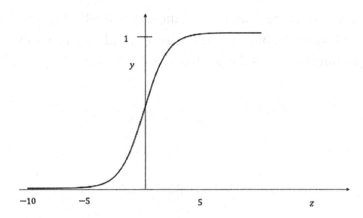

Figure 2-11. *Sigmoid activation function*

Figure 2-11 illustrates the input–output relationship of a sigmoid activation function. The output of a neuron that has a sigmoid activation function is very smooth and gives nice continuous derivatives, which works well when training a neural network. The output of a sigmoid activation function ranges between 0 and 1. Because of its capability to provide continuous values in the range of 0 to 1, the sigmoid function is generally used to output probability with respect to a given class for a binary classification. The sigmoid activation functions in the hidden layers introduce nonlinearity so that the model can learn more complex features.

SoftMax Activation Function

The SoftMax activation function is a generalization of the sigmoid function and is best suited for multiclass classification problems. If there are k output classes and the weight vector for the ith class is w(i), then the predicted probability for the ith class given the input vector $x \in R^{n \times 1}$ is given by the following:

$$P(y_i = 1 / x) = \frac{e^{w^{(i)T}x + b^{(i)}}}{\sum_{j=1}^{k} e^{w^{(j)T}x + b^{(j)}}}$$

where b(i) is the bias term for each output unit of the SoftMax.

Let's try to see the connection between a sigmoid function and a two-class SoftMax function.

Let's say the two classes are y1 and y2 and the corresponding weight vectors for them are w(1) and w(2). Also, let the biases for them be b(1) and b(2), respectively. Let's say the class corresponding to $y_1 = 1$ is the positive class.

$$P(y_1 = 1/x) = \frac{e^{w^{(1)T}x + b^{(1)}}}{e^{w^{(1)T}x + b^{(1)}} + e^{w^{(2)T}x + b^{(2)}}}$$

$$= \frac{1}{1 + e^{-\left(w^{(1)} - w^{(2)}\right)^T x - \left(b^{(1)} - b^{(2)}\right)}}$$

We can see from the preceding expression that the probability of the positive class for a two-class SoftMax has the same expression as that of a sigmoid activation function, the only difference being that in the sigmoid we only use one set of weights, while in the two-class SoftMax, there are two sets of weights. In sigmoid activation functions, we don't use different sets of weights for the two different classes, and the set of weight taken is generally the weight of the positive class with respect to the negative class. In SoftMax activation functions, we explicitly take different sets of weights for different classes.

The loss function for the SoftMax layer as represented by Figure 2-12 is called categorical cross-entropy and is given by the following:

$$C = \sum_{i=1}^{k} -y_i \log P(y_i = 1/x)$$

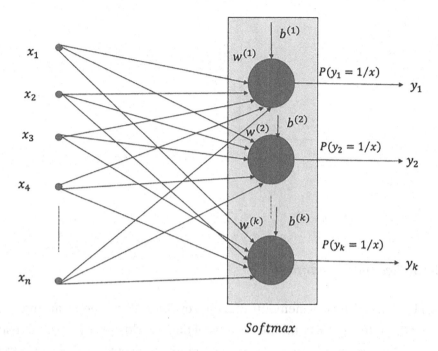

Softmax

Figure 2-12. *SoftMax activation function*

Rectified Linear Unit (ReLU) Activation Function

In a rectified linear unit, as shown in Figure 2-13, the output equals the net input to the neuron if the overall input is greater than 0; however, if the overall input is less than or equal to 0, the neuron outputs a 0.

The output for a *ReLU* unit can be represented as follows:

$$y = \max\left(0, w^T x + b\right)$$

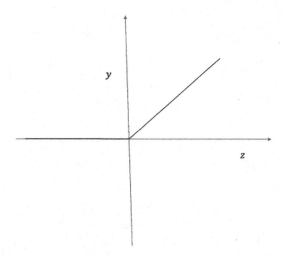

Figure 2-13. *Rectified linear unit*

The ReLU is one of the key elements that has revolutionized deep learning. They are easier to compute. ReLUs combine the best of both worlds—they have a constant gradient, while the net input is positive and zero gradient elsewhere. If we take the sigmoid activation function, for instance, the gradient of the same is almost zero for very large positive and negative values, and hence the neural network might suffer from a vanishing-gradient problem. This constant gradient for positive net input ensures the gradient-descent algorithm doesn't stop learning because of a vanishing gradient. At the same time, the zero output for a non-positive net input renders nonlinearity.

There are several versions of rectified linear unit activation functions such as parametric rectified linear unit (PReLU) and leaky rectified linear unit.

For a normal ReLU activation function, the output and the gradients are zero for non-positive input values, and so the training can stop because of the zero gradient. For the model to have a non-zero gradient even while the input is negative, PReLU can be useful. The input–output relationship for a PReLU activation function is given by the following:

$$y = \max(0,z) + \beta \min(0,z)$$

where $z = w^T x + b$ is the net input to the PReLU activation function and β is the parameter learned through training.

When β is set to -1, then $y = |z|$ and the activation function is called absolute value ReLU. When β is set to some small positive value, typically around 0.01, then the activation function is called leaky ReLU.

Tanh Activation Function

The input–output relationship for a tanh activation function (see Figure 2-14) is expressed as follows:

$$y = \frac{e^z - e^{-z}}{e^z + e^{-z}}$$

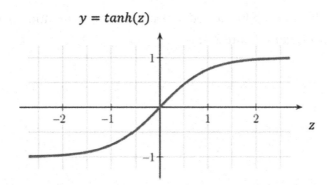

Figure 2-14. *Tanh activation function*

where $z = w^T x + b$ is the net input to the tanh activation function.

- When the net input z is a positive large number $e^{-z} \gg 0$ and so $y \gg 1$.
- When the net input z is a negative large number $e^z \gg 0$ and so $y \gg$ -1.
- When the net input z is 0, then $e^{-z} = 1$ and so $y = 0$.

As we can see, tanh activation functions can output values between -1 and +1.

The sigmoid activation function saturates at around output 0. While training the network, if the outputs in the layer are close to zero, the gradient vanishes and training stops. The tanh activation function saturates at -1 and + 1 values for the output and has well-defined gradients at around the 0 value of the output. So, with tanh activation functions such vanishing-gradient problems can be avoided around output 0.

SoftPlus Activation Function

The SoftPlus activation function on an input z is defined as follows:

$$f(z) = \log_e\left(1 + e^z\right)$$

It is not difficult to see that the gradient of the activation function with respect to the input is nothing but the sigmoid activation on the input as shown in the following:

$$f'(z) = \frac{1}{\left(1 + e^{-z}\right)}$$

The SoftPlus activation can be viewed as a smooth approximation of the ReLU function as can be seen from Figure 2-15.

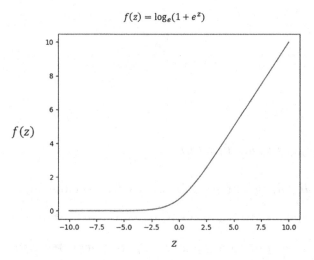

Figure 2-15. *SoftPlus activation function*

The SoftPlus function can lead to instability when z is close to zero or when z is too large and hence is not often used. A modified version of the SoftPlus as shown in the following helps avoid such issues.

$$f(z) = \max\left(0, z\right) + \log_e\left(1 + e^{-|x|}\right)$$

The SoftPlus function much like ReLU activation provides nonnegative outputs. For this reason, the SoftPlus activation is often used to output standard deviation in neural networks.

Swish Activation Function

The SoftPlus activation function on an input z is defined as the product of the input and the sigmoid activation on the input(see below):

$$f(z) = z * sigmoid(z) = \frac{z}{\left(1 + e^{-z}\right)}$$

One way Swish is different from the existing activation function is that Swish is non-monotonic in nature as can be seen from Figure 2-16.

$$f(z) = \frac{z}{1 + e^{-z}}$$

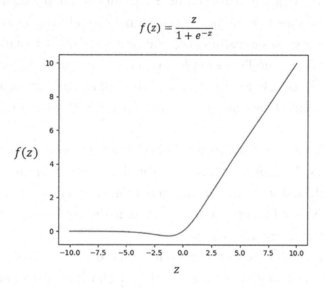

Figure 2-16. *Swish activation function*

One of the potential advantages of Swish is that it has non-zero gradient for negative values of z which ensures that we do not have the dead neuron problem we face with ReLU activation. The sigmoid in the Swish activation acts like a gate attenuating the linear activation z for z < 0 yet keeping the gradient alive.

Swish as a replacement for ReLU suggests an improvement in classification accuracy on ImageNet by 0.9% and around 0.6% for Inception ResNet v2.

Learning Rule for Multi-layer Perceptron Network

In an earlier section, we saw that the Perceptron learning rule can only learn linear decision boundaries. Nonlinear complex decision boundaries can be modeled through multi-layer Perceptrons; however, such a model cannot be learned through the Perceptron learning rule. Hence, one would need a different learning algorithm.

In the Perceptron learning rule, the goal is to keep updating the weights of the model till all the training data points have been correctly classified. If there is no such feasible weight vector that classifies all the points correctly, the algorithm does not converge. In such cases, the algorithm can be stopped by predefining the number of passes (iterations) to train or by defining a threshold for the number of correctly classified training data points after which to stop the training.

For multi-layer Perceptrons and for most of the deep-learning training networks, the best way to train the model is to compute a cost function based on the error of misclassification and then to minimize the cost function with respect to the parameters of the model. Since cost-based learning algorithms minimize the cost function, for binary classifications—generally the log-loss cost function—the negative of the log likelihood function is used. For reference, how the log-loss cost function is derived from maximum likelihood methods has been illustrated in Chapter 1 under "Logistic Regression."

A multi-layer Perceptron network would have hidden layers, and to learn nonlinear decision boundaries, the activation functions should be nonlinear themselves, such as sigmoid, ReLU, tanh, and so forth. The output neuron for binary classification should have a sigmoid activation function in order to cater to the log-loss cost function and to output probability values for the classes.

Now, with the preceding considerations, let us try to solve the XOR function by building a log-loss cost function and then minimizing it with respect to the weight and bias parameters of the model. All the neurons in the network are taken to have sigmoid activation functions.

Referring to Figure 2-7, let the input and output at hidden unit h_1 be i_1 and z_1, respectively. Similarly, let the input and output at hidden unit h_2 be i_2 and z_2, respectively. Finally, let the input and output at output layer p_1 be i_3 and z_3, respectively.

$$i_1 = w_{11}x_1 + w_{21}x_2 + b_1$$

$$i_2 = w_{12}x_1 + w_{22}x_2 + b_2$$

$$z_1 = 1/\left(1 + e^{-i_1}\right)$$

$$z_2 = 1/\left(1 + e^{-i_2}\right)$$

$$i_3 = w_1 z_1 + w_2 z_2 + b_3$$

$$z_3 = 1/\left(1 + e^{-i_3}\right)$$

Considering the log-loss cost function, the total cost function for the XOR problem can be defined as follows:

$$C = \sum_{i=1}^{4} -y^{(i)} log z_3^{(i)} - \left(1 - y^{(i)}\right) log\left(1 - z_3^{(i)}\right)$$

If all the weights and biases put together can be thought of as a parameter vector θ, we can learn the model by minimizing the cost function $C(\theta)$:

$$\theta^* = \underset{\theta}{Arg\ Min} C(\theta)$$

For minima, the gradient of the cost function $C(\theta)$ with respect to θ (i.e., $\nabla C(\theta)$) should be zero. The minima can be reached through gradient-descent methods. The update rule for gradient descent is $\theta^{(t+1)} = \theta^{(t)} - \eta \nabla C(\theta^{(t)})$, where η is the learning rate and $\theta^{(t+1)}$ and $\theta^{(t)}$ are the parameter vectors at iterations $t + 1$ and t, respectively.

If we consider individual weight within the parameter vector, the gradient-descent update rule becomes the following:

$$w_k^{(t+1)} = w_k^{(t)} - \eta \frac{\partial C\left(w_k^{(t)}\right)}{\partial w_k} \quad \forall w_k \in \theta$$

The gradient vector would not be as easy to compute as it would be in linear or logistic regression since in neural networks the weights follow a hierarchical order. However, the chain rule of derivatives provides for some simplification to methodically compute the partial derivatives with respect to the weights (including biases).

The method is called *backpropagation*, and it provides simplification in gradient computation.

Backpropagation for Gradient Computation

Backpropagation is a useful method of propagating the error at the output layer backward so that the gradients at the preceding layers can be computed easily using the chain rule of derivatives.

Let us consider one training example and work through the backpropagation, taking the XOR network structure into account (see Figure 2-8). Let the input be $x = [x_1\ x_2]^T$ and the corresponding class be y. So, the cost function for the single record becomes the following:

$$C = -y log z_3 - (1-y) log (1-z_3)$$

$$\frac{\partial C}{\partial w_1} = \frac{dC}{dz_3} \frac{dz_3}{di_3} \frac{\partial i_3}{\partial w_1}$$

$$\frac{dC}{dz_3} = \frac{(z_3 - y)}{z_3 (1 - z_3)}$$

$$z_3 = 1/(1 + e^{-z_3})$$

$$\frac{dz_3}{di_3} = z_3 (1 - z_3)$$

$$\frac{dC}{di_3} = \frac{dC}{dz_3} \frac{dz_3}{di_3} = \frac{(z_3 - y)}{z_3 (1 - z_3)} z_3 (1 - z_3) = (z_3 - y)$$

As we can see, the derivative of the cost function with respect to the net input in the final layer is nothing but the error in estimating the output $(z_3 - y)$:

$$\frac{\partial i_3}{\partial w_1} = z_1$$

$$\frac{\partial C}{\partial w_1} = \frac{dC}{dz_3} \frac{dz_3}{di_3} \frac{\partial i_3}{\partial w_1} = (z_3 - y) z_1$$

Similarly,

$$\frac{\partial C}{\partial w_2} = \frac{dC}{dz_3}\frac{dz_3}{di_3}\frac{\partial i_3}{\partial w_2} = (z_3 - y)z_2$$

$$\frac{\partial C}{\partial b_3} = \frac{dC}{dz_3}\frac{dz_3}{di_3}\frac{\partial i_3}{\partial b_3} = (z_3 - y)$$

Now, let us compute the partial derivatives of the cost function with respect to the weights in the previous layer:

$$\frac{\partial C}{\partial z_1} = \frac{dC}{dz_3}\frac{dz_3}{di_3}\frac{\partial i_3}{\partial z_1} = (z_3 - y)w_1$$

$\frac{\partial C}{\partial z_1}$ can be treated as the error with respect to the output of the hidden-layer unit h_1. The error is propagated in proportion to the weight that is joining the output unit to the hidden-layer unit. If there were multiple output units, then $\frac{\partial C}{\partial z_1}$ would have a contribution from each of the output units. We will see this in detail in the next section.

Similarly,

$$\frac{\partial C}{\partial i_1} = \frac{dC}{dz_3}\frac{dz_3}{di_3}\frac{\partial i_3}{\partial z_1}\frac{dz_1}{di_1} = (z_3 - y)w_1 z_1 (1 - z_1)$$

$\frac{\partial C}{\partial i_1}$ can be considered the error with respect to the net input of the hidden-layer unit h_1. It can be computed by just multiplying the $z_1(1 - z_1)$ factor by $\frac{\partial C}{\partial z_1}$:

$$\frac{\partial C}{\partial w_{11}} = \frac{dC}{dz_3}\frac{dz_3}{di_3}\frac{\partial i_3}{\partial z_1}\frac{dz_1}{di_1}\frac{\partial i_1}{\partial w_{11}} = (z_3 - y)w_1 z_1 (1 - z_1)x_1$$

$$\frac{\partial C}{\partial w_{21}} = \frac{dC}{dz_3}\frac{dz_3}{di_3}\frac{\partial i_3}{\partial z_1}\frac{dz_1}{di_1}\frac{\partial i_1}{\partial w_{21}} = (z_3 - y)w_1 z_1 (1 - z_1)x_2$$

$$\frac{\partial C}{\partial b_1} = \frac{dC}{dz_3}\frac{dz_3}{di_3}\frac{\partial i_3}{\partial z_1}\frac{dz_1}{di_1}\frac{\partial i_1}{\partial w_{21}} = (z_3 - y)w_1 z_1 (1 - z_1)$$

Once we have the partial derivative of the cost function with respect to the input in each neuron unit, we can compute the partial derivative of the cost function with respect to the weight contributing to the input—we just need to multiply the input coming through that weight.

Generalizing the Backpropagation Method for Gradient Computation

In this section, we try to generalize the backpropagation method through a more complicated network. We assume the final output layer is composed of three independent sigmoid output units, as depicted in Figure 2-17. Also, we assume that the network has a single record for ease of notation and for simplification in learning.

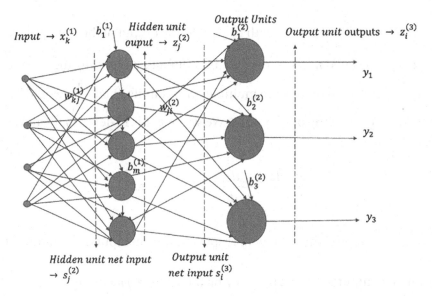

Figure 2-17. *Network to illustrate backpropagation for independent sigmoid output layers*

The cost function for a single input record is given by the following:

$$C = \sum_{i=1}^{3} -y_i \log P\big(y_i = 1\big) - \big(1 - y_i\big) \log\big(1 - P\big(y_i = 1\big)\big)$$

$$= \sum_{i=1}^{3} -y_i \log z_i^{(3)} - \big(1 - y_i\big) \log\big(1 - z_i^{(3)}\big)$$

In the preceding expression, $y_i \in \{0, 1\}$, depending on whether the event specific to y_i is active or not.

$P(y_i = 1) = z_i^{(3)}$ denotes the predicted probability of the *ith* class.

Let us compute the partial derivative of the cost function with respect to the weight $w_{ji}^{(2)}$. The weight would only impact the output of the *ith* output unit of the network.

$$\frac{\partial C}{\partial w_{ji}^{(2)}} = \frac{\partial C}{\partial z_i^{(3)}} \frac{\partial z_i^{(3)}}{\partial s_i^{(3)}} \frac{\partial s_i^{(3)}}{\partial w_{ji}^{(2)}}$$

$$\frac{\partial C}{\partial z_i^{(3)}} = \frac{\left(z_i^{(3)} - y_i\right)}{z_i^{(3)}\left(1 - z_i^{(3)}\right)}$$

$$P(y_i = 1) = z_i^{(3)} = 1 / \left(1 + e^{-s_i^{(3)}}\right)$$

$$\frac{\partial z_i^{(3)}}{\partial s_i^{(3)}} = z_i^{(3)}\left(1 - z_i^{(3)}\right)$$

$$\frac{\partial C}{\partial s_i^{(3)}} = \frac{\partial C}{\partial z_i^{(3)}} \frac{\partial z_i^{(3)}}{\partial s_i^{(3)}} = \frac{\left(z_i^{(3)} - y_i\right)}{z_i^{(3)}\left(1 - z_i^{(3)}\right)} z_i^{(3)}\left(1 - z_i^{(3)}\right) = \left(z_i^{(3)} - y_i\right)$$

So, as before, the partial derivative of the cost function with respect to the net input for the *ith* output unit is $\left(z_i^{(3)} - y_i\right)$, which is nothing but the error in prediction at the *ith* output unit.

$$\frac{\partial s_i^{(3)}}{\partial w_{ji}^{(2)}} = z_j^{(2)}$$

Combining $\dfrac{\partial C}{\partial s_i^{(3)}}$ and $\dfrac{\partial s_i^{(3)}}{\partial w_{ji}^{(2)}}$, we get the following:

$$\frac{\partial C}{\partial w_{ji}^{(2)}} = \left(z_i^{(3)} - y_i\right) z_j^{(2)}$$

$$\frac{\partial C}{\partial b_i^{(2)}} = \left(z_i^{(3)} - y_i\right)$$

The preceding gives a generalized expression for partial derivatives of the cost function with respect to weights and biases in the last layer of the network. Next, let us compute the partial derivative of the weights and biases in the lower layers. Things get a little more involved but still follow a generalized trend. Let us compute the partial derivative of the cost function with respect to the weight $w_{kj}^{(1)}$. The weight would be impacted by the errors at all three output units. Basically, the error at the output of the *jth* unit in the hidden layer would have an error contribution from all output units, scaled by the weights connecting the output layers to the *jth* hidden unit. Let us compute the partial derivative just by the chain rule and see if it lives up to what we have proclaimed:

$$\frac{\partial C}{\partial w_{kj}^{(1)}} = \frac{\partial C}{\partial z_j^{(2)}} \frac{\partial z_j^{(2)}}{\partial s_j^{(2)}} \frac{\partial s_j^{(2)}}{\partial w_{kj}^{(1)}}$$

$$\frac{\partial s_j^{(2)}}{\partial w_{kj}^{(1)}} = z_k^{(1)}$$

$$\frac{\partial z_j^{(2)}}{\partial s_j^{(2)}} = z_j^{(2)}\left(1 - z_j^{(2)}\right)$$

Now, $\frac{\partial C}{\partial z_j^{(2)}}$ is the tricky computation since $z_j^{(2)}$ influences all three output units:

$$\frac{\partial C}{\partial z_j^{(2)}} = \sum_{i=1}^{3} \frac{\partial C}{\partial z_i^{(3)}} \frac{\partial z_i^{(3)}}{\partial s_i^{(3)}} \frac{\partial s_i^{(3)}}{\partial z_j^{(2)}}$$

$$= \sum_{i=1}^{3} \left(z_i^{(3)} - y_i\right)w_{ji}^{(2)}$$

Combining the expressions for $\frac{\partial s_j^{(2)}}{\partial w_{kj}^{(1)}}, \frac{\partial z_j^{(2)}}{\partial s_j^{(2)}}$ and $\frac{\partial C}{\partial z_j^{(2)}}$, we have the following:

$$\frac{\partial C}{\partial w_{kj}^{(1)}} = \sum_{i=1}^{3} \left(z_i^{(3)} - y_i\right)w_{ji}^{(2)}z_j^{(2)}\left(1 - z_j^{(2)}\right)x_k^{(1)}$$

In general, for a multi-layer neural network to compute the partial derivative of the cost function C with respect to a specific weight w contributing to the net input s in a neuron unit, we need to compute the partial derivative of the cost function with respect to the net input (i.e., $\frac{\partial C}{\partial s}$) and then multiply the input x associated with the weight w, as follows:

$$\frac{\partial C}{\partial w} = \frac{\partial C}{\partial s}\frac{\partial s}{\partial w} = \frac{\partial C}{\partial s}x$$

$\frac{\partial C}{\partial s}$ can be thought of as the error at the neural unit and can be computed iteratively by passing the error at the output layer to the neural units in the lower layers. Another point to note is that an error in a higher-layer neural unit is distributed to the output of the preceding layer's neural units in proportion to the weight connections between them. Also, the partial derivative of the cost function with respect to the net input to a sigmoid activation neuron $\frac{\partial C}{\partial s}$ can be computed from the partial derivative of the cost function with respect to the output z of a neuron (i.e., $\frac{\partial C}{\partial z}$) by multiplying $\frac{\partial C}{\partial z}$ by $z(1 - z)$. For linear neurons, this multiplying factor becomes 1.

All these properties of neural networks make computing gradients easy. This is how a neural network learns in each iteration through backpropagation.

Each iteration is composed of a forward pass and a backward pass, or backpropagation. In the forward pass, the net input and output at each neuron unit in each layer are computed. Based on the predicted output and the actual target values, the error is computed in the output layers. The error is backpropagated by combining it with the neuron outputs computed in the forward pass and with existing weights. Through backpropagation, the gradients get computed iteratively. Once the gradients are computed, the weights are updated by gradient-descent methods.

Please note that the deductions shown are valid for sigmoid activation functions. For other activation functions, while the approach remains the same, changes specific to the activation functions are required in the implementation.

The cost function for a SoftMax function is different from that for the independent multiclass classification.

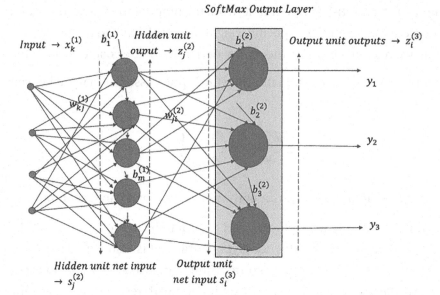

Figure 2-18. *Network to illustrate backpropagation for SoftMax output layer*

The cross-entropy cost for the SoftMax activation layer in the network represented in Figure 2-18 is given by the following:

$$C = \sum_{i=1}^{3} -y_i \log P(y_i = 1) = \sum_{i=1}^{3} -y_i \log z_i^{(3)}$$

Let us compute the partial derivative of the cost function with respect to the weight $w_{ji}^{(2)}$. Now, the weight would impact the net input $s_i^{(3)}$ to the *ith* SoftMax unit. However, unlike the independent binary activations in the earlier network, here all three SoftMax output units $z_k^{(3)} \forall k \in \{1,2,3\}$ would be influenced by $s_i^{(3)}$ since

$$z_k^{(3)} = \frac{e^{s_k^{(3)}}}{\sum_{l=1}^{3} e^{s_l^{(3)}}} = \frac{e^{s_k^{(3)}}}{\sum_{l \neq i} e^{s_l^{(3)}} + e^{s_i^{(3)}}}$$

Hence, the derivative $\dfrac{\partial C}{\partial w_{ji}^{(2)}}$ can be written as follows:

$$\frac{\partial C}{\partial w_{ji}^{(2)}} = \frac{\partial C}{\partial s_i^{(3)}} \frac{\partial s_i^{(3)}}{\partial w_{ji}^{(2)}}$$

Now, as just stated, since $s_i^{(3)}$ influences all the outputs $z_k^{(3)}$ in the SoftMax layer,

$$\frac{\partial C}{\partial s_i^{(3)}} = \sum_{k=1}^{3} \frac{\partial C}{\partial z_k^{(3)}} \frac{\partial z_k^{(3)}}{\partial s_i^{(3)}}$$

The individual components of the partial derivative are as follows:

$$\frac{\partial C}{\partial z_k^{(3)}} = \frac{-y_k}{z_k^{(3)}}$$

For $k = i$, $\dfrac{\partial z_k^{(3)}}{\partial s_i^{(3)}} = z_i^{(3)} \left(1 - z_i^{(3)} \right)$

For $k \neq i$, $\dfrac{\partial z_k^{(3)}}{\partial s_i^{(3)}} = -z_i^{(3)} z_k^{(3)}$

$$\frac{\partial s_i^{(3)}}{\partial w_{ji}^{(2)}} = z_j^{(2)}$$

$$\frac{\partial C}{\partial s_i^{(3)}} = \sum_{k=1}^{3} \frac{\partial C}{\partial z_k^{(3)}} \frac{\partial z_k^{(3)}}{\partial s_i^{(3)}} = \sum_{k=i} \frac{\partial C}{\partial z_k^{(3)}} \frac{\partial z_k^{(3)}}{\partial s_i^{(3)}} + \sum_{k \neq i} \frac{\partial C}{\partial z_k^{(3)}} \frac{\partial z_k^{(3)}}{\partial s_i^{(3)}}$$

$$= \frac{-y_i}{z_i^{(3)}} z_i^{(3)} \left(1 - z_i^{(3)} \right) + \sum_{k \neq i} \frac{-y_k}{z_k^{(3)}} \left(-z_i^{(3)} z_k^{(3)} \right)$$

$$= -y_i \left(1 - z_i^{(3)} \right) + z_i^{(3)} \sum_{k \neq i} y_k$$

$$= -y_i + y_i z_i^{(3)} + z_i^{(3)} \sum_{k \neq i} y_k$$

$$= -y_i + z_i^{(3)} \sum_{k} y_k$$

143

Since $y_k = 1$ for only one value of k, hence $\sum_k y_k = 1$. Hence,

$$\frac{\partial C}{\partial s_i^{(3)}} = -y_i + z_i^{(3)}$$

$$= z_i^{(3)} - y_i$$

As it turns out, the cost derivative with respect to the net input to the *ith* SoftMax unit is the error in predicting the output at the *ith* SoftMax output unit. Combining $\frac{\partial C}{\partial s_i^{(3)}}$ and $\frac{\partial s_i^{(3)}}{\partial w_{ji}^{(2)}}$, we get the following:

$$\frac{\partial C}{\partial w_{ji}^{(2)}} = \frac{\partial C}{\partial s_i^{(3)}} \frac{\partial s_i^{(3)}}{\partial w_{ji}^{(2)}} = \left(z_i^{(3)} - y_i\right) z_j^{(2)}$$

Similarly, for the bias term to the *ith* SoftMax output unit, we have the following:

$$\frac{\partial C}{\partial b_i^{(2)}} = \left(z_i^{(3)} - y_i\right)$$

Computing the partial derivative of the cost function with respect to the weight $w_{kj}^{(1)}$ in the previous layer, i.e., $\frac{\partial C}{\partial w_{kj}^{(1)}}$, would have the same form as that in the case of a network with independent binary classes. This is obvious since the networks only differ in terms of the output units' activation functions, and even then, the expressions that we get for $\frac{\partial C}{\partial s_i^{(3)}}$ and $\frac{\partial s_i^{(3)}}{\partial w_{ji}^{(2)}}$ remain the same. As an exercise, interested readers can verify whether $\frac{\partial C}{\partial w_{kj}^{(1)}} = \sum_{i=1}^{3}\left(z_i^{(3)} - y_i\right) w_{ji}^{(2)} z_j^{(2)}\left(1 - z_j^{(2)}\right) x_k^{(1)}$ still holds true.

Deep Learning vs. Traditional Methods

In this book, we will use TensorFlow from Google as the deep-learning library since it has several advantages. Before moving on to TensorFlow, let us look at some of the key advantages of deep learning and a few of its shortcomings if it is not used in the right place.

- Deep learning outperforms traditional machine-learning methods by a huge margin in several domains, especially in the fields of computer vision, speech recognition, natural language processing, and time series.

- With deep learning, more and more complex features can be learned as the layers in the deep-learning neural network increase. Because of this automatic feature-learning property, deep learning reduces the feature-engineering time, which is a time-consuming activity in traditional machine-learning approaches.

- Deep learning works best for unstructured data, and there is a plethora of unstructured data in the form of images, text, speech, sensor data, and so forth, which when analyzed would revolutionize different domains, such as healthcare, manufacturing, banking, aviation, e-commerce, and so on.

A few limitations of deep learning are as follows:

- Deep-learning networks generally tend to have a lot of parameters, and for such implementations, there should be a sufficiently large volume of data to train. If there are not enough data, deep-learning approaches will not work well since the model will suffer from overfitting.

- The complex features learned by the deep-learning network are often hard to interpret.

- Deep-learning networks require a lot of computational power to train because of the large number of weights in the model as well as the data volume.

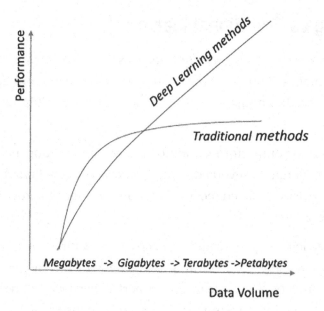

Figure 2-19. *Performance comparison of traditional methods vs. deep-learning methods*

When the data volume is less, traditional methods tend to perform better that deep-learning ones. However, when the data volume is huge, the deep-learning method wins over traditional methods by a huge margin, which has been roughly depicted in Figure 2-19.

TensorFlow

TensorFlow from Google is an open source library that primarily focuses on deep learning. It uses computational data-flow graphs to represent complicated neural network architecture. The nodes in the graph denote mathematical computations, also called ops (operations), whereas the edges denote the data tensors transferred between them. Also, the relevant gradients are stored at each node of the computational graph, and during backpropagation, these are combined to get the gradients with respect to each weight. Tensors are multidimensional data arrays used by TensorFlow.

Common Deep-Learning Packages

The common deep-learning packages are as follows:

- *Pytorch*: Pytorch is a deep-learning open source framework from Facebook based on Torch Library. It was primarily built with the intention of expediting research prototyping to deployment. Pytorch uses a C++ frontend with Python interface. Reputed organizations such as Facebook AI Research and IBM use Pytorch. Pytorch can utilize GPU for fast computation.

- *Theano*: Theano is a deep-learning package in Python that is primarily used for computationally intensive research-oriented activities. It is tightly integrated with Numpy array and has efficient symbolic differentiators. It also provides transparent use of GPU for much faster computation.

- *Caffe*: Caffe is a deep-learning framework developed by Berkeley AI Research (BAIR). Speed makes Caffe perfect for research experiments and industry deployment. Caffe implementation can use GPU very efficiently.

- *CuDNN*: CuDNN stands for CUDA Deep Neural Network library. It provides a library of primitives for GPU implementation of deep neural networks.

- *TensorFlow*: TensorFlow is an open source deep-learning framework from Google inspired by Theano. TensorFlow is slowly becoming the preferred library for deep learning in research-oriented work as well as for production implementation. Also, for distributed production implementation over the cloud, TensorFlow is becoming the go-to library.

- *MxNet*: MxNet is an open source deep-learning framework that can scale to multiple GPUs and machines and is supported by major cloud providers such as AWS and Azure. Popular machine-learning library GraphLab has good deep-learning implementation using MxNet.

- ***deeplearning4j***: deeplearning4j is an open source distributed deep-learning framework for Java virtual machines.

- ***Sonnet***: Sonnet is a high-level library for building complex neural network architectures using TensorFlow as the backend. Sonnet was developed at DeepMind.

- ***ONNX***: ONNX stands for Open Neural Network Exchange and is jointly developed by Microsoft and Facebook. ONNX as the name suggests helps in the seamless transfer of models from one framework to another. For example, a model trained on Pytorch can be seamless setup for inference in TensorFlow using ONNX.

TensorFlow and Pytorch are the most popular and widely used deep-learning frameworks across academia and industry. The following are some of the salient features of the two frameworks.

- Python is the high-level language of choice for TensorFlow and Pytorch. Both TensorFlow and Pytorch provide the right amount of abstractions to speed up model development.

- TensorFlow much like Theano, MxNet, and Caffe uses automatic differentiators, while Torch uses AutoGrad. Automatic differentiators are different from symbolic differentiation and numeric differentiation. Automatic differentiators are very efficient when used in neural networks because of the backpropagation method of learning that utilizes the chain rule of differentiation.

- For production implementation on the cloud, TensorFlow is on its way to becoming the go-to platform for applications targeting large distributed systems.

- TensorFlow has better visualization capabilities in the form of TensorBoard which enables one to track and debug training issues much more conveniently.

TensorFlow Installation

TensorFlow can be installed with ease in Linux-, Mac OS–, and Windows-based machines. It is always preferable to create separate environments for TensorFlow. In this book, we will be using the capabilities of TensorFlow 2 which requires your Python version to be greater than or equal to 3.7. The details of installation for TensorFlow 2 are documented well on the official website for TensorFlow: `www.tensorflow.org/install`.

TensorFlow Basics for Development

TensorFlow has its own format of commands to define and manipulate tensors. TensorFlow 1 worked on the principle of executing the logic using computational graphs within activated sessions. However, TensorFlow 2 looks at eager execution as its default. We discuss on both eager execution and computation graph approaches in detail in the following and then compare their salient features.

Eager Execution

Eager execution provides an execution environment to execute operations immediately. It does not work on the principle of building computation graphs to be executed later. Eager execution simplifies the model development effort in TensorFlow as one can see the results of the operations immediately. Eager execution makes debugging easier and is a flexible tool for research and fast prototyping.

Graph-Based Execution

Although Eager execution has several advantages as discussed in the earlier section, it is in general slower than the graph execution. Eager execution cannot take advantage of acceleration opportunities that exist as it runs the TensorFlow operations one by one. In contrast to eager execution, graph execution extracts the TensorFlow-specific computation operations from Python and builds an efficient graph that can take advantage of potential acceleration opportunities. Also, graphs provide cross-platform flexibility as the graphs can be run, stored, and restored without the need for the original Python code. Hence, graph-based execution turns out to be important for models in devices such as mobile where Python code might not be available.

The following are some of the key features of eager execution and graph-based execution.

Eager execution	Graph-based execution
Highly intuitive and easy to debug Simplifies rapid model development	Less intuitive and in general hard to debug than eager execution
Slower than graph-based execution as it executes the TensorFlow operations one by one	Generally faster than eager execution as graph can be built to take advantage of acceleration opportunities prior to execution
Better for beginners.	Ideal for large-scale training
Support for GPU and TPU acceleration	Support for GPU and TPU acceleration

The Best Strategy for TensorFlow 2: Eager Execution vs. Graph Execution

Although TensorFlow 2 has prioritized eager execution, one can build models in an eager execution fashion and then execute it using graph execution. One can wrap eager execution operations with just **tf.function()** to get the benefits of the graph-based execution without having to create a graph and then run the graph within a session through **session.run().** Use this mixed scheme in a way we are adopting the intuitive way of coding as in eager execution but eventually executing the code in a graph-based way with minimal changes to the eager execution code by wrapping it around with **tf. function().**

Listings 2-1 to 2-15 are a few of the basic TensorFlow commands used to define tensors and TensorFlow variables and to execute TensorFlow computational graphs within sessions. The idea is to emphasize both eager execution and the graph-based execution methods as part of the following code listings.

Listing 2-1. Import TensorFlow and Numpy Library and to Check if Eager Execution Is Active by Default

```
from platform import python_version
import tensorflow as tf
import numpy as np
import os
```

```
print(f"Python version: {python_version()}")
print(f"Tensorflow version: {tf.__version__}")
print("Eager executive active:", tf.executing_eagerly())

-- output --
Python version: 3.9.5
Tensorflow version: 2.4.1
"Eager executive active:", True
```

Listing 2-2. Defining Zeros and Ones Tensors

```
a = tf.zeros((2,2))
print('a:',a)
b = tf.ones((2,2))
print('b:',b)

-- output --
a: tf.Tensor(
[[0. 0.]
 [0. 0.]], shape=(2, 2), dtype=float32)
b: tf.Tensor(
[[1. 1.]
 [1. 1.]], shape=(2, 2), dtype=float32)
```

True

With eager execution as the default, we will be able to see the values of the tensor immediately.

Listing 2-3. Sum the Elements of the Matrix (2D Tensor) Across the Horizontal Axis

```
out = tf.math.reduce_sum(b,axis=1)
print(out)

-- output --
tf.Tensor([2. 2.], shape=(2,), dtype=float32)
```

Listing 2-4. Check the Shape of the Tensor

```
a.get_shape()
```

```
-- output --
tf.Tensor([2. 2.], shape=(2,), dtype=float32)
```

Listing 2-5. Reshaping a Tensor

```
a_ = tf.reshape(a,(1,4))
print(a_)
```

```
-- output -
tf.Tensor([[0. 0. 0. 0.]], shape=(1, 4), dtype=float32)
```

Listing 2-6. Convert a Tensor to Numpy

```
ta = a.numpy()
print(ta)
```

```
--output–
[[0. 0.]
 [0. 0.]]
```

Listing 2-7. Define TensorFlow Constants

```
# Tensorflow constants are immutable
a = tf.constant(2)
b = tf.constant(5)
c= a*b
print(c)
--output–

tf.Tensor([10], shape=(1,), dtype=int32)
```

Listing 2-8. Illustration of Between Eager Execution and Graph-Based Execution

```
import timeit
# Eager function
def func_eager(a,b):
    return a*b
# Graph function using tf.function on eager func
@tf.function
def graph_func(a,b):
    return a*b

a = tf.constant([2])
b = tf.constant([5])

# Eager execution
print("Eager execution:",func_eager(a,b))
# Function with graph execution
print("Graph execution:",graph_func(a,b))

--output--

Eager execution: tf.Tensor([10], shape=(1,), dtype=int32)
Graph execution: tf.Tensor([10], shape=(1,), dtype=int32)
```

Listing 2-9. Execution Time Comparison of Eager Execution vs. Graph Execution in Simple Operation

```
import timeit
# Eager function
def func_eager(a,b):
    return a*b
# Graph function using tf.function on eager func
@tf.function
def graph_func(a,b):
    return a*b

a = tf.constant([2])
b = tf.constant([5])
```

```
# Eager execution
print("Eager execution:",timeit.timeit(lambda:func_eager(a,b),number=100))
# Function with graph execution
print("Graph execution:",timeit.timeit(lambda: graph_func(a,b),number=100))
print("For simple operations Graph execution takes more time..")
```

--output--

```
Eager execution: 0.0020395979954628274
Graph execution: 0.038001397988409735
```

From Listing 2-9, we can observe that for simple operations, graph execution takes more time than eager execution. Next we will see how the inference time looks when the model is heavy in parameters.

Listing 2-10. Execution Time Comparison of Eager Execution vs. Graph Execution in a Model Inference

```
# TensorFlow imports
from tensorflow.keras import Input, Model
from tensorflow.keras.layers import Flatten, Dense

# Define the model (Inspired by mnist inputs)
model = tf.keras.Sequential()
model.add(tf.keras.Input(shape=(28,28,)))
model.add(Flatten())
model.add(Dense(256,"relu"))
model.add(Dense(128,"relu"))
model.add(Dense(256,"relu"))
model.add(Dense(10,"softmax"))
# Dummy data with MNIST image sizes
X = tf.random.uniform([1000, 28, 28])

# Eager Execution to do inference (Model untrained as we are evaluating
speed of inference)
eager_model = model
print("Eager time:", timeit.timeit(lambda: eager_model(X,training=False),
number=10000))
```

```
#Graph Execution to do inference (Model untrained as we are evaluating
speed of inference)
graph_model = tf.function(eager_model) # Wrap the model with tf.function
print("Graph time:", timeit.timeit(lambda: graph_model(X,training=False),
number=10000))
--output-

Eager time: 7.980951177989482
Graph time: 1.995524710000609
```

We can see from Listing 2-10 that in case of a model with several parameters, graph execution method scores over eager execution.

Listing 2-11. Defining TensorFlow Variables

```
w = tf.Variable([5.,10])

print('Intial value of Variable w =', w.numpy())
w.assign([2.,2.])
print('New assigned value of Variable w =', w.numpy())

--output-
Intial value of Variable w = [ 5. 10.]
New assigned value of Variable w = [2. 2.]
```

TensorFlow variable method tf.Variable produces mutable tensors which can only be changed using assign.

Listing 2-12. Converting a Numpy Array to Tensor

```
nt = np.random.randn(5,3)
nt_tensor = tf.convert_to_tensor(nt)
print(nt_tensor)
--output-
tf.Tensor(
[[ 1.21409834  0.17042089 -0.3132248 ]
 [ 0.58964541  0.42423984  1.00614624]
 [ 0.9511394   1.80499692  0.36418302]
 [ 0.93088843  0.68589623  1.43379157]
 [-1.5732957  -0.06314358  1.36723688]], shape=(5, 3), dtype=float64)
```

Listing 2-13. Computing Gradient

```
x = tf.Variable(2.0)

with tf.GradientTape() as tape:
    y = x**3

dy_dx = tape.gradient(y,x) # Compute gradient of y wrt to x at x =2.

print(dy_dx.numpy()) # dy/dx = ( 3(x^2) at x = 2 ) == 3*(2^2) = 12.0

--output–
12.0
```

TensorFlow needs to remember the order of operation in forward pass so that during the backward pass (backpropagation), it can traverse the list of operations in reverse order to compute gradients.

tf.GradientTape() is precisely the method that provides a way of recording relevant operations executed within its scope so that the information can be used to compute the gradient.

Listing 2-14. Gradient with Respect to Model

```
# TensorFlow imports
from tensorflow.keras import Input, Model
from tensorflow.keras.layers import Flatten, Dense

# Define the model
model = tf.keras.Sequential()
model.add(tf.keras.Input(shape=(2,)))
model.add(Dense(5,'relu'))
model.add(Dense(1))
print(model.summary())

X = tf.constant([[2,2],[1,1]])
y = tf.constant([3.4,4.7])
with tf.GradientTape() as tape:
    # Forward pass
    y_hat = model(X,training=True)
    loss = tf.reduce_mean((y- y_hat)**2) # Made up loss

grad_ = tape.gradient(loss,model.trainable_variables)
```

```
# Print the gradient tensors shape in each layer
for var, g in zip(model.trainable_variables, grad_):
  print(f'{var.name}, shape: {g.shape}')
```

-- output -

Layer (type)	Output Shape	Param #
dense_31 (Dense)	(None, 5)	15
dense_32 (Dense)	(None, 1)	6

```
Total params: 21
Trainable params: 21
Non-trainable params: 0
```

```
None
dense_31/kernel:0, shape: (2, 5)
dense_31/bias:0, shape: (5,)
dense_32/kernel:0, shape: (5, 1)
dense_32/bias:0, shape: (1,)
```

We can see from the output that the gradient of loss with respect to each of the units in each layer is stored by TensorFlow. TensorFlow is able to compute the same because it is tracking the variables and parameters of interest in the forward pass using the tf.GradientTape() method.

Gradient-Descent Optimization Methods from a Deep-Learning Perspective

Before we dive into the TensorFlow optimizers, it is important to understand a few key points regarding full-batch gradient descent and stochastic gradient descent, including their shortcomings, so that one can appreciate the need to come up with variants of these gradient-based optimizers.

Elliptical Contours

The cost function for a linear neuron with a least square error is quadratic. When the cost function is quadratic, the direction of the gradient resulting from the full-batch gradient-descent method gives the best direction for cost reduction in a linear sense, but it does not point to the minimum unless the different elliptical contours of the cost function are circles. In cases of long elliptical contours, the gradient components might be large in directions where less change is required and small in directions where more change is required to move to the minimum point.

As we can see in Figure 2-20, the gradient at S does not point to the direction of the minimum, i.e., point M. The problem with this condition is that if we take small steps by making the learning rate small, then the gradient descent would take a while to converge, whereas if we were to use a big learning rate, the gradients will change direction rapidly in directions where the cost function had curvature, leading to oscillations. The cost function for a multi-layer neural network is not quadratic but rather is mostly a smooth function. Locally, such non-quadratic cost functions can be approximated by quadratic functions, and so the problems of gradient descent inherent to elliptical contours still prevail for non-quadratic cost functions.

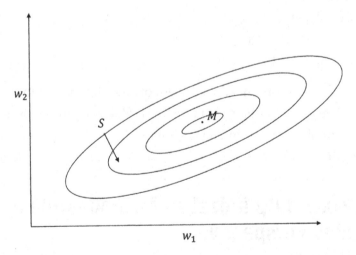

Figure 2-20. *Contour plot for a quadratic cost function with elliptical contours*

The best way to get around this problem is to take larger steps in those directions in which the gradients are small but consistent and take smaller steps in those directions that have big but inconsistent gradients. This can be achieved if, instead of having a fixed learning rate for all dimensions, we have a separate learning rate for each dimension.

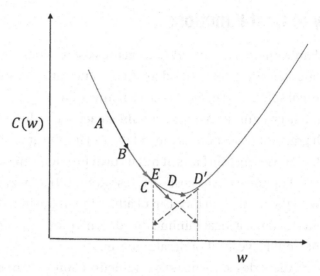

Figure 2-21. *Gradient descent for a cost function with one variable*

In Figure 2-21, the cost function between *A* and *C* is almost linear, and so gradient descent works well. However, from point *C*, the curvature of the cost function takes over, and so the gradient at *C* is not able to keep up with the direction of the change in cost function. Based on the gradient, if we take a small learning rate at *C*, we will end up at *D*, which is reasonable enough since it does not overshoot the point of minima. However, a larger step size at *C* will get us to *D'*, which is not desirable, because it is on the other side of the minima. Again, a large step size at *D'* would get us to *E*, and if the learning rate is not reduced, the algorithm tends to toggle between points on either side of the minima, leading to oscillations. When this happens, one way to stop it and achieve convergence is to look at the sign of the gradient $\frac{\partial C}{\partial w}$ or $\frac{dC}{dw}$ in successive iterations, and if they have opposite signs, reduce the learning rate so that the oscillations are reduced. Similarly, if the successive gradients have the same sign, then the learning rate can be increased accordingly. When the cost function is a function of multiple weights, the cost function might have curvatures in some dimensions of the weights, while it might be linear along other dimensions. Hence, for multivariate cost functions, the partial derivative of the cost function with respect to each weight $\left(\frac{\partial C}{\partial w_i} \right)$ can be similarly analyzed to update the learning rate for each weight or dimension of the cost function.

Non-convexity of Cost Functions

The other big problem with neural networks is that the cost functions are mostly non-convex, and so the gradient-descent method might get stuck at local minimum points, leading to a suboptimal solution. The non-convex nature of the neural network is the result of the hidden-layer units that have nonlinear activation functions, such as sigmoid. Full-batch gradient descent uses the full dataset for the gradient computation. While this is good for convex cost surfaces, it has its own problems in cases of non-convex cost functions. For non-convex cost surfaces with full-batch gradients, the model is going to end up with the minima in its basin of attraction. If the initialized parameters are in the basin of attraction of a local minima that doesn't provide good generalization, a full-batch gradient would give a suboptimal solution.

With stochastic gradient descent, the noisy gradients computed may force the model out of the basin of attraction of the bad local minima—one that does not provide good generalization—and place it in a more optimal region. Stochastic gradient descent with single data points produces very random and noisy gradients. Gradients with mini-batches tend to produce much more stable estimates of gradients when compared to gradients of single data points, but they are still noisier than those produced by the full batches. Ideally, the mini-batch size should be carefully chosen such that the gradients are noisy enough to avoid or escape bad local minima points but stable enough to converge at global minima or a local minimum that provides good generalization.

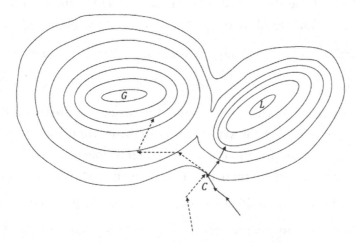

Figure 2-22. *Contour plot showing basins of attraction for global and local minima and traversal of paths for gradient descent and stochastic gradient descent*

In Figure 2-22, the dotted arrows correspond to the path taken by stochastic gradient descent (SGD), while the continuous arrows correspond to the path taken by full-batch gradient descent. Full-batch gradient descent computes the actual gradient at a point, and if it is in the basin of attraction of a poor local minimum, gradient descent almost surely ensures that the local minima L is reached. However, in the case of stochastic gradient descent, because the gradient is based on only portion of the data and not on the full batch, the gradient direction is only a rough estimate. Since the noisy rough estimate does not always point to the actual gradient at the point C, stochastic gradient descent may escape the basin of attraction of the local minima and fortunately land in the basin of a global minima. Stochastic gradient descent may escape the global minima basin of attraction too, but generally if the basin of attraction is large and the mini-batch size is carefully chosen so that the gradients it produces are moderately noisy, stochastic gradient descent is most likely to reach the global minima G (as in this case) or some other optimal minima that has a large basin of attraction. For non-convex optimization, there are other heuristics as well, such as momentum, which when adopted along with stochastic gradient descent increases the chances of the SGD's avoiding shallow local minima. Momentum generally keeps track of the previous gradients through the velocity component. So, if the gradients are steadily pointing toward a good local minimum that has a large basin of attraction, the velocity component would be high in the direction of the good local minimum. If the new gradient is noisy and points toward a bad local minimum, the velocity component would provide momentum to continue in the same direction and not get influenced by the new gradient too much.

Saddle Points in the High-Dimensional Cost Functions

Another impediment to optimizing non-convex cost functions is the presence of saddle points. The number of saddle points increases exponentially with the dimensionality increase of the parameter space of a cost function. Saddle points are stationary points (i.e., points where the gradient is zero) but are neither a local minimum nor a local maximum point. Since the saddle points are associated with a long plateau of points with the same cost as that of the saddle point, the gradient in the plateau region is either zero or very close to zero. Because of this near-zero gradient in all directions, gradient-based optimizers have a hard time coming out of these saddle points. Mathematically, to determine whether a point is a saddle point, the Eigen values of the Hessian matrix of the cost function must be computed at the given point. If there are both positive and negative Eigen values, then it is a saddle point. Just to refresh our memory of local and global minima tests, if all the Eigen values of the Hessian matrix are positive at a stationary point, then the point is a global minimum, whereas if all the Eigen values of the Hessian matrix are negative at the stationary point, then the point is a global maximum. The Eigen vectors of the Hessian matrix for a cost function give the direction of change in the curvature of the cost function, whereas the Eigen values denote the magnitude of the curvature changes along those directions. Also, for cost functions with continuous second derivatives, the Hessian matrix is symmetrical and hence would always produce an orthogonal set of Eigen vectors, thus giving mutually orthogonal directions for cost curvature changes. If in all such directions given by Eigen vectors the values of the curvature changes (Eigen values) are positive, then the point must be a local minimum, whereas if all the values of curvature changes are negative, then the point is a local maximum. This generalization works for cost functions with any input dimensionality, whereas the determinant rules for determining extremum points vary with the dimensionality of the input to the cost function. Coming back to saddle points, since the Eigen values are positive for some directions but negative for other directions, the curvature of the cost function increases in the direction of positive Eigen values while decreasing in the direction of Eigen vectors with negative coefficients. This nature of the cost surface around a saddle point generally leads to a region of long plateau with a near-zero gradient and makes it tough for gradient-descent methods to escape the plateau of this low gradient. The point $(0, 0)$ is a saddle point for the function $f(x, y) = x^2 - y^2$ as we can see from the following evaluation:

$$\nabla f(x, y) = 0 => \frac{\partial f}{\partial x} = 0 \quad \text{and} \quad \frac{\partial f}{\partial y} = 0$$

$$\frac{\partial f}{\partial x} = 2x = 0 => x = 0$$

$$\frac{\partial f}{\partial y} = -2y = 0 => y = 0$$

So, $(x, y) = (0, 0)$ is a stationary point. The next thing to do is to compute the Hessian matrix and evaluate its Eigen values at $(x, y) = (0, 0)$. The Hessian matrix $Hf(x, y)$ is as follows:

$$Hf(x,y) = \begin{bmatrix} \dfrac{\partial^2 f}{\partial x^2} & \dfrac{\partial^2 f}{\partial x \partial y} \\[2ex] \dfrac{\partial^2 f}{\partial x \partial y} & \dfrac{\partial^2 f}{\partial y^2} \end{bmatrix} = \begin{bmatrix} 2 & 0 \\ 0 & -2 \end{bmatrix}$$

So, the Hessian $Hf(x, y)$ at all points including $(x, y) = (0, 0)$ is $\begin{bmatrix} 2 & 0 \\ 0 & -2 \end{bmatrix}$.

The two Eigen values of the $Hf(x, y)$ are 2 and -2, corresponding to the Eigen vectors $\begin{bmatrix} 1 \\ 0 \end{bmatrix}$ and $\begin{bmatrix} 0 \\ 1 \end{bmatrix}$, which are nothing but the directions along the X and Y axes. Since one Eigen value is positive and the other negative, $(x, y) = (0, 0)$ is a saddle point.

The non-convex function $f(x, y) = x^2 - y^2$ is plotted in Figure 2-23, where S is the saddle point at $x, y = (0, 0)$.

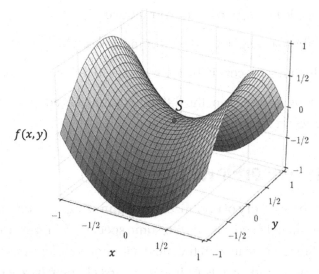

Figure 2-23. *Plot of f(x, y) = x² − y²*

Learning Rate in Mini-Batch Approach to Stochastic Gradient Descent

When there is high redundancy in the dataset, the gradient computed on a mini-batch of data points is almost the same as the gradient computed on the whole dataset, provided the mini-batch is a good representation of the entire dataset. In such cases, computing the gradient on the whole dataset can be avoided, and instead the gradient on the mini-batch of data points can be used as the approximate gradient for the whole dataset. This is the mini-batch approach to gradient descent, which is also called mini-batch stochastic gradient descent. When, instead of using a mini-batch, the gradients are approximated by one data point, it is called online learning or stochastic gradient descent. However, it is always better to use the mini-batch version of stochastic gradient descent over online learning since the gradients for the mini-batch method are less noisy compared to the online mode of learning. Learning rate plays a vital role in the convergence of mini-batch stochastic gradient descent. The following approach tends to provide good convergence:

- Start with an initial learning rate.

- Increase the learning rate if the error reduces.

- Decrease the learning rate if the error increases.

- Stop the learning process if the error ceases to reduce.

As we will see in the next section, the different optimizers adopt an adaptive learning-rate approach in their implementations.

Optimizers in TensorFlow

TensorFlow has a rich inventory of optimizers for optimizing cost functions. The optimizers are all gradient based, along with some special optimizers to handle local minima problems. Since we dealt with the most common gradient-based optimizers used in machine learning and deep learning in the first chapter, here we will stress the customizations added in TensorFlow to the base algorithms.

GradientDescentOptimizer

`GradientDescentOptimizer` implements the fundamental full-batch gradient-descent algorithm and takes the learning rate as an input. The gradient-descent algorithm will not loop over the iterations automatically, so such logic must be specified by the developer.

Usage

```
optimizer = tf.compat.v1.train.GradientDescentOptimizer(learning_
rate=0.001)
```

The optimizer object for full-batch gradient descent can be defined as above where `learning_rate` is the constant learning rate to be used for the entire duration of training.

AdagradOptimizer

`AdagradOptimizer` is a first-order optimizer like gradient descent but with some modifications. Instead of having a global learning rate, the learning rate is normalized for each dimension on which the cost function is dependent. The learning rate in each iteration is the global learning rate divided by the l^2 norm of the prior gradients up to the current iteration for each dimension.

If we have a cost function $C(\theta)$ where $\theta = [\theta_1\theta_2. . \theta_n]^T \in R^{n \times 1}$, then the update rule for θ_i is as follows:

$$\theta_i^{(t+1)} - \frac{\eta}{\sqrt{\sum_{\tau=1}^{t}\theta_i^{(\tau)2}+\epsilon}}\frac{\partial C^{(t)}}{\partial \theta_i}$$

where η is the learning rate and $\theta_i^{(t)}$ and $\theta_i^{(t+1)}$ are the values for the *ith* parameter at iterations t and $t + 1$, respectively.

In a matrix format, the parameter update for the vector θ can be represented by the following:

$$\theta^{(t+1)} = \theta^{(t)} - \eta G_{(t)}^{-1}\nabla C\left(\theta^{(t)}\right)$$

where $G_{(t)}$ is the diagonal matrix containing the l^2 norm of the past gradients till iteration t for each dimension. The matrix $G_{(t)}$ would be of the following form:

$$G_{(t)} = \begin{bmatrix} \sqrt{\sum_{\tau=1}^{t} \theta_1^{(\tau)2} + \epsilon} & \cdots & 0 \\ \cdots & \sqrt{\sum_{\tau=1}^{t} \theta_i^{(\tau)2} + \epsilon} & \cdots \\ 0 & \cdots & \sqrt{\sum_{\tau=1}^{t} \theta_n^{(\tau)2} + \epsilon} \end{bmatrix}$$

Sometimes sparse features that don't show up much in the data can be very useful to an optimization problem. However, with basic gradient descent or stochastic gradient descent, the learning rate gives equal importance to all the features in each iteration. Since the learning rate is the same, the overall contribution of non-sparse features would be much more than that of sparse features. Hence, we end up losing critical information from the sparse features. With Adagrad, each parameter is updated with a different learning rate. The sparser the feature is, the higher its parameter update would be in an iteration. This is because for sparse features, the quantity $\sqrt{\sum_{\tau=1}^{t} \theta_i^{(\tau)2} + \epsilon}$ would be less, and hence the overall learning rate would be high.

This is a good optimizer to use in applications with natural language processing and image processing where the data is sparse.

Usage

```
optimizer = tf.keras.optimizers.Adagrad(learning_rate=0.001, initial_
accumulator_value=0.1)
```

where `learning_rate` represents η and `initial_accumulator_value` represents the initial non-zero normalizing factor for each weight.

RMSprop

RMSprop is the mini-batch version of the resilient backpropagation (Rprop) optimization technique that works best for full-batch learning. Rprop solves the issue of gradients not pointing to the minimum in cases where the cost function contours are elliptical. As we discussed earlier, in such cases, instead of a global learning rule, a separate adaptive update rule for each weight would lead to better convergence. The special thing with Rprop is that it doesn't use the magnitude of the gradients of the weight but only the signs in determining how to update each weight. The following is the logic by which Rprop works:

- Start with the same magnitude of weight update for all weights, i.e., $\Delta_{ij}^{(t=0)} = \Delta_{ij}^{(0)} = \Delta$. Also, set the maximum and minimum allowable weight updates to Δ_{max} and Δ_{min}, respectively.

- At each iteration, check the sign of both the previous and the current gradient components, i.e., the partial derivatives of the cost function with respect to the different weights.

- If the signs of the current and previous gradient components for a weight connection are the same—i.e., $sign\left(\dfrac{\partial C^{(t)}}{\partial w_{ij}} \dfrac{\partial C^{(t-1)}}{\partial w_{ij}}\right) = +ve$ — then increase the learning by a factor $\eta_+ = 1.2$. The update rule becomes as follows:

$$\Delta_{ij}^{(t+1)} = \min\left(\eta_+ \Delta_{ij}^{(t)}, \Delta_{max}\right)$$

$$w_{ij}^{(t+1)} = w_{ij}^{(t)} - \text{sign}\left(\dfrac{\partial C^{(t)}}{\partial w_{ij}}\right).\Delta_{ij}^{(t+1)}$$

- If the signs of the current and previous gradient components for a dimension are different—i.e., $\text{sign}\left(\dfrac{\partial C^{(t)}}{\partial w_{ij}} \dfrac{\partial C^{(t-1)}}{\partial w_{ij}}\right) = -ve$ —then reduce the learning rate by a factor $\eta_- = 0.5$. The update rule becomes as follows:

$$\Delta_{ij}^{(t+1)} = \max\left(\eta_- \Delta_{ij}^{(t)}, \Delta_{min}\right)$$

$$w_{ij}^{(t+1)} = w_{ij}^{(t)} - \text{sign}\left(\dfrac{\partial C^{(t)}}{\partial w_{ij}}\right).\Delta_{ij}^{(t+1)}$$

167

- If $\dfrac{\partial C^{(t)}}{\partial w_{ij}} \dfrac{\partial C^{(t-1)}}{\partial w_{ij}} = 0$, the update rule is as follows:

$$\varDelta_{ij}^{(t+1)} = \varDelta_{ij}^{(t)}$$

$$w_{ij}^{(t+1)} = w_{ij}^{(t)} - \operatorname{sign}\left(\frac{\partial C^{(t)}}{\partial w_{ij}}\right).\varDelta_{ij}^{(t+1)}$$

The dimensions along which the gradients are not changing sign at a specific interval during gradient descent are the dimensions along which the weight changes are consistent. Hence, increasing the learning rate would lead to faster convergence of those weights to their final value.

The dimensions along which the gradients are changing sign indicate that along those dimensions, the weight changes are inconsistent, and so by decreasing the learning rate, one would avoid oscillations and better catch up with the curvatures. For a convex function, gradient sign change generally occurs when there is curvature in the cost function surface and the learning rate is set high. Since the gradient does not have the curvature information, a large learning rate takes the updated parameter value beyond the minima point, and the phenomena keep on repeating on either side of the minima point.

`Rprop` works well with full batches but does not do well when stochastic gradient descent is involved. When the learning rate is very small, gradients from different mini-batches average out in cases of stochastic gradient descent. If through stochastic gradient descent for a cost function the gradients for a weight are +0.2 each for nine mini-batches and -0.18 for the tenth mini-batch when the learning rate is small, then the effective gradient effect for stochastic gradient descent is almost zero, and the weight remains almost at the same position, which is the desired outcome.

However, with `Rprop` the learning rate will increase about nine times and decrease only once, and hence the effective weight would be much larger than zero. This is undesirable.

To combine the qualities of `Rprop`'s adaptive learning rule for each weight with the efficiency of stochastic gradient descent, `RMSprop` came into the picture. In `Rprop` we do not use the magnitude but rather just the sign of the gradient for each weight. The sign of the gradient for each weight can be thought of as dividing the gradient for the weight by its magnitude. The problem with stochastic gradient descent is that with each mini-batch, the cost function keeps on changing and hence so do the gradients. So, the idea is to get a magnitude of gradient for a weight that would not fluctuate much over nearby mini-batches. What would work well is a root mean of the squared gradients for each weight over the recent mini-batches to normalize the gradient.

$$g_{ij}^{(t)} = \alpha g_{ij}^{(t-1)} + (1-\alpha)\left(\frac{\partial C^{(t)}}{\partial w_{ij}}\right)^2$$

$$w_{ij}^{(t+1)} = w_{ij}^{(t)} - \frac{\eta}{\sqrt{g_{ij}^{(t)} + \epsilon}}\frac{\partial C^{(t)}}{\partial w_{ij}}$$

where $g^{(t)}$ is the root mean square of the gradients for the weight w_{ij} at iteration t and α is the decay rate for the root mean square gradient for each weight w_{ij}.

Usage

```
optimizer = tf.keras.optimizers.RMSprop(learning_rate=0.001, decay=0.9,
momentum=0.0,epsilon=1e-10)
```

where decay represents α, epsilon represents ϵ, and η represents the learning rate.

AdadeltaOptimizer

AdadeltaOptimizer is a variant of AdagradOptimizer that is less aggressive in reducing the learning rate. For each weight connection, AdagradOptimizer scales the learning rate constant in an iteration by dividing it by the root mean square of all past gradients for that weight till that iteration. So, the effective learning rate for each weight is a monotonically decreasing function of the iteration number, and after a considerable number of iterations, the learning rate becomes infinitesimally small. AdagradOptimizer overcomes this problem by taking the mean of the exponentially decaying squared gradients for each weight or dimension. Hence, the effective learning rate in AdadeltaOptimizer remains more of a local estimate of its current gradients and does not shrink as fast as the AdagradOptimizer method. This ensures that learning continues even after a considerable number of iterations or epochs. The learning rule for Adadelta can be summed up as follows:

$$g_{ij}^{(t)} = \gamma g_{ij}^{(t-1)} + (1-\gamma) \left(\frac{\partial C^{(t)}}{\partial w_{ij}} \right)^2$$

$$w_{ij}^{(t+1)} = w_{ij}^{(t)} - \frac{\eta}{\sqrt{g_{ij}^{(t)} + \epsilon}} \frac{\partial C^{(t)}}{\partial w_{ij}}$$

where γ is the exponential decay constant, η is a learning-rate constant, and $g_{ij}^{(t)}$ represents the effective mean square gradient at iteration t. We can denote the term $\sqrt{g_{ij}^{(t)} + \epsilon}$ as $RMS(g_{ij}^{(t)})$, which gives the update rule as follows:

$$w_{ij}^{(t+1)} = w_{ij}^{(t)} - \frac{\eta}{RMS\left(g_{ij}^{(t)}\right)} \frac{\partial C^{(t)}}{\partial w_{ij}}$$

If we observe carefully, the unit for the change in weight doesn't have the unit of the weight. The units for $\frac{\partial C^{(t)}}{\partial w_{ij}}$ and $RMS(g_{ij}^{(t)})$ are the same—i.e., the unit of gradient (cost function change/per unit weight change)—and hence they cancel each other out. Therefore, the unit of the weight change is the unit of the learning-rate constant. `Adadelta` solves this problem by replacing the learning-rate constant η with a square root of the mean of the exponentially decaying squared-weight updates up to the current iteration. Let $h_{ij}^{(t)}$ be the mean of the square of the weight updates up to iteration t, β be the decaying constant, and $\Delta w_{ij}^{(t)}$ be the weight update in iteration t. Then, the update rule for $h_{ij}^{(t)}$ and the final weight update rule for `Adadelta` can be expressed as follows:

$$h_{ij}^{(t)} = \beta h_{ij}^{(t-1)} + (1-\beta)\left(\Delta w_{ij}^{(t)}\right)^2$$

$$w_{ij}^{(t+1)} = w_{ij}^{(t)} - \frac{\sqrt{h_{ij}^{(t)} + \epsilon}}{RMS\left(g_{ij}^{(t)}\right)} \frac{\partial C^{(t)}}{\partial w_{ij}}$$

If we denote $\sqrt{h_{ij}^{(t)} + \epsilon}$ as $RMS(h_{ij}^{(t)})$, then the update rule becomes as follows:

$$w_{ij}^{(t+1)} = w_{ij}^{(t)} - \frac{RMS\left(h_{ij}^{(t)}\right)}{RMS\left(g_{ij}^{(t)}\right)} \frac{\partial C^{(t)}}{\partial w_{ij}}$$

Usage

```
optimizer = tf.keras.optimizers.Adadelta(learning_rate=0.001, rho=0.95,
epsilon=1e-08)
```

where `rho` represents γ, `epsilon` represents ϵ, and η represents the learning rate.

One significant advantage of `Adadelta` is that it eliminates the learning-rate constant altogether. If we compare `Adadelta` and `RMSprop,` both are the same if we leave aside the learning-rate constant elimination. `Adadelta` and `RMSprop` were both developed independently around the same time to resolve the fast learning-rate decay problem of `Adagrad`.

AdamOptimizer

Adam, or Adaptive Moment Estimator, is another optimization technique that, much like `RMSprop` or `Adagrad`, has an adaptive learning rate for each parameter or weight. `Adam` not only keeps a running mean of squared gradients but also keeps a running mean of past gradients.

Let the decay rate of the mean of gradients $m_{ij}^{(t)}$ and the mean of the square of gradients $v_{ij}^{(t)}$ for each weight w_{ij} be β_1 and β_2, respectively. Also, let η be the constant learning-rate factor. Then, the update rules for `Adam` are as follows:

$$m_{ij}^{(t)} = \beta_1 m_{ij}^{(t-1)} + \left(1 - \beta_1\right) \frac{\partial C^{(t)}}{\partial w_{ij}}$$

$$v_{ij}^{(t)} = \beta_2 v_{ij}^{(t-1)} + \left(1 - \beta_2\right) \left(\frac{\partial C^{(t)}}{\partial w_{ij}}\right)^2$$

If we expand the expression for the moving average of the gradient $m_{ij}^{(t)}$ and the square of the gradient $v_{ij}^{(t)}$, we would see their expectations are biased. For instance, the gradient iteration t can be simplified as follows:

$$m_{ij}^{(t)} = \beta_1 m_{ij}^{(t-1)} + (1-\beta_1)\frac{\partial C^{(t)}}{\partial w_{ij}}$$

$$= \beta_1\left(\beta_1 m_{ij}^{(t-2)} + (1-\beta_1)\frac{\partial C^{(t-1)}}{\partial w_{ij}}\right) + (1-\beta_1)\frac{\partial C^{(t)}}{\partial w_{ij}}$$

$$= \beta_1^2 m_{ij}^{(t-2)} + \beta_1(1-\beta_1)\frac{\partial C^{(t-1)}}{\partial w_{ij}} + (1-\beta_1)\frac{\partial C^{(t)}}{\partial w_{ij}}$$

$$= \beta_1^t m_{ij}^{(0)} + \beta_1^{t-1}(1-\beta_1)\frac{\partial C^{(1)}}{\partial w_{ij}} + \ldots + (1-\beta_1)\frac{\partial C^{(t)}}{\partial w_{ij}}$$

Taking expectation of the $E\left[m_{ij}^{(t)}\right]$ and assuming that $E\left[\dfrac{\partial C^{(i)}}{\partial w_{ij}}\right] = g$ is the actual gradient of the full dataset, we have the following:

$$E\left[m_{ij}^{(t)}\right] = E\left[\beta_1^t m_{ij}^{(0)} + \beta_1^{t-1}(1-\beta_1)\frac{\partial C^{(1)}}{\partial w_{ij}} + \ldots + (1-\beta_1)\frac{\partial C^{(t)}}{\partial w_{ij}}\right]$$

$$= E\left[\beta_1^t m_{ij}^{(0)}\right] + (1-\beta_1)g\left[\beta_1^{t-1} + \ldots + 1\right]$$

$$= E\left[\beta_1^t m_{ij}^{(0)}\right] + \frac{(1-\beta_1)g(1-\beta_1^t)}{(1-\beta_1)}$$

If we take the initial estimate of the moving average of gradient $m_{ij}^{(0)}$ to be zero, then we have from above:

$$E\left[m_{ij}^{(t)}\right] = E\left[\beta_1 m_{ij}^{(t-1)} + (1-\beta_1)\frac{\partial C^{(t)}}{\partial w_{ij}}\right] = g(1-\beta_1^t)$$

Hence to make the moving average of the gradient unbiased estimate of the true gradient g of the entire dataset at iteration t, we define the normalized moving average of gradient as $\hat{m}_{ij}^{(t)}$ as follows:

$$\hat{m}_{ij}^{(t)} = \frac{m_{ij}^{(t)}}{\left(1 - \beta_1^t\right)}$$

Similarly, we define the normalized moving average of the square of the gradient $v_{ij}^{(t)}$ as follows:

$$\hat{v}_{ij}^{(t)} = \frac{v_{ij}^{(t)}}{\left(1 - \beta_2^t\right)}$$

The final update rule for each weight w_{ij} is as follows:

$$w_{ij}^{(t+1)} = w_{ij}^{(t)} - \frac{\eta}{\sqrt{\widehat{v_{ij}^{(t)}}} + \in} \widehat{m_{ij}^{(t)}}$$

$$w_{ij}^{(t+1)} = w_{ij}^{(t)} - \frac{\eta \, \hat{m}_{ij}^{(t)}}{\sqrt{\hat{v}_{ij}^{(t)}} + \epsilon}$$

Usage

```
optimizer =tf.keras.optimizers.Adam(learning_rate=0.001,beta_1=0.9,beta_2=
0.999,epsilon=1e-08)
```

where `learning_rate` is the constant learning rate η and `cost` C is the cost function that needs to be minimized through `AdamOptimizer`. The parameters `beta1` and `beta2` correspond to β_1 and β_2, respectively, whereas `epsilon` represents $\in$.

MomentumOptimizer and Nesterov Algorithm

The momentum-based optimizers have evolved to take care of non-convex optimizations. Whenever we are working with neural networks, the cost functions that we generally get are non-convex in nature, and thus the gradient-based optimization methods might get caught up in bad local minima. As discussed earlier, this is highly undesirable since in such cases we get a suboptimal solution to the optimization problem—and likely a suboptimal model. Also, gradient descent follows the slope at each point and makes small advances toward the local minima, but it can be terribly slow. Momentum-based methods introduce a component called velocity v that dampens the parameter update when the gradient computed changes sign, whereas it accelerates the parameter update when the gradient is in the same direction of velocity. This introduces faster convergence as well as fewer oscillations around the global minima or around a local minimum that provides good generalization. The update rule for momentum-based optimizers is as follows:

$$v_i^{(t+1)} = \alpha v_i^{(t)} - \eta \frac{\partial C}{\partial w_i}\left(w_i^{(t)}\right)$$

$$w_i^{(t+1)} = w_i^{(t)} + v_i^{(t+1)}$$

where α is the momentum parameter and η is the learning rate. The terms $v_i^{(t)}$ and $v_i^{(t+1)}$ represent the velocity at iterations t and $(t+1)$, respectively, for the *ith* parameter. Similarly, $w_i^{(t)}$ and $w_i^{(t+1)}$ represent the weight of the *ith* parameter at iterations t and $t+1$, respectively.

Imagine that while optimizing a cost function the optimization algorithm reaches a local minimum where $\frac{\partial C}{\partial w_i}\left(w_i^{(t)}\right) \to 0 \forall i \in \{1,2,..n\}$. In normal gradient-descent methods that do not take momentum into consideration, the parameter update would stop at that local minimum or the saddle point. However, in momentum-based optimization, the prior velocity would drive the algorithm out of the local minima, considering the local minima has a small basin of attraction, as $v_i^{(t+1)}$ would be non-zero because of the non-zero velocity from prior gradients. Also, if the prior gradients consistently pointed toward a global minimum or a local minimum with good generalization and a reasonably large basin of attraction, the velocity or the momentum of gradient descent would be in that direction. So, even if there were a bad local minimum with a small basin of attraction, the momentum component would not only drive the algorithm out of the bad local minima but also would continue the gradient descent toward the global minima or the good local minima.

If the weights are part of the parameter vector θ, the vectorized update rule for momentum-based optimizers would be as follows (refer to Figure 2-24 for the vector-based illustration):

$$v^{(t+1)} = \alpha v^{(t)} - \eta \nabla C\left(\theta = \theta^{(t)}\right)$$

$$\theta^{(t+1)} = \theta^{(t)} + v^{(t+1)}$$

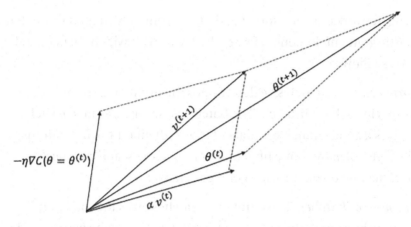

Figure 2-24. *Parameter vector update in momentum-based gradient-descent optimizer*

A specific variant of momentum-based optimizers is the Nesterov accelerated gradient technique. This method utilizes the existing velocity $v^{(t)}$ to make an update to the parameter vector. Since it is an intermediate update to the parameter vector, it's convenient to denote it by $\theta^{\left(t+\frac{1}{2}\right)}$. The gradient of the cost function is evaluated at $\theta^{\left(t+\frac{1}{2}\right)}$, and the same is used to update the new velocity. Finally, the new parameter vector is the sum of the parameter vector at the previous iteration and the new velocity.

$$\theta^{\left(t+\frac{1}{2}\right)} = \theta^{(t)} + \alpha v^{(t)}$$

$$v^{(t+1)} = \alpha v^{(t)} - \eta \nabla C\left(\theta = \theta^{\left(t+\frac{1}{2}\right)}\right)$$

$$\theta^{(t+1)} = \theta^{(t)} + v^{(t+1)}$$

Usage

```
optimizer = tf.keras.optimizers.SGD( learning_rate=0.001, momentum=0.9,
nesterov=True)
```

where `learning_rate` represents η, `momentum` represents α, and `use_nesterov` determines whether to use the Nesterov version of momentum.

Epoch, Number of Batches, and Batch Size

Deep-learning networks, as mentioned earlier, are generally trained through mini-batch stochastic gradient descent. A few of the terms with which we need to familiarize ourselves are as follows:

- *Batch size*: The batch size determines the number of training data points in each mini-batch. The batch size should be chosen such that it gives a good enough estimate of the gradient for the full training dataset and at the same time noisy enough to escape bad local minima that do not provide good generalization.

- *Number of batches*: The number of batches gives the total number of mini-batches in the entire training dataset. It can be computed by dividing the count of the total training data points by the batch size. Please note that the last mini-batch might have a smaller number of data points than the batch size.

- *Epochs*: One epoch consists of one full pass of training over the entire dataset. To be more specific, one epoch is equivalent to a forward pass plus one backpropagation over the entire training dataset. So, one epoch would consist of n number of forward pass + backpropagation where n denotes the number of batches.

XOR Implementation Using TensorFlow

Now that we have a decent idea of the components and training methods involved with an artificial neural network; we will implement a XOR network using sigmoid activation functions in the hidden layers as well as in the output. The detailed implementation has been outlined in Listing 2-15.

Listing 2-15. XOR Implementation with Hidden Layers That Have Sigmoid Activation Functions

```python
# Import the required packages
import tensorflow as tf
import numpy as np
from tensorflow.keras import Model, layers
print( 'tensorflow version',tf.__version__ )

# Model with one hidden layer of 2 units with sigmoid activation to render
non-linearity
class MLP(Model):
    # Set layers.
    def __init__(self,n_hidden=2,n_out=1):
        super(MLP, self).__init__()
        # Fully-connected hidden layer.
        self.fc1 = layers.Dense(n_hidden, activation=tf.nn.sigmoid,
        name="hidden")
        # Output layer
        self.out = layers.Dense(n_out,activation=tf.nn.sigmoid,name="out")

    # Forward pass through the MLP

    def call(self, x):
        x = self.fc1(x)
        x = self.out(x)
        return x

# Define Model by instantiating the MLP Class
model = MLP(n_hidden=2,n_out=1)

# Wrap the model with tf.function to create Graph execution for the Model
model_graph = tf.function(model)

# Learning rate
learning_rate = 0.01

# Define Optimizer
optimizer = tf.compat.v1.train.GradientDescentOptimizer(learning_rate)
```

```python
# Define Crossentropy loss. Since we have set the output layer activation
to be sigmoid from_logits is set to False
loss_fn = tf.keras.losses.BinaryCrossentropy(from_logits=False)

# Define the XOR specific datapoints
XOR_X = [[0,0],[0,1],[1,0],[1,1]]
XOR_Y = [[0],[1],[1],[0]]

# Convert the data to Constant Tensors
x_ = tf.constant(np.array(XOR_X))
y_ = tf.constant(np.array(XOR_Y))

num_epochs = 100000

for i in range(num_epochs):
 # Track variables/operations along with their order within the
tf.GradientTape scope so as to use them for Gradient
# Computation
    with tf.GradientTape() as tape:
        y_pred = model_graph(x_)
        loss = loss_fn(y_,y_pred)

    # Compute gradient
    gradients = tape.gradient(loss, model.trainable_variables)

    # update the parameters
    optimizer.apply_gradients(zip(gradients, model.trainable_variables))

    if i % 10000 == 0:
        print(f"Epoch: {i}, loss: {loss.numpy()}")

print('Final Prediction:', model_graph(x_).numpy())

--output -

Final Prediction: [[0.02323644]
 [0.9814549 ]
 [0.9816729 ]
 [0.02066191]]
```

In Listing 2-15, the XOR logic has been implemented using TensorFlow 2. The hidden-layer units have sigmoid activation functions to introduce nonlinearity. The output activation function has a sigmoid activation function to give probability outputs. We are using the gradient-descent optimizer with a learning rate of 0.01 and total iterations of around 100,000. If we see the final prediction, the first and fourth training samples have a near-zero value for probabilities, while the second and fourth training samples have probabilities near 1. So, the network can predict the classes accurately corresponding to the XOR labels with high precision. Any reasonable threshold on the output probabilities would classify the data points correctly into the XOR classes.

TensorFlow Computation Graph for XOR Network

In Figure 2-25, the computation graph for the preceding implemented XOR network is illustrated. The computation graph summary is written to the log files that can we defined as in the following anywhere in the code before the actual training or execution of the function of interest using tf.summary.create_file_writer() function. Here we choose to define the log prior to the MLP class instantiation for the model creation.

```
stamp = datetime.now().strftime("%Y%m%d-%H%M%S")
logdir = 'logs/func1/%s' % stamp
writer = tf.summary.create_file_writer(logdir)

        # Define Model by instantiating the MLP Class

model = MLP(n_hidden=2,n_out=1)
```

Once we have defined the log writer, we need to set the summary trace on by using tf.summary.trace_on() function immediately before the function call or the model training of interest to us. For graph information, we need to set **graph=True**, while for profile information such as Memory and CPU time, we need to set **profile=True** while setting the summary trace on. Also, we need to set the trace_export functionality to write the traces to the log just after the computation logic (training, function call, etc.) we are trying to trace ends.

In fact, the trace_on and trace_export functionalities should sandwidth the execution logic that we plan to trace. For the XOR example, we place the trace_on and trace_export functionality to trace the training of the model as in the following:

```
tf.summary.trace_on(graph=True, profiler=True)

for i in range(num_epochs):
 #  Track variables/operations along with their order within the
tf.GradientTape scope so as to use them for Gradient
# Computation
    with tf.GradientTape() as tape:
        y_pred = model_graph(x_)
        loss = loss_fn(y_,y_pred)

    # Compute gradient
    gradients = tape.gradient(loss, model.trainable_variables)

    # update the parameters
    optimizer.apply_gradients(zip(gradients, model.trainable_variables))

    if i % 10000 == 0:
        print(f"Epoch: {i}, loss: {loss.numpy()}")

with writer.as_default():
    tf.summary.trace_export(
      name="my_func_trace",
      step=0,
      profiler_outdir=logdir)

print('Final Prediction:', model_graph(x_).numpy())
```

Once the training has completed and output has been written to the log, we can view the computation graph by loading the log into TensorBoard. The following is the code that can be used in Jupyter notebook for the same.

```
%load_ext tensorboard
import tensorboard
%tensorboard -logdir logs/func1
```

This would start the TensorBoard session where we can observe the computation graph. Figure 2-25 shows the computation graph for the XOR network in Listing 2-15.

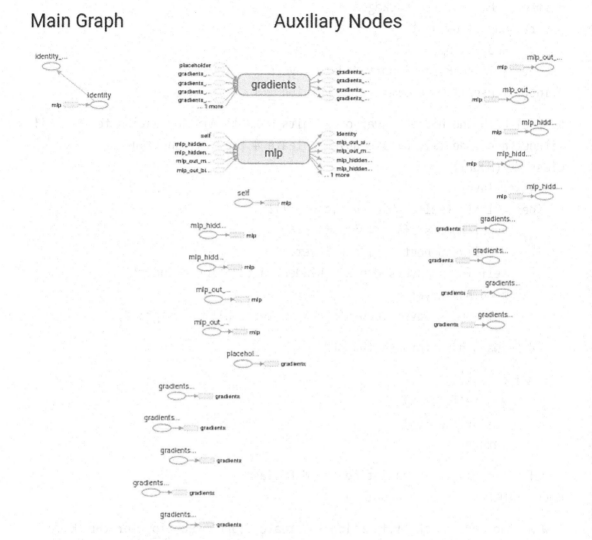

Figure 2-25. *Computation graph for the XOR network*

Now, we implement the XOR logic again, using linear activation functions in the hidden layer and keeping the rest of the network as it is. Listing 2-16 shows the TensorFlow implementation.

Listing 2-16. XOR Implementation with Linear Activation Functions in Hidden Layer

```python
# Import the required packages
import tensorflow as tf
import numpy as np
from tensorflow.keras import Model, layers
print( 'tensorflow version',tf.__version__ )

# Model with one hidden layer of 2 units with linear activation to check if
without non linearity we are able to learn # the XOR function
class MLP(Model):
    # Set layers.
    def __init__(self,n_hidden=2,n_out=1):
        super(MLP, self).__init__()
        # Fully-connected hidden layer.
        self.fc1 = layers.Dense(n_hidden, activation='linear')
        # Output layer
        self.out = layers.Dense(n_out,activation=tf.nn.sigmoid)

    # Forward pass through the MLP

    def call(self, x):
        x = self.fc1(x)
        x = self.out(x)
        return x

# Define Model by instantiating the MLP Class
model = MLP(n_hidden=2,n_out=1)

# Wrap the model with tf.function to create Graph execution for the Model
model_graph = tf.function(model)

# Learning rate
learning_rate = 0.01

# Define Optimizer
optimizer = tf.compat.v1.train.GradientDescentOptimizer(learning_rate)
```

```
# Define Crossentropy loss. Since we have set the output layer activation
to be sigmoid from_logits is set to False
loss_fn = tf.keras.losses.BinaryCrossentropy(from_logits=False)

# Define the XOR specific datapoints
XOR_X = [[0,0],[0,1],[1,0],[1,1]]
XOR_Y = [[0],[1],[1],[0]]

# Convert the data to Constant Tensors
x_ = tf.constant(np.array(XOR_X))
y_ = tf.constant(np.array(XOR_Y))

num_epochs = 100000

for i in range(num_epochs):
 #  Track variables/operations along with their order within the
tf.GradientTape scope so as to use them for Gradient
# Computation
    with tf.GradientTape() as tape:
        y_pred = model_graph(x_)
        loss = loss_fn(y_,y_pred)

    # Compute gradient
    gradients = tape.gradient(loss, model.trainable_variables)

    # update the parameters
    optimizer.apply_gradients(zip(gradients, model.trainable_variables))

    if i % 10000 == 0:
        print(f"Epoch: {i}, loss: {loss.numpy()}")

print('Final Prediction:', model_graph(x_).numpy())

--output -

Final Prediction: [[0.5006326 ]
 [0.5000254 ]
 [0.49998236]
 [0.49937516]]
```

The final predictions as shown in Listing 2-16 are all near 0.5, which means that the implemented XOR logic is not able to do a good job in discriminating the positive class from the negative one. When we have linear activation functions in the hidden layer, the network primarily remains linear, as we have seen previously, and hence the model is not able to do well where nonlinear decision boundaries are required to separate classes.

Linear Regression in TensorFlow

Linear regression can be expressed as a single-neuron regression problem. The mean of the square of the errors in prediction is taken as the cost function to be optimized with respect to the coefficient of the model. Listing 2-17 shows a TensorFlow implementation of linear regression with the Boston housing price dataset.

Listing 2-17. Linear Regression Implementation in TensorFlow

```
# Importing TensorFlow, Numpy and the Boston Housing price dataset

import tensorflow as tf
print('tensorflow version',tf.__version__)
import numpy as np
import sklearn
from sklearn.datasets import load_boston

# Function to load the Boston data set
def read_infile():
    data = load_boston()
    features = np.array(data.data)
    target = np.array(data.target)
    return features,target

# Normalize the features by Z scaling i.e. subract form each feature value
its mean and then divide by its
# standard deviation. Accelerates Gradient Descent.

def feature_normalize(data):
    mu = np.mean(data,axis=0)
    std = np.std(data,axis=0)
    return (data - mu)/std
```

```python
# Execute the functions to read and normalize  the data

features,target = read_infile()
z_features = feature_normalize(features)
num_features = z_features.shape[1]

X = tf.constant( z_features , dtype=tf.float32 )
Y = tf.constant( target , dtype=tf.float32 )

# Create Tensorflow linear Model
model = tf.keras.models.Sequential()
model.add(tf.keras.Input(shape=(num_features,)))
model.add(tf.keras.layers.Dense(1, use_bias=True,activation='linear'))

#Learning rate
learning_rate = 0.01
# Define optimizer
optimizer = tf.keras.optimizers.SGD(learning_rate)
num_epochs = 1000
cost_trace = []
loss_fn = tf.keras.losses.MeanSquaredError()
# Execute the gradient descent learning

for i in range(num_epochs):
    with tf.GradientTape() as tape:
        y_pred = model(X, training=True)
        loss = loss_fn(Y,y_pred)
    # compute gradient
    gradients = tape.gradient(loss, model.trainable_variables)
    # update the parameters
    optimizer.apply_gradients(zip(gradients, model.trainable_variables))
    cost_trace.append(loss.numpy())
    if i % 100 == 0:
        print(f"Epoch: {i}, loss: {loss.numpy()}")
print(f'Final Prediction..\n')
#print(model(X,training=False).numpy())
print('MSE in training:',cost_trace[-1])
```

```
-- output --

tensorflow version 2.4.1
Epoch: 0, loss: 596.2260131835938
Epoch: 100, loss: 32.83129119873047
Epoch: 200, loss: 22.925716400146484
Epoch: 300, loss: 22.40866470336914
Epoch: 400, loss: 22.21877670288086
Epoch: 500, loss: 22.1116943359375
Epoch: 600, loss: 22.04631996154785
Epoch: 700, loss: 22.00408935546875
Epoch: 800, loss: 21.975479125976562
Epoch: 900, loss: 21.955341339111328
Final Prediction..

MSE in training: 21.940863
```

Listing 2-17a. Linear Regression Cost Plot over Epochs or Iterations

```
import matplotlib.pyplot as plt
%matplotlib inline
plt.plot(cost_trace)
```

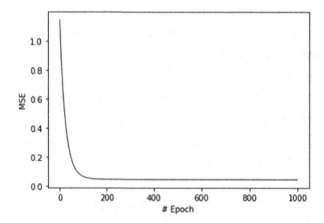

Figure 2-26. *Cost (MSE) vs. epochs while training*

Listing 2-17b. Linear Regression Actual House Price vs. Predicted House Price

```
# Plot the Predicted house Prices vs the Actual House Prices

fig, ax = plt.subplots()
plt.scatter(target,y_pred.numpy())
ax.set_xlabel('Actual House price')
ax.set_ylabel('Predicted House price')
```

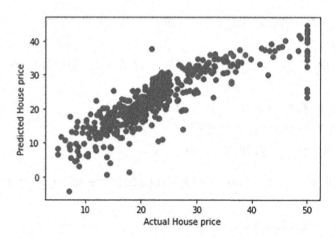

Figure 2-27. *Predicted house price vs. actual house price*

Figure 2-26 illustrates the cost progression against the epochs, and Figure 2-27 illustrates the predicted house price vs. the actual house price after training. As we can see, the linear regression using TensorFlow does a decent job of predicting the housing prices.

Multiclass Classification with SoftMax Function Using Full-Batch Gradient Descent

In this section, we illustrate a multiclass classification problem using full-batch gradient descent. The MNIST dataset has been used because there are ten output classes corresponding to the ten integers. The detailed implementation is provided in Listing 2-18. A SoftMax has been used as the output layer.

Listing 2-18. Multiclass Classification with SoftMax Function Using Full-Batch Gradient Descent

```
# Load the required packages
import tensorflow as tf
print('tensorflow version', tf.__version__)
import numpy as np
from sklearn import datasets

# Function to Read the MNIST dataset along with the labels
def read_infile():
    (train_X, train_Y), (test_X, test_Y) = tf.keras.datasets.mnist.
    load_data()
    train_X = train_X.reshape(-1,28*28)
    test_X = test_X.reshape(-1,28*28)
    return train_X, train_Y,test_X, test_Y

# Define the Model class to have a MLP architecture with just one
hidden layer.
# The model is linear classification Model
class MLP(Model):
    # Set layers.
    def __init__(self,n_classes=10):
        super(MLP, self).__init__()
        # Fully-connected hidden layer.
        self.out = layers.Dense(n_classes,activation='linear')

    # Forward pass.
    def call(self, x):
        x = self.out(x)
        return x

# Define the Categorical Cross Entropy that does a softmax on the final
layer output logits
loss_fn = tf.keras.losses.SparseCategoricalCrossentropy(from_logits=True)

#Learning rate
learning_rate = 0.01
```

```
# Define optimizer for Full Gradient Descent
optimizer = tf.compat.v1.train.GradientDescentOptimizer(learning_rate)

X_train, y_train, X_test, y_test = read_infile()
num_train_recs, num_test_recs = X_train.shape[0], X_test.shape[0]

# Build the model by instantiating the MLP Class
model = MLP(n_classes=max(y_train) +1)

# Wrap the model with tf.function to create Graph execution for the model
model_graph = tf.function(model)

# Defining the train and test input outputs as tensoflow constants
X_train = tf.constant(X_train, dtype=tf.float32 )
X_test = tf.constant(X_test, dtype=tf.float32 )
y_train = tf.constant(y_train)
y_test = tf.constant(y_test)

epochs = 1000
loss_trace = []
accuracy_trace = []

# Execute the training
for i in range(epochs):
    #  Track variables/operations along with their order within the
       tf.GradientTape scope so as to use them for
    # Gradient  Computation
    with tf.GradientTape() as tape:
        y_pred = model_graph(X_train)
        loss = loss_fn(y_train,y_pred)

    # compute gradient
    gradients = tape.gradient(loss, model.trainable_variables)

    # update the parameters
    optimizer.apply_gradients(zip(gradients, model.trainable_variables))

    # Compute accurcy
    accuracy_ = np.mean(y_train.numpy() == np.argmax(y_pred.
    numpy(),axis=1))
```

```
        loss_trace.append(loss.numpy()/num_train_recs)
        accuracy_trace.append(accuracy_)
        if (((i+1) >= 100) and ((i+1) % 100 == 0 )) :
            loss_ = np.round((loss.numpy()/num_recs),4)
            print(f"Epoch {i+1} : loss: {loss_} ,accuracy:{np.
            round(accuracy_,4)}\n")

y_pred_test = model_graph(X_test)
loss_test = loss_fn(y_test,y_pred_test).numpy()/num_test_recs
accuracy_test = np.mean(y_test.numpy() == np.argmax(y_pred_test.
numpy(),axis=1))
print('Results on Test Dataset:','loss:',np.round(loss_
test,4),'accuracy:',np.round(accuracy_test,4))
```

-- output -

tensorflow version 2.4.1

Epoch 100 : loss: 0.1058 ,accuracy:0.903

Epoch 200 : loss: 0.0965 ,accuracy:0.9086

Epoch 300 : loss: 0.218 ,accuracy:0.8239

Epoch 400 : loss: 0.0984 ,accuracy:0.9154

Epoch 500 : loss: 0.082 ,accuracy:0.921

Epoch 600 : loss: 0.0884 ,accuracy:0.9201

Epoch 700 : loss: 0.0896 ,accuracy:0.9204

Epoch 800 : loss: 0.081 ,accuracy:0.9228

Epoch 900 : loss: 0.1502 ,accuracy:0.8858

Epoch 1000 : loss: 0.0944 ,accuracy:0.8995

Results on Test Dataset: loss: 0.0084 accuracy: 0.861

Listing 2-18a. Display of the Actual Digits vs. the Predicted Digits Along with the Images of the Actual Digits

```
import matplotlib.pyplot as plt
%matplotlib inline
f, a = plt.subplots(1, 10, figsize=(10, 2))
print('Actual digits: ', y_test[0:10].numpy())
print('Predicted digits:',np.argmax(y_pred_test[0:10],axis=1))
print('Actual images of the digits follow:')
for i in range(10):
        a[i].imshow(np.reshape(X_test[i],(28, 28)))

-- output --

Actual digits:     [7 2 1 0 4 1 4 9 5 9]
Predicted digits:  [7 2 1 0 4 1 4 9 6 9]
Actual images of the digits follow:
```

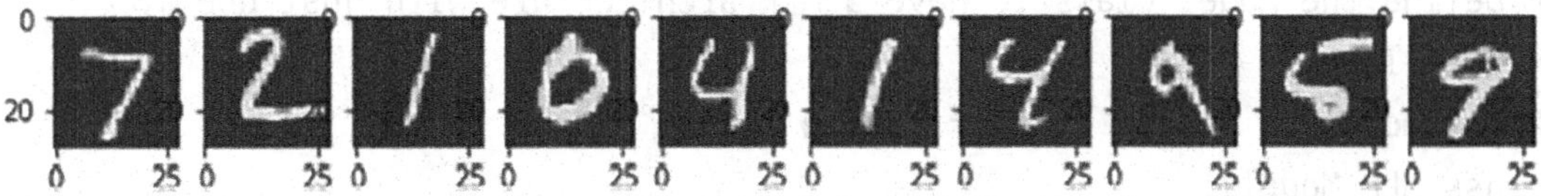

Figure 2-28. *Actual digits vs. predicted digits for SoftMax classification through gradient descent*

Figure 2-28 displays the actual digits vs. the predicted digits for SoftMax classification of the validation dataset samples after training through gradient-descent full-batch learning.

Multiclass Classification with SoftMax Function Using Stochastic Gradient Descent

We now perform the same classification task, but instead of using full-batch learning, we resort to stochastic gradient descent with a batch size of 1000. The detailed implementation has been outlined in Listing 2-19.

Listing 2-19. Multiclass Classification with SoftMax Function Using Stochastic Gradient Descent

```
# Load the required packages
import tensorflow as tf
print('tensorflow version', tf.__version__)
import numpy as np
from sklearn import datasets

# Function to Read the MNIST dataset along with the labels
def read_infile():
    (train_X, train_Y), (test_X, test_Y) = tf.keras.datasets.mnist.
    load_data()
    train_X = train_X.reshape(-1,28*28)
    test_X = test_X.reshape(-1,28*28)
    return train_X, train_Y,test_X, test_Y

# Define the Model class to have a MLP architecture with just one
hidden layer.
# The model is linear classification Model
class MLP(Model):
    # Set layers.
    def __init__(self,n_classes=10):
        super(MLP, self).__init__()
        # Fully-connected hidden layer.
        self.out = layers.Dense(n_classes,activation='linear')

    # Forward pass.
    def call(self, x):
        x = self.out(x)
        return x

# Define the Categorical Cross Entropy that does a softmax on the final
layer output logits
loss_fn = tf.keras.losses.SparseCategoricalCrossentropy(from_logits=True)

#Learning rate
learning_rate = 0.01
```

```
# Define  the Stochastic Gradient Descent Optimizer for mini-batch based
training
optimizer = tf.keras.optimizers.SGD(learning_rate)

X_train, y_train, X_test, y_test = read_infile()

# Build the model by instantiating the MLP class
model = MLP(n_classes=max(y_train) +1)

# Wrap the model with tf.function to create Graph execution for the model
model_graph = tf.function(model)

X_test = tf.constant(X_test, dtype=tf.float32 )
y_test = tf.constant(y_test)

epochs = 1000
loss_trace = []
accuracy_trace = []
batch_size = 1000

num_train_recs,num_test_recs = X_train.shape[0], X_test.shape[0]
num_batches = num_train_recs // batch_size
order_ = np.arange(num_train_recs)

# Invoke the training
for i in range(epochs):
    loss, accuracy = 0,0
  # Randomize the order of the training data
    np.random.shuffle(order_)
    X_train,y_train = X_train[order_], y_train[order_]
    # Interate of the mini batches
    for j in range(num_batches):
        X_train_batch = tf.constant(X_train[j*batch_size:(j+1)*batch_
        size],dtype=tf.float32)
        y_train_batch = tf.constant(y_train[j*batch_size:(j+1)*batch_size])

        #  Track variables/operations along with their order within the
           tf.GradientTape scope so as to use them for
        # Gradient  Computation
```

```python
    with tf.GradientTape() as tape:
        y_pred_batch = model_graph(X_train_batch)
        loss_ = loss_fn(y_train_batch,y_pred_batch)

    # Compute gradient
    gradients = tape.gradient(loss_, model.trainable_variables)

    # Update the parameters
    optimizer.apply_gradients(zip(gradients, model.trainable_
    variables))

    accuracy += np.sum(y_train_batch.numpy() == np.argmax(y_pred_batch.
    numpy(),axis=1))
    loss += loss_.numpy()

loss /= num_train_recs
accuracy /= num_train_recs
loss_trace.append(loss)
accuracy_trace.append(accuracy)

if (((i+1) >= 100) and ((i+1) % 100 == 0 )) :
    print(f"Epoch {i+1} : loss: {np.round(loss,4)} ,accuracy:{np.
    round(accuracy,4)}\n")
y_pred_test = model_graph(X_test)
loss_test = loss_fn(y_test,y_pred_test).numpy()/num_test_recs
accuracy_test = np.mean(y_test.numpy() == np.argmax(y_pred_test.numpy(),
axis=1))
print('Results on Test Dataset:','loss:',np.round(loss_test,4),
'accuracy:',np.round(accuracy_test,4))

-- output -
tensorflow version 2.4.1

Epoch 100 : loss: 0.1215 ,accuracy:0.862

Epoch 200 : loss: 0.082 ,accuracy:0.8949

Epoch 300 : loss: 0.0686 ,accuracy:0.9046

Epoch 400 : loss: 0.0774 ,accuracy:0.8947
```

Epoch 500 : loss: 0.0698 ,accuracy:0.901

Epoch 600 : loss: 0.0628 ,accuracy:0.9081

Epoch 700 : loss: 0.0813 ,accuracy:0.8912

Epoch 800 : loss: 0.0837 ,accuracy:0.8887

Epoch 900 : loss: 0.068 ,accuracy:0.9056

Epoch 1000 : loss: 0.0709 ,accuracy:0.902

Results on Test Dataset: loss: 0.0065 accuracy: 0.9105

Listing 2-19a. Actual Digits vs. Predicted Digits for SoftMax Classification Through Stochastic Gradient Descent

```
import matplotlib.pyplot as plt
%matplotlib inline
f, a = plt.subplots(1, 10, figsize=(10, 2))
print('Actual digits: ', y_test[0:10].numpy())
print('Predicted digits:',np.argmax(y_pred_test[0:10],axis=1))
print('Actual images of the digits follow:')
for i in range(10):
        a[i].imshow(np.reshape(X_test[i],(28, 28)))

--output --
```

```
Actual digits:    [7 2 1 0 4 1 4 9 5 9]
Predicted digits: [7 2 1 0 4 1 4 9 6 9]
Actual images of the digits follow:
```

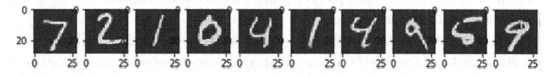

Figure 2-29. *Actual digits vs. predicted digits for SoftMax classification through stochastic gradient descent*

Figure 2-29 displays the actual digits vs. predicted digits for SoftMax classification of the validation dataset samples after training through stochastic gradient descent.

GPU

Before we end this chapter, we want to talk a little about GPU, which has revolutionized the deep-learning world. GPU stands for graphical processing unit, which was initially used for gaming purposes to display more screens per second for better gaming resolution. Deep-learning networks use a lot of matrix multiplication, especially convolution, for both the forward pass and for backpropagation. GPUs are good at matrix-to-matrix multiplication; hence, several thousand cores of GPU are utilized to process data in parallel. These speed up the deep-learning training.

Common GPUs available in the market are as follows:

- NVIDIA RTX 30 Series

- NVIDIA GTX TITAN X

- NVIDIA GeForce GTX 1080

- NVIDIA GeForce GTX 1070

- NVIDIA Tesla V100

One limitation of GPU is that they are designed as general-purpose processor that can support a wide variety of applications. In doing so, the GPU must access registers and shared memory to read and store results. A large amount of energy is expended by the GPU in accessing memory which increases the footprint of the GPU.

TPU

An alternative to GPU is the Tensor Processing Units or TPUs from Google which is designed to accelerate the deep-learning and machine-learning applications. Unlike the CPU and GPU which are general-purpose processors, TPUs are designed solely as matrix processor specialized for neural network workloads.

In TPUs, the entire matrix computation is done once the data and the parameters of the model are loaded into TPU and post that there is no memory access during the entire process of the massive computation. Hence the power expended in accessing memory is much less in TPUs than in GPUs.

TPUs were used by Google internally since 2015, and it has been made publicly available to others since 2018.

Summary

In this chapter, we have covered how deep learning has evolved from artificial neural networks over the years. Also, we discussed the Perceptron method of learning, its limitations, and the current method of training neural networks. Problems pertaining to non-convex cost functions, elliptical localized cost contours, and saddle points were discussed in some detail, along with the need for different optimizers to tackle such problems. Also, in the second half of the chapter, we caught up on TensorFlow basics and how to execute simple models pertaining to linear regression, multiclass SoftMax, and XOR classification through TensorFlow. In the next chapter, the emphasis is going to be on convolutional neural networks for images.

## CHAPTER 3

Convolutional Neural Networks

Artificial neural networks have flourished in recent years in the processing of unstructured data, especially images, text, audio, and speech. Convolutional neural networks (CNNs) work best for such unstructured data. Whenever there is a topology associated with the data, convolutional neural networks do a good job of extracting the important features out of the data. From an architectural perspective, CNNs are inspired by multi-layer Perceptrons. By imposing local connectivity constraints between neurons of adjacent layers, CNN exploits local spatial correlation.

The core element of convolutional neural networks is the processing of data through the convolution operation. Convolution of any signal with another signal produces a third signal that may reveal more information about the signal than the original signal itself. Let us go into detail about convolution before we dive into convolutional neural networks.

Convolution Operation

The convolution of a temporal or spatial signal with another signal produces a modified version of the initial signal. The modified signal may have better feature representation than the original signal suitable for a specific task. For example, by convolving a grayscale image as a 2D signal with another signal, generally called a filter or kernel, an output signal can be obtained that contains the edges of the original image. Edges in an image can correspond to object boundaries, changes in illumination, changes in material property, discontinuities in depth, and so on, which may be useful for several applications. Knowledge about the linear time invariance or shift invariance properties of systems helps one appreciate the convolution of signals better. We will discuss this first before moving on to convolution itself.

199

© Santanu Pattanayak 2023
S. Pattanayak, *Pro Deep Learning with TensorFlow 2.0*, https://doi.org/10.1007/978-1-4842-8931-0_3

Linear Time Invariant (LTI)/Linear Shift Invariant (LSI) Systems

A system works on an input signal in some way to produce an output signal. If an input signal $x(t)$ produces an output, $y(t)$, then $y(t)$ can be expressed as follows:

$$y(t) = f(x(t))$$

For the system to be linear, the following properties for scaling and superposition should hold true:

$$\textbf{Scaling} : f(\alpha x(t)) = \alpha f(x(t))$$

$$\textbf{Superposition} : f(\alpha x_1(t) + \beta x_2(t)) = \alpha f(x(t)) + \beta f(x_2(t))$$

Similarly, for the system to be time invariant or in general shift invariant,

$$f(x(t-\tau)) = y(t-\tau)$$

Such systems that have properties of linearity and shift invariance are termed linear shift invariant (LSI) systems in general. When such systems work on time signals, they are referred to as linear time invariant (LTI) systems. For the rest of the chapter, we will refer to such systems as LSI systems without any loss of generality. See Figure 3-1.

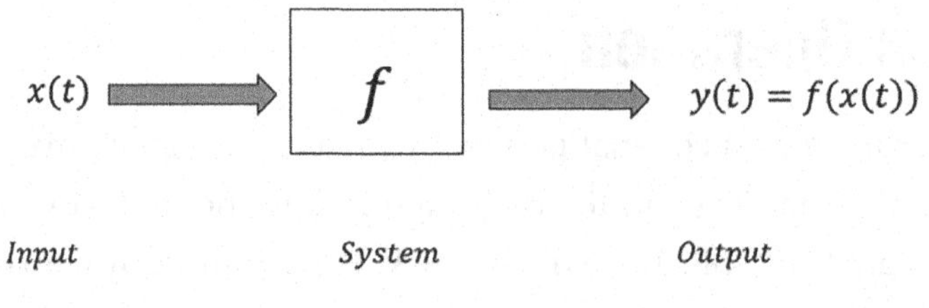

$x(t) \implies \boxed{f} \implies y(t) = f(x(t))$

Input   System   Output

Figure 3-1. *Input–output system*

The key feature of an LSI system is that if one knows the output of the system to an impulse response, then one can compute the output response to any signal.

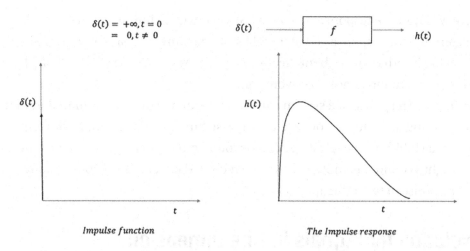

Figure 3-2a. *Response of an LSI system to an impulse (Dirac Delta) function*

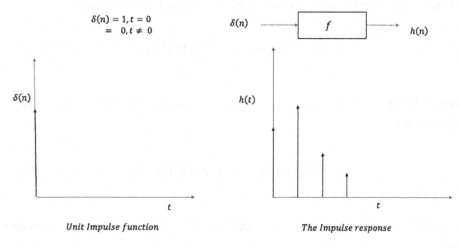

Figure 3-2b. *Response of an LTI system to a unit step impulse*

In Figures 3-2a and 3-2b, we illustrate the impulse response of the systems to different kinds of impulse functions. Figure 3-2a shows the continuous impulse response of the system to a Dirac Delta impulse, whereas Figure 3-2b shows the discrete impulse response of the system to a step impulse function. The system in Figure 3-2a is a continuous LTI system, and hence a Dirac Delta is required to determine its impulse response. On the other hand, the system in Figure 3-2b is a discrete LTI system, and so a unit step impulse is needed to determine its impulse response.

Once we know the response $h(t)$ of an LSI system to an impulse function $\delta(t)$, we can compute the response $y(t)$ of the LTI system to any arbitrary input signal $x(t)$ by convolving it with $h(t)$. Mathematically, it can be expressed as $y(t) = x(t)(^{*})h(t)$, where the $(^{*})$ operation denotes convolution.

The impulse response of a system can either be known or be determined from the system by noting down its response to an impulse function. For example, the impulse response of a Hubble space telescope can be found out by focusing it on a distant star in the dark night sky and then noting down the recorded image. The recorded image is the impulse response of the telescope.

Convolution for Signals in One Dimension

Intuitively, convolution measures the degree of overlap between one function and the reversed and translated version of another function. In the discrete case,

$$y(t)=x(t)\left(^{*}\right)h(t)=\sum_{\tau=-\infty}^{+\infty}x(\tau)h(t-\tau)$$

Similarly, in the continuous domain, the convolution of two functions can be expressed as follows:

$$y(t)=x(t)\left(^{*}\right)h(t)=\int_{\tau=-\infty}^{+\infty}x(\tau)h(t-\tau)d\tau$$

Let's perform convolution of two discrete signals to better interpret this operation. See Figures 3-3a to 3-3c.

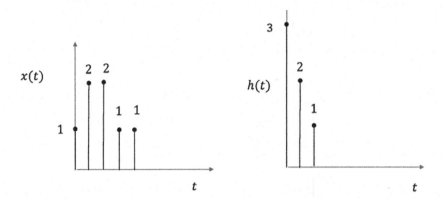

Figure 3-3a. *Input signals*

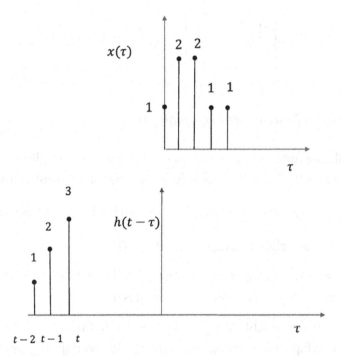

Figure 3-3b. *Functions for computing convolution operation*

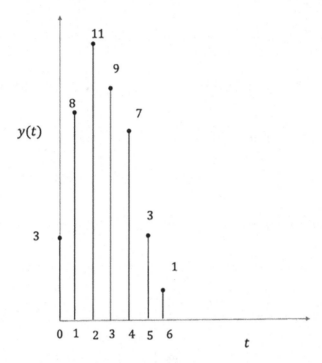

Figure 3-3c. *Output function from convolution*

In Figure 3-3b, the function $h(t - \tau)$ needs to be computed for different values of t by sliding it across the horizontal axis. At each value of t, the convolution sum $\sum_{\tau=-\infty}^{+\infty} x(\tau)h(t-\tau)$ needs to be computed. The sum can be thought of as a weighted average of $x(\tau)$ with the weights being provided by $h(t - \tau)$.

- When $t = -1$, the weights are given by $h(1 - \tau)$, but the weights do not overlap with $x(\tau)$ and hence the sum is 0.

- When $t = 0$, the weights are given by $h(-\tau)$, and the only element of $x(\tau)$ in overlap with the weights is $x(\tau = 0)$, the overlapping weight being $h(0)$. Hence, the convolving sum is $x(\tau = 0) * h(0) = 1*3 = 3$. Thus, $y(0) = 3$.

- When $t = 1$, the weights are given by $h(1 - \tau)$. The elements $x(0)$ and $x(1)$ are in overlap with the weights $h(1)$ and $h(0)$, respectively. Hence, the convolving sum is $x(0)*h(1) + x(1)*h(0) = 1*2 + 2*3 = 8$.

- When $t = 2$, the weights are given by $h(2 - \tau)$. The elements $x(0)$, $x(1)$, and $x(2)$ are in overlap with the weights $h(2)$, $h(1)$, and $h(0)$, respectively. Hence, the convolving sum is elements $x(0)^*h(2) + x(1)^*h(1) + x(2)^*h(0) = 1^*1 + 2^*2 + 2^*3 = 11$. The overlap of the two functions for $t = 2$ is illustrated in Figure 3-3d.

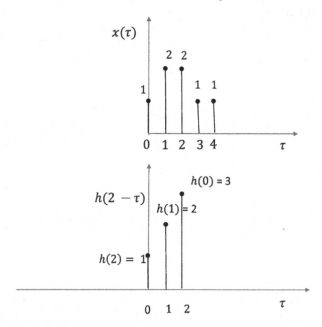

Figure 3-3d. *Overlap of the functions in convolution at t = 2*

Analog and Digital Signals

In general, any quantity of interest that shows variation in time and/or space represents a signal. Hence, a signal is a function of time and/or space. For instance, the stock market prices of a specific stock over a period of a week represent a signal.

Signals can be analogous or digital in nature. However, a computer can't process analogous continuous signals, so the signal is made into a digital signal for processing. For example, speech is an acoustic signal in time where both time and the amplitude of the speech energy are continuous signals. When the speech is transmitted through a microphone, this acoustic continuous signal is converted into an electrical continuous signal. If we want to process the analog electrical signal through a digital computer, we need to convert the analog continuous signal into a discrete signal. This is done through sampling and quantization of the analog signal.

Sampling refers to taking the signal amplitudes only at fixed spatial or time intervals. This has been illustrated in Figure 3-4a.

Not all possible continuous values of the signal amplitude are generally noted, but the signal amplitude is generally quantized to some fixed discrete values, as shown in Figure 3-4b. Through sampling and quantization, some information is lost from the analog continuous signal.

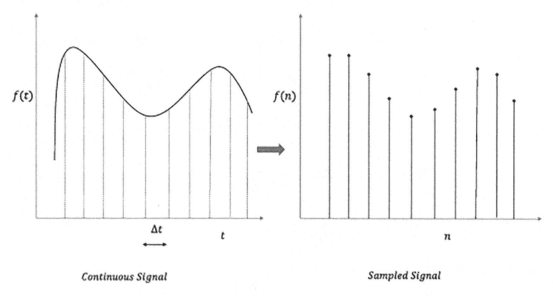

Figure 3-4a. Sampling of a signal

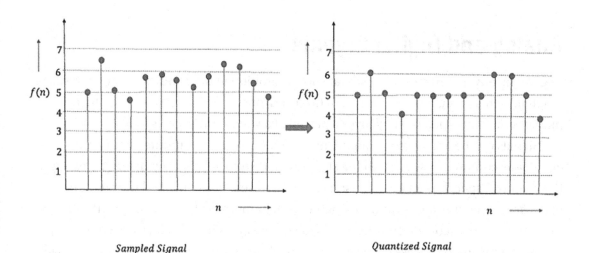

Figure 3-4b. Quantization of signal at discrete amplitude values

The activities of sampling and quantization convert an analog signal to a digital one.

A digital image can be expressed as a digital signal in the two-dimensional spatial domain. The colored RGB image has three channels: red, green, and blue. Each of the channels can be considered a signal in the spatial domain such that at each spatial location, the signal is represented by a pixel intensity. Each pixel can be represented by 8 bits, which in binary allows for 256 pixel intensities from 0 to 255. The color at any location is determined by the vector of pixel intensities at that location corresponding to the three channels. So, to represent a specific color, 24 bits of information is used. For a grayscale image, there is only 1 channel, and the pixel intensities range from 0 to 255. 255 represents the color white, while 0 represents the color black.

A video is a sequence of images with a temporal dimension. A black and white video can be expressed as a signal of its spatial and temporal coordinates (x, y, t). A colored video can be expressed as a combination of three signals, with the spatial and temporal coordinates corresponding to the three color channels—red, green, and blue.

So, a grayscale $n \times m$ image can be expressed as function $I(x, y)$, where I denotes the intensity of the pixel at the x, y coordinate. For a digital image, x, y are sampled coordinates and take discrete values. Similarly, the pixel intensity is quantized between 0 and 255.

2D and 3D Signals

A grayscale image of dimension $N \times M$ can be expressed as a scalar 2D signal of its spatial coordinates. The signal can be represented as follows:

$$x(n_1, n_2), 0 < n_1 < M - 1, 0 < n_2 < N - 1$$

where n_1 and n_2 are the discrete spatial coordinates along the horizontal and vertical axes, respectively, and $x(n_1, n_2)$ denotes the pixel intensity at the spatial coordinates. The pixel intensities take up values from 0 to 255.

A colored RGB image is a vector 2D signal since there is a vector of pixel intensities at each spatial coordinate. For an RGB image of dimensions $N \times M \times 3$, the signal can be expressed as follows:

$$x(n_1, n_2) = \left[x_R(n_1, n_2), x_G(n_1, n_2), x_B(n_1, n_2) \right], 0 < n_1 < M - 1, 0 < n_2 < N - 1$$

where x_R, x_G, and x_B denote the pixel intensities along the red, green, and blue color channels. See Figures 3-5a and 3-5b.

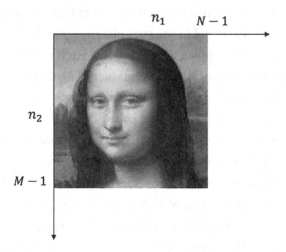

Figure 3-5a. *Grayscale image as a 2D discrete signal*

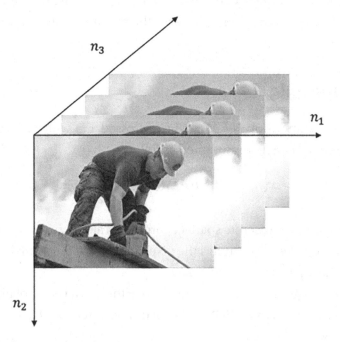

Figure 3-5b. *Video as a 3D object*

2D Convolution

Now that we have expressed grayscale images as 2D signals, we would like to process those signals through 2D convolution. The images can be convolved with the impulse response of an image-processing system to achieve different objectives, such as the following:

- Remove the visible noise in the image through noise-reduction filters. For white noise, we can use a Gaussian filter. For salt and pepper noise, a median filter can be used.

- For detecting edges, we need filters that extract high-frequency components from an image.

The image-processing filters can be thought of as image-processing systems that are linear and shift invariant. Before we go to image processing, it's worthwhile to know the different impulse functions.

Two-Dimensional Unit Step Function

A two-dimensional unit step function $\delta(n_1, n_2)$, where n_1 and n_2 are the horizontal and vertical coordinates, can be expressed as follows:

$$\delta(n_1,n_2)=1 \ when \ n_1=0 \ and \ n_2=0$$
$$=0 \ elsewhere$$

Similarly, a shifted unit step function can be expressed as follows:

$$\delta(n_1-k_1,n_2-k_2)=1 \ when \ n_1=k_1 \ and \ n_2=k_2$$
$$=0 \ elsewhere$$

This has been illustrated in Figure 3-6.

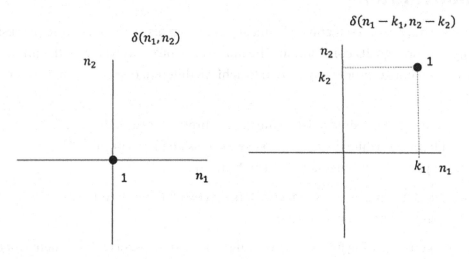

Figure 3-6. *Unit step functions*

Any discrete two-dimensional signal can be expressed as the weighted sum of unit step functions at different coordinates. Let us consider the signal $x(n_1, n_2)$ as shown in Figure 3-7.

$$x(n_1,n_2)=1\, when\, n_1 =0\, and\, n_2 =0$$
$$=2\, when\, n_1 =0\, and\, n_2 =1$$
$$=3\, when\, n_1 =1\, and\, n_2 =1$$
$$=0\, elsewhere$$

$$x(n_1,n_2)=x(0,0)^*\delta(n_1,n_2)+x(0,1)^*\delta(n_1,n_2 -1)+x(1,1)^*\delta(n_1 -1,n_2 -1)$$
$$=1^*\delta(n_1,n_2)+2^*\delta(n_1,n_2 -1)+3^*\delta(n_1 -1,n_2 -1)$$

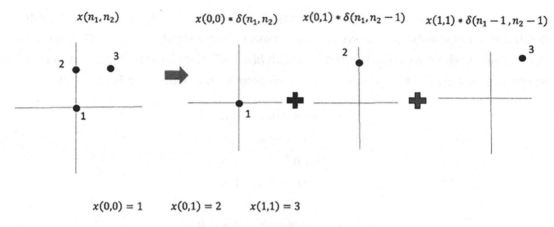

$x(0,0) = 1$ $x(0,1) = 2$ $x(1,1) = 3$

Figure 3-7. *Representing a 2D discrete signal as the weighted sum of unit step functions*

So, in general, any discrete 2D signal can be written as follows:

$$x(n_1,n_2) = \sum_{k_2=-\infty}^{+\infty} \sum_{k_1=-\infty}^{+\infty} x(k_1,k_2)\delta(n_1-k_1,n_2-k_2)$$

2D Convolution of a Signal with an LSI System Unit Step Response

When any discrete 2D signal as expressed above signal passes through an LSI system with transformation f, then, because of the linearity property of LSI systems,

$$f(x(n_1,n_2)) = \sum_{k_2=-\infty}^{+\infty} \sum_{k_1=-\infty}^{+\infty} x(k_1,k_2)f(\delta(n_1-k_1,n_2-k_2))$$

Now, the unit step response of an LSI system $f(\delta(n_1, n_2) = h(n_1, n_2)$, and since an LSI system is shift invariant, $f(\delta(n_1 - k_1, n_2 - k_2) = h(n_1 - k_1, n_2 - k_2)$.

Hence, $f(x(n_1, n_2))$ can be expressed as follows:

$$f(x(n_1,n_2)) = \sum_{k_2=-\infty}^{+\infty} \sum_{k_1=-\infty}^{+\infty} x(k_1,k_2)h(n_1-k_1,n_2-k_2) \qquad (1)$$

The preceding expression denotes the expression for 2D convolution of a signal with the unit step response of an LSI system. To illustrate 2D convolution, let's walk through an example in which we convolve $x(n_1, n_2)$ with $h(n_1, n_2)$. The signal and the unit step response signal are defined as follows and have also been illustrated in Figure 3-8:

$$x(n_1, n_2) = 4 \; when \; n_1 = 0, n_2 = 0$$
$$= 5 \; when \; n_1 = 1, n_2 = 0$$
$$= 2 \; when \; n_1 = 0, n_2 = 1$$
$$= 3 \; when \; n_1 = 1, n_2 = 1$$
$$= 0 \; elsewhere$$
$$h(n_1, n_2) = 1 \; when \; n_1 = 0, n_2 = 0$$
$$= 2 \; when \; n_1 = 1, n_2 = 0$$
$$= 3 \; when \; n_1 = 0, n_2 = 1$$
$$= 4 \; when \; n_1 = 1, n_2 = 1$$
$$= 0 \; elsewhere$$

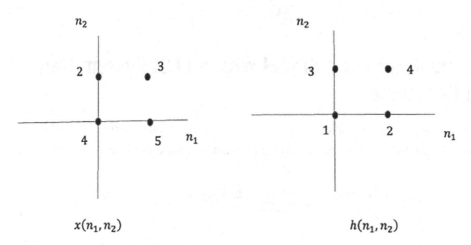

$$x(n_1, n_2) \qquad h(n_1, n_2)$$

Figure 3-8. *2D signal and unit step response of LSI system*

To compute the convolution, we need to plot the signals on a different set of coordinate points. We chose k_1 and k_2 on the horizontal and vertical axes, respectively. Also, we reverse the impulse response $h(k_1, k_2)$ to $h(-k_1, -k_2)$ as plotted in Figure 3-9b. We then place the reversed function $h(-k_1, -k_2)$ at different offset values for n_1 and n_2. The generalized reversed function can be expressed as $h(n_1 - k_1, n_2 - k_2)$. For computing the output $y(n_1, n_2)$ of convolution at a specific value of n_1 and n_2, we see the points at which $h(n_1 - k_1, n_2 - k_2)$ overlaps with $x(k_1, k_2)$ and take the total sum of the coordinate-wise product of the signal and impulse response values as the output.

As we can see in Figure 3-9c, for the $(n_1 = 0, n_2 = 0)$ offset, the only point of overlap is $(k_1 = 0, k_2 = 0)$ and so $y(0,0) = x(0,0)*h(0,0) = 4*1 = 4$.

Similarly, for offset $(n_1 = 1, n_2 = 0)$, the points of overlap are the points $(k_1 = 0, k_2 = 0)$ and $(k_1 = 1, k_2 = 0)$, as shown in Figure 3-9d.

$$
\begin{aligned}
y(1,0) &= x(0,0)^* h(1-0,0-0) + x(1,0)^* h(1-1,0-0) \\
&= x(0,0)^* h(1,0) + x(1,0)^* h(0,0) \\
&= 4^*2 + 5^*1 = 13
\end{aligned}
$$

For offset $(n_1 = 1, n_2 = 1)$, the points of overlap are the points $(k_1 = 1, k_2 = 0)$, as shown in Figure 3-9e.

$$
\begin{aligned}
y(2,0) &= x(1,0)^* h(2-1,0-0) \\
&= x(1,0)^* h(1,0) \\
&= 5^*2 = 10
\end{aligned}
$$

Following this approach of shifting the unit step response signal by altering n_1 and n_2, the entire function $y(n_1, n_2)$ can be computed.

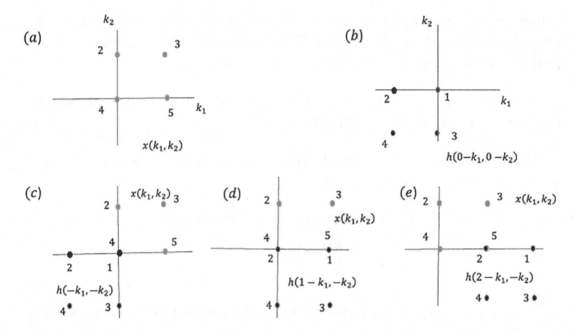

Figure 3-9. *Convolution at different coordinate points*

2D Convolution of an Image to Different LSI System Responses

Any image can be convolved with an LSI system's unit step response. Those LSI system unit step responses are called filters or kernels. For example, when we try to take an image through a camera and the image gets blurred because of shaking of hands, the blur introduced can be treated as an LSI system with a specific unit step response. This unit step response convolves the actual image and produces the blurred image as output. Any image that we take through the camera gets convolved with the unit step response of the camera. So, the camera can be treated as an LSI system with a specific unit step response.

Any digital image is a 2D discrete signal. The convolution of an $N \times M$ 2D image $x(n_1, n_2)$ with a 2D image-processing filter $h(n_1, n_2)$ is given by the following:

$$y(n_1, n_2) = \sum_{k_2=0}^{N-1} \sum_{k_1=0}^{M-1} x(k_1, k_2) h(n_1 - k_1, n_2 - k_2)$$

$$0 \leq n_1 \leq N-1, 0 \leq n_2 \leq M-1.$$

The image-processing filters work on a grayscale image's (2D) signal to produce another image (2D signal). In cases of multichannel images, generally 2D image-processing filters are used for image processing, which means one must process each image channel as a 2D signal or convert the image into a grayscale image.

Now that we have gone through the concepts of convolution, we know to term any unit step response of an LSI system with which we convolve an image as a filter or kernel.

An example of 2D convolution is illustrated in Figure 3-10a.

1	2	3	4	5	6	7
8	9	10	11	12	13	14
15	16	17	18	19	20	21
22	23	24	25	26	27	28
29	30	31	32	33	34	35
36	37	38	39	40	41	42
43	44	45	46	47	48	49

Image

-1	1	-1
-2	3	1
2	-4	0

Filter kernel

0	0	0	0	0	0	0	0	0
0	1	2	3	4	5	6	7	0
0	8	9	10	11	12	13	14	0
0	15	16	17	18	19	20	21	0
0	22	23	24	25	26	27	28	0
0	29	30	31	32	33	34	35	0
0	36	37	38	39	40	41	42	0
0	43	44	45	46	47	48	49	0
0	0	0	0	0	0	0	0	0

Image padded with zeroes at the boundaries

0	-4	2
1	3	-2
-1	1	-1

Flipped Filter kernel

Figure 3-10a. *Example of 2D convolution of images*

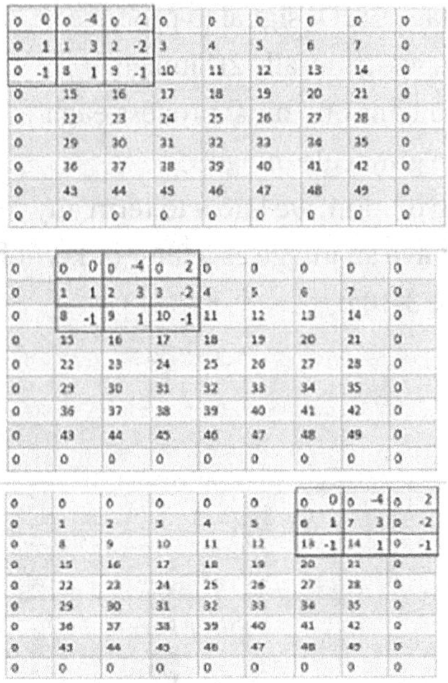

0	0	0	-4	0	2	0	0	0	0	0	0
0	1	1	3	2	-2	3	4	5	6	7	0
0	-1	8	1	9	-1	10	11	12	13	14	0
0		15	16	17	18	19	20	21			0
0		22	23	24	25	26	27	28			0
0		29	30	31	32	33	34	35			0
0		36	37	38	39	40	41	42			0
0		43	44	45	46	47	48	49			0
0		0	0	0	0	0	0	0			0

$$I[0,0] = 1 \times 3 + 2 \times -2 + 8 \times 1 + 9 \times -1 = -2$$

0	0	0	0	-4	0	2	0	0	0	0	0
0		1	1	2	3	3	-2	4	5	6	7	0
0		8	-1	9	1	10	-1	11	12	13	14	0
0		15		16	17	18	19	20	21	0
0		22		23	24	25	26	27	28	0
0		29		30	31	32	33	34	35	0
0		36		37	38	39	40	41	42	0
0		43		44	45	46	47	48	49	0
0		0		0	0	0	0	0	0	0

$$I[0,1] = 1 \times 1 + 2 \times 3 + 3 \times -2 + 8 \times -1 + 9 \times 1 + 10 \times -1 = -8$$

0	0	0	0	0	0	0	0	-4	0	2
0	1	2	3	4	5	6	1	7	3	0	-2
0	8	9	10	11	12	13	-1	14	1	0	-1
0	15	16	17	18	19	20	21	0
0	22	23	24	25	26	27	28	0
0	29	30	31	32	33	34	35	0
0	36	37	38	39	40	41	42	0
0	43	44	45	46	47	48	49	0
0	0	0	0	0	0	0	0	0

$$I[0,6] = 6 \times 1 + 7 \times 3 + 13 \times -1 + 14 \times 1 = 28$$

Figure 3-10b.

To keep the length of the output image the same as that of the input image, the original image has been zero padded. As we can see, the flipped filter or kernel is slid over various areas of the original image, and the convolution sum is computed at each coordinate point. Please note that the indexes in the intensity $I[i, j]$ as mentioned in Figure 3-10b denote the matrix coordinates. The same example problem is worked out through scipy 2D convolution as well as through basic logic in Listing 3-1. In both cases, the results are the same.

Listing 3-1. Illustrate 2D convolution of Images through a Toy example¶

```
## Illustrate 2D convolution of Images through a Toy example
import scipy
import scipy.signal
import numpy as np
print(f"scipy version: {scipy.__version__}")
print(f"numpy version: {np.__version__}")
print('\n')
```

```
# Take a 7x7 image as example

image = np.array([[1, 2, 3, 4, 5, 6, 7],
                  [8, 9, 10, 11, 12, 13, 14],
                  [15, 16, 17, 18, 19, 20, 21],
                  [22, 23, 24, 25, 26, 27, 28],
                  [29, 30, 31, 32, 33, 34, 35],
                  [36, 37, 38, 39, 40, 41, 42],
                  [43, 44, 45, 46, 47, 48, 49]])

# Defined an image processing kernel

filter_kernel = np.array([[-1, 1, -1],
                          [-2, 3, 1],
                          [2, -4, 0]])

# Convolve the image with the filter kernel through scipy 2D convolution to
produce an output image of same dimension as that of the input

I = scipy.signal.convolve2d(image, filter_kernel,mode='same',
boundary='fill', fillvalue=0)
print(f'Scipy convolve2d output\n')
print(I)
print('\n')

# We replicate the same logic of a Scipy 2D convolution by following the
below steps
# a) The boundaries need to be extended in both directions for the image
and padded with zeroes.
#     For convolving the 7x7 image by 3x3 kernel the dimensions needs to be
      extended by (3-1)/2 i.e 1
#     on either size for each dimension. So a skeleton image of 9x9 image
      would be created
#     in which the boundaries of 1 pixel are pre-filled with zero.
# b) The kernel needs to be flipped i.e rotated by 180 degrees
# c) The flipped kernel needs to placed at each cordinate location for the
image and then the sum of
```

```
#     cordinatewise product with the image intensities need to be computed.
      These sum for each co-ordinate would give
#     the intensities for the output image.

row,col=7,7

## Rotate the filter kernel twice by 90 degree to get 180 rotation

filter_kernel_flipped = np.rot90(filter_kernel,2)

## Pad the boundaries of the image with zeroes and fill the rest from the
original image

image1 = np.zeros((9,9))

for i in range(row):
    for j in range(col):
        image1[i+1,j+1] = image[i,j]

#print(image1)

## Define the output image

image_out = np.zeros((row,col))

## Dynamic shifting of the flipped filter at each image cordinate and then
computing the convolved sum.

for i in range(1,1+row):
    for j in range(1,1+col):
        arr_chunk = np.zeros((3,3))
        for k,k1 in zip(range(i-1,i+2),range(3)):
            for l,l1 in zip(range(j-1,j+2),range(3)):
                arr_chunk[k1,l1] = image1[k,l]

        image_out[i-1,j-1] = np.sum(np.multiply(arr_chunk,filter_kernel_
        flipped))
print(f"2D convolution implementation\n")
print(image_out)
```

```
[[ -2  -8  -7  -6  -5  -4  28]          [[ -2.  -8.  -7.  -6.  -5.  -4.  28.]
 [  5  -3  -4  -5  -6  -7  28]           [  5.  -3.  -4.  -5.  -6.  -7.  28.]
 [ -2 -10 -11 -12 -13 -14  28]           [ -2. -10. -11. -12. -13. -14.  28.]
 [ -9 -17 -18 -19 -20 -21  28]           [ -9. -17. -18. -19. -20. -21.  28.]
 [-16 -24 -25 -26 -27 -28  28]           [-16. -24. -25. -26. -27. -28.  28.]
 [-23 -31 -32 -33 -34 -35  28]           [-23. -31. -32. -33. -34. -35.  28.]
 [-29  13  13  13  13  13  27]]          [-29.  13.  13.  13.  13.  13.  27.]]
```

Scipy convolve2d output   *2D convolution implementation*

Figure 3-11.

As we can see in Figure 3-11 the output of the scipy based convolution matches our 2D convolution implementation in Listing 3-1.

Based on the choice of image-processing filter, the nature of the output images will vary. For example, a Gaussian filter would create an output image that would be a blurred version of the input image, whereas a Sobel filter would detect the edges in an image and produce an output image that contains the edges of the input image.

Common Image-Processing Filters

Let us discuss image-processing filters commonly used on 2D images. Make sure to be clear with notations since the natural way of indexing an image does not align well with how one would prefer to define the x and y axes. Whenever we represent an image-processing filter or an image in the coordinate space, n_1 and n_2 are the discrete coordinates for the x and y directions. The column index of the image in Numpy matrix form coincides nicely with the x axis, whereas the row index moves in the opposite direction of the y axis. Also, it does not matter which pixel location one chooses as the origin for the image signal while doing convolution. Based on whether zero padding is used or not, one can handle the edges accordingly. Since the filter kernel is of a smaller size, we generally flip the filter kernel and then slide it over the image and not the other way around.

Mean Filter

The mean filter or average filter is a low-pass filter that computes the local average of the pixel intensities at any specific point. The impulse response of the mean filter can be any of the form seen here (see Figure 3-12):

$$\begin{bmatrix} 1/9 & 1/9 & 1/9 \\ 1/9 & 1/9 & 1/9 \\ 1/9 & 1/9 & 1/9 \end{bmatrix}$$

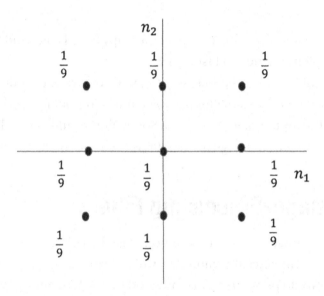

Figure 3-12. *Impulse response of a mean filter*

Here, the matrix entry h_{22} corresponds to the entry at the origin. So, at any given point, the convolution will represent the average of the pixel intensities at that point. The code in Listing 3-2 illustrates how one can convolve an image with an image-processing filter such as the mean filter.

Please note that in many Python implementations, we would be using OpenCV to perform basic operations on the image, such as reading the image, converting the image from RGB format to grayscale format, and so on. OpenCV is an open source image-processing package that has a rich set of methodologies for image processing. Readers are advised to explore OpenCV or any other image-processing toolbox to get accustomed to the basic image-processing functions.

Listing 3-2. Convolution of an Image with Mean Filter

```
import cv2
print("Opencv version",cv2.__version__)
import matplotlib.pyplot as plt
from scipy.signal import convolve2d
%matplotlib inline

img = cv2.imread('monalisa.jpeg')
gray = cv2.cvtColor(img, cv2.COLOR_BGR2GRAY)
plt.imshow(gray,cmap='gray')
mean = 0
var = 100
sigma = var**0.5
row,col = np.shape(gray)
gauss = np.random.normal(mean,sigma,(row,col))
gauss = gauss.reshape(row,col)
gray_noisy = gray + gauss
print(f"Image after applying Gaussian Noise")
plt.imshow(gray_noisy,cmap='gray')
```

In Listing 3-2, we read an image of the *Mona Lisa* and then introduce some Gaussian white noise into the image. The Gaussian noise has a mean of 0 and a variance of 100. We then convolve the noisy image with a mean filter to reduce the white noise. The noisy image and the image after convolution has been plotted are shown in Figure 3-13.

Noisy Image with Gaussian Noise   *Image after Convolving with Mean Filter*

Figure 3-13. *Mean filter processing on Mona Lisa image*

The mean filter is mostly used to reduce the noise in an image. If there is some white Gaussian noise present in the image, then the mean filter will reduce the noise since it averages over its neighborhood, and hence the white noise of the zero mean will be suppressed. As we can see from Figure 3-13, the Gaussian white noise is reduced once the image has been convolved with the mean filter. The new image has fewer high-frequency components and thus is relatively less sharp than the image before convolution, but the filter has done a good job of reducing the white noise.

Median Filter

A 2D median filter replaces each pixel in a neighborhood with the median pixel intensity in that neighborhood based on the filter size. The median filter is good for removing salt and pepper noise. This type of noise presents itself in the images in the form of black and white pixels and is generally caused by sudden disturbances while capturing the images. Listing 3-3 illustrates how salt and pepper noise can be added to an image and then how the noise can be suppressed using a median filter.

Listing 3-3. Median Filter Illustration

```
#----------------------------------------------------------------------
# First create an image with Salt and Pepper Noise
#----------------------------------------------------------------------
# Generate random integers from 0 to 20
# If the value is zero we will replace the image pixel with a low value of
0 that corresponds to a black pixel
# If the value is 20 we will replace the image pixel with a high value of
255 that correspondsa to a white pixel
# Since we have taken 20 intergers and out of which we will only tag
integers 1 and 20 as salt and pepper noise
# hence we would have approximately 10% of the overall pixels as salt and
pepper noise. If we want to reduce it
# to 5 % we can taken integers from 0 to 40 and then treat 0 as indicator
for black pixel and 40 as an indicator
# for white pixel.

np.random.seed(0)

gray_sp = gray*1
sp_indices = np.random.randint(0,21,[row,col])

for i in range(row):
    for j in range(col):
        if sp_indices[i,j] == 0:
            gray_sp[i,j] = 0
        if sp_indices[i,j] == 20:
            gray_sp[i,j] = 255
print(f"Image after applying Salt and Pepper Noise\n")
      plt.imshow(gray_sp,cmap='gray')
      #----------------------------------------------------------------------
      # Remove the Salt and Pepper Noise
      #----------------------------------------------------------------------
      # Now we want to remove the salt and pepper noise through a
        median filter.
      # Using the opencv Median Filter for the same
```

```
gray_sp_removed = cv2.medianBlur(gray_sp,3)
print(f"Removing Salt and Pepper Noise with OpenCV Median
Filter\n")
plt.imshow(gray_sp_removed,cmap='gray')

# Implementation of the 3x3 Median Filter without using openc
gray_sp_removed_exp = gray*1

for i in range(row):
    for j in range(col):
local_arr = []
for k in range(np.max([0,i-1]),np.min([i+2,row])):
    for l in range(np.max([0,j-1]),np.min([j+2,col])):
        local_arr.append(gray_sp[k,l])

                gray_sp_removed_exp[i,j] = np.median(local_arr)
        print(f"Image produced by applying Median Filter Logic\n")
        plt.imshow(gray_sp_removed_exp,cmap='gray')
```

Figure 3-14. *Median filter processing*

As we can see in Figure 3-14, the salt and pepper noise has been removed by the median filter.

Gaussian Filter

The Gaussian filter is a modified version of the mean filter where the weights of the impulse function are distributed normally around the origin. Weight is highest at the center of the filter and falls normally away from the center. A Gaussian filter can be created with the code in Listing 3-4. As we can see, the intensity falls in a Gaussian fashion away from the origin. The Gaussian filter, when displayed as an image, has the highest intensity at the origin and then diminishes for pixels away from the center. Gaussian filters are used to reduce noise by suppressing the high-frequency components. However, in its pursuit of suppressing the high-frequency components, it ends up producing a blurred image, called Gaussian blur.

In Figure 3-15, the original image is convolved with the Gaussian filter to produce an image that has Gaussian blur. We then subtract the blurred image from the original image to get the high-frequency component of the image. A small portion of the high-frequency image is added to the original image to improve the sharpness of the image.

Listing 3-4. Illustration of Gaussian Filter

```
# Creating the Gaussian Filter
Hg = np.zeros((20,20))

for i in range(20):
    for j in range(20):
        Hg[i,j] = np.exp(-((i-10)**2 + (j-10)**2)/10)
print(f"Gaussian Blur Filter\n")
plt.imshow(Hg,cmap='gray')

gray_blur = convolve2d(gray,Hg,mode='same')
print(f"Image after convolving with  Gaussian Blur Filter Created above\n")
plt.imshow(gray_blur,cmap='gray')

gray_high = gray - gray_blur
print(f"High Frequency Component of Image\n")
plt.imshow(gray_high,cmap='gray')
gray_enhanced = gray + 0.025*gray_high
print(f"Enhanced Image with some portion of High Frequency Component
added\n")
plt.imshow(gray_enhanced,cmap='gray')
```

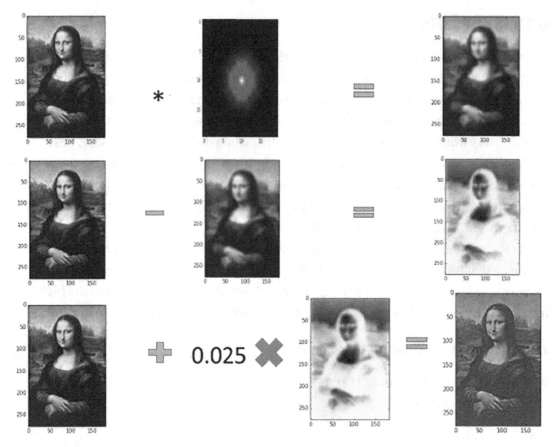

Figure 3-15. *Various activities with Gaussian filter kernel*

Gradient-Based Filters

To review, the gradient of a two-dimensional function $I(x, y)$ is given by the following:

$$\nabla I(x,y) = \left[\frac{\partial I(x,y)}{\partial x} \frac{\partial I(x,y)}{\partial y} \right]^{T}$$

where the gradient along the horizontal direction is given by - $\frac{\partial I(x,y)}{\partial x} =$ $\lim_{h \to 0} \frac{I(x+h,y)-I(x,y)}{h}$ or $\lim_{h \to 0} \frac{I(x+h,y)-I(x-h,y)}{2h}$ based on convenience and the problem at hand.

For discrete coordinates, we can take $h = 1$ and approximate the gradient along the horizontal as follows:

$$\frac{\partial I(x,y)}{\partial x} = I(x+1,y) - I(x,y)$$

This derivative of a signal can be achieved by convolving the signal with the filter kernel $\begin{bmatrix} 0 & 0 & 0 \\ 0 & 1 & -1 \\ 0 & 0 & 0 \end{bmatrix}$.

Similarly,

$$\frac{\partial I(x,y)}{\partial x} \propto I(x+1,y) - I(x-1,y)$$

from the second representation.

This form of derivative can be achieved by convolving the signal with the filter kernel $\begin{bmatrix} 0 & 0 & 0 \\ 1 & 0 & -1 \\ 0 & 0 & 0 \end{bmatrix}$.

For the vertical direction, the gradient component in the discrete case can be expressed as follows:

$$\frac{\partial I(x,y)}{\partial y} = I(x,y+1) - I(x,y) \text{ or by } \frac{\partial I(x,y)}{\partial y} \propto I(x,y+1) - I(x,y-1)$$

The corresponding filter kernels to compute gradients through convolution are $\begin{bmatrix} 0 & -1 & 0 \\ 0 & 1 & 0 \\ 0 & 0 & 0 \end{bmatrix}$ and $\begin{bmatrix} 0 & -1 & 0 \\ 0 & 0 & 0 \\ 0 & 1 & 0 \end{bmatrix}$, respectively.

Do note that these filters take the direction of the x axis and y axis, as shown in Figure 3-16. The direction of x agrees with the matrix index n_2 increment, whereas the direction of y is opposite to that of the matrix index n_1 increment.

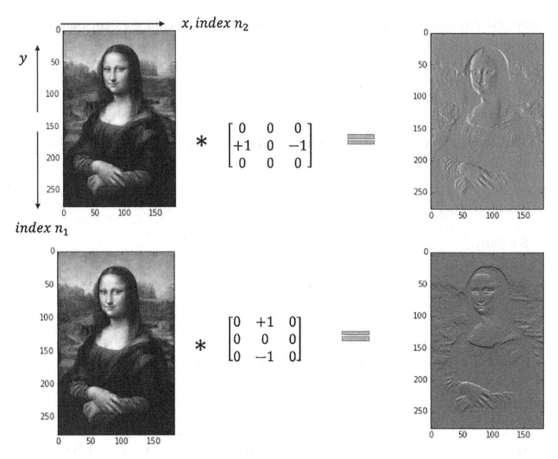

Figure 3-16. *Vertical and horizontal gradient filters*

Figure 3-16 illustrates the convolution of the *Mona Lisa* image with the horizontal and vertical gradient filters.

Sobel Edge-Detection Filter

The impulse response of a Sobel edge detector along the horizontal and vertical axes can be expressed by the following H_x and H_y matrices, respectively. The Sobel detectors are extensions of the horizontal and vertical gradient filters just illustrated. Instead of only taking the gradient at the point, it also takes the sum of the gradients at the points on either side of it. Also, it gives double weight to the point of interest. See Figure 3-17.

$$Hx = \begin{bmatrix} 1 & 0 & -1 \\ 2 & 0 & -2 \\ 1 & 0 & -1 \end{bmatrix} = \begin{bmatrix} 1 \\ 2 \\ 1 \end{bmatrix} \begin{bmatrix} 1 & 0 & -1 \end{bmatrix}$$

$$Hy = \begin{bmatrix} -1 & -2 & -1 \\ 0 & 0 & 0 \\ 1 & 2 & 1 \end{bmatrix}$$

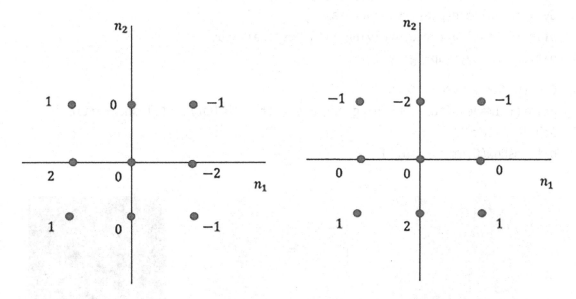

Horizontal Sobel Filter   *Vertical Sobel Filter*

Figure 3-17. *Sobel filter impulse response*

The convolution of the image with the Sobel filters is illustrated in Listing 3-5.

Listing 3-5. Convolution Using a Sobel Filter

```
Hx = np.array([[ 1,0, -1],[2,0,-2],[1,0,-1]],dtype=np.float32)
Gx = convolve2d(gray,Hx,mode='same')
print(f"Image after convolving with Horizontal Sobel Filter\n")
plt.imshow(Gx,cmap='gray')

Hy = np.array([[ -1,-2, -1],[0,0,0],[1,2,1]],dtype=np.float32)
Gy = convolve2d(gray,Hy,mode='same')
print(f"Image after convolving with Vertical Sobel Filter\n")
plt.imshow(Gy,cmap='gray')

G = (Gx*Gx + Gy*Gy)**0.5
print(f'Image after combining outputs from both Horizontal and Vertical
Sobel Filters')
plt.imshow(G,cmap='gray')
```

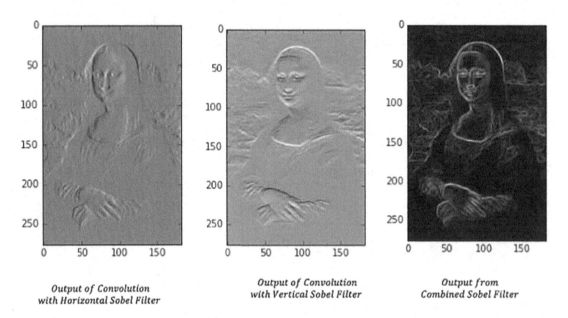

Output of Convolution
with Horizontal Sobel Filter

Output of Convolution
with Vertical Sobel Filter

Output from
Combined Sobel Filter

Figure 3-18. *Output of various Sobel filters*

Listing 3-5 has the logic required to convolve the image with the Sobel filters. The horizontal Sobel filter detects edges in the horizontal direction, whereas the vertical Sobel filter detects edges in the vertical direction. Both are high-pass filters since they attenuate the low frequencies from the signals and capture only the high-frequency components within the image. Edges are important features for an image and help one detect local changes within an image. Edges are generally present on the boundary between two regions in an image and are often the first step in retrieving information from images. We see the output of Listing 3-5 in Figure 3-18. The pixel values obtained for the images with horizontal and vertical Sobel filters for each location can be thought of as a vector $I'(x, y) = [I_x(x, y)I_y(x, y)]^T$, where $I_x(x, y)$ denotes the pixel intensity of the image obtained through the horizontal Sobel filter and $I_y(x, y)$ denotes the pixel intensity of the image obtained through the vertical Sobel filter. The magnitude of the vector $I'(x, y)$ can be used as the pixel intensity of the combined Sobel filter.

$$C(x,y) = \sqrt{\left(I_x(x,y)\right)^2 + \left(I_y(x,y)\right)^2}$$, where $C(x, y)$ denotes the pixel intensity function for the combined Sobel filter.

Identity Transform

The filter for identity transform through convolution is as follows:

$$\begin{bmatrix} 0 & 0 & 0 \\ 0 & 1 & 0 \\ 0 & 0 & 0 \end{bmatrix}$$

Figure 3-19 illustrates a unity transform through convolution.

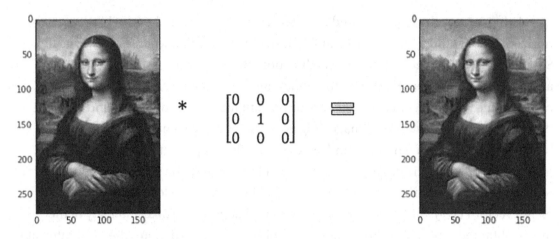

Figure 3-19. *Identity transform through convolution*

Table 3-1 lists several useful image-processing filters and their uses.

Table 3-1. *Image-Processing Filters and Their Uses*

Filter	Use
Mean filter	Reduce Gaussian noise; smooth the image after upsampling
Median filter	Reduce salt and pepper noise
Sobel filter	Detect edges in an image
Gaussian filter	Reduce noise in an image
Canny filter	Detect edges in an image
Weiner filter	Reduce additive noise and blurring

Convolution Neural Networks

Convolution neural networks (CNNs) are based on the convolution of images and detect features based on filters that are learned by the CNN through training. For example, we do not apply any known filter, such as the ones for the detection of edges or for removing the Gaussian noise, but through the training of the convolutional neural network, the algorithm learns image-processing filters on its own that might be very different from normal image-processing filters. For supervised training, the filters are learned in such a way that the overall cost function is reduced as much as possible. Generally, the first convolution layer learns to detect edges, while the second may learn to detect more complex shapes that can be formed by combining different edges, such as circles and rectangles, and so on. The third layer and beyond learn much more complicated features based on the features generated in the previous layer.

The good thing about convolutional neural networks is the sparse connectivity that results from weight sharing, which greatly reduces the number of parameters to learn. The same filter can learn to detect the same edge in any given portion of the image through its equivariance property, which is a great property of convolution useful for feature detection.

Components of Convolution Neural Networks

The following are the typical components of a convolutional neural network:

> *Input layer* will hold the pixel intensity of the image. For example, an input image with width 64, height 64, and depth 3 for the red, green, and blue color channels (RGB) would have input dimensions of 64 × 64 × 3.

> *Convolution layer* will take images from the preceding layers and convolve with them the specified number of filters to create images called *output feature maps*. The number of output feature maps is equal to the specified number of filters. Till now, CNNs in TensorFlow have used mostly 2D filters; however, recently 3D convolution filters have been introduced.

> *Activation function* for CNNs are generally ReLUs, which we discussed in Chapter 2. The output dimension is the same as the input after passing through the ReLU activation layers. The ReLU layer adds nonlinearity in the network and at the same time provides non-saturating gradients for positive net inputs.

> *Pooling layer* will downsample the 2D activation maps along the height and width dimensions. The depth or the number of activation maps is not compromised and remains the same.

Fully connected layers contain traditional neurons that receive different sets of weights from the preceding layers; there is no weight sharing between them as is typical for convolution operations. Each neuron in this layer will be connected either to all the neurons in the previous layer or to all the coordinate-wise outputs in the output maps through separate weights. For classification, the class output neurons receive inputs from the final fully connected layers.

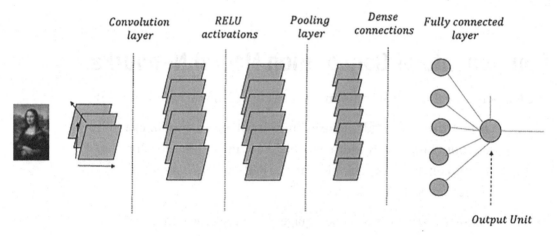

Figure 3-20. *Basic flow diagram of a convolutional neural network*

Figure 3-20 illustrates a basic convolutional neural network (CNN) that uses one convolutional layer, one ReLU layer, and one pooling layer followed by a fully connected layer and finally the output classification layer. The network tries to discern the *Mona Lisa* images from the non–*Mona Lisa* images. The output unit can be taken to have a sigmoid activation function since it's a binary classification problem for images. Generally, for most of the CNN architectures, a few to several convolutional layer-ReLU layer-pooling layer combinations are stacked one after another before the fully connected layers. We will discuss the different architectures at a later point in time. For now, let's look at the different layers in much more detail.

Input Layer

The input to this layer is images. Generally, the images are fed in batches as four-dimensional tensors where the first dimension is specific to the image index, second and third dimensions are specific to the height and width of the image, and the fourth dimension corresponds to the different channels. For a colored image, generally we have the red (R), green (G), and blue (B) channels, while for grayscale images we have only one channel. The number of images in a batch would be determined by the mini-batch size chosen for the mini-batch stochastic gradient descent. The batch size is one for stochastic gradient descent.

The inputs are fed to the input layer through mini-batches.

Convolution Layer

Convolution is the heart of any CNN network. TensorFlow supports both 2D and 3D convolutions. However, 2D convolutions are more common since 3D convolutions are computationally memory intensive. The input images or intermediate images in the form of output feature maps are 2D convolved with 2D filters of the size specified. 2D convolution happens along the spatial dimensions, while there is no convolution along the depth channel of the image volume. For each depth channel, the same number of feature maps is generated, and then they are summed together along the depth dimension before they pass through the ReLU activations. These filters help to detect features in the images. The deeper the convolutional layer is in the network, the more complicated features it learns. For instance, the initial convolutional layer might learn to detect edges in an image, while the second convolutional layer might learn to connect the edges to form geometric shapes such as circles and rectangles. The even deeper convolutional layers might learn to detect more complicated features; for example, in cat vs. dog classification, it might learn to detect the eyes, nose, or other body parts of the animals.

In a CNN, only the size of the filters is specified; the weights are initialized to arbitrary values before the start of training. The weights of the filters are learned through the CNN training process, and hence they might not represent the traditional image-processing filters such as Sobel, Gaussian, mean, median, or other kind of filters. Instead, the learned filters would be such that the overall loss function defined is minimized or a good generalization is achieved based on the validation. Although it might not learn the traditional edge-detection filter, it would learn several filters that detect edges in some form since edges are good feature detectors for images.

Some of the terms with which one should be familiar while defining the convolution layer are as follows:

> **Filter size**: Filter size defines the height and width of the filter kernel. A filter kernel of size 3 × 3 would have nine weights. Generally, these filters are initialized and slide over the input image for convolution without flipping these filters. Technically, when convolution is performed without flipping the filter kernel, it is called cross-correlation and not convolution. However, it does not matter, as we can consider the filters learned as a flipped version of image-processing filters.
>
> **Stride**: The stride determines the number of pixels to move in each spatial direction while performing convolution. In normal convolution of signals, we generally don't skip any pixels and instead compute the convolution sum at each pixel location, and hence we have a stride of 1 along both spatial directions for 2D signals. However, one may choose to skip every alternate pixel location while convolving and thus chose a stride of 2. If a stride of 2 is chosen along both the height and the width of the image, then after convolving, the output image would be approximately $\frac{1}{4}$ of the input image size. Why it is *approximately* $\frac{1}{4}$ and not *exactly* $\frac{1}{4}$ of the original image or feature-map size will be covered in our next topic of discussion.
>
> **Padding**: When we convolve an image of a specific size by a filter, the resulting image is generally smaller than the original image. For example, if we convolve a 5 × 5 2D image by a filter of size 3 × 3, the resulting image is 3 × 3.

Padding is an approach that appends zeroes to the boundary of an image to control the size of the output of convolution. The convolved output image length L' along a specific spatial dimension is given by the following:

$$L' = \frac{L - K + 2P}{S} + 1$$

where

$L \rightarrow$ Length of the input image in a specific dimension

$K \rightarrow$ Length of the kernel/filter in a specific dimension

$P \rightarrow$ Zeroes padded along a dimension in either end

$S \rightarrow$ Stride of the convolution

In general, for a stride of 1, the image size along each dimension is reduced by $(K-1)/2$ on either end, where K is the length of the filter kernel along that dimension. So, to keep the output image the same as that of the input image, a pad length of $\frac{K-1}{2}$ would be required.

Whether a specific stride size is possible can be found out from the output image length along a specific direction. For example, if $L = 12$, $K = 3$, and $P = 0$, stride $S = 2$ is not possible, since it would produce an output length along the spatial dimension as $\frac{(12-3)}{2} = 4.5$, which is not an integer value.

In TensorFlow, padding can be chosen as either `"VALID"` or `"SAME"`. `"SAME"` ensures that the output spatial dimensions of the image are the same as those of the input spatial dimensions in cases where a stride of 1 is chosen. It uses zero padding to achieve this. It tries to keep the zero-pad length even on both sides of a dimension, but if the total pad length for that dimension is odd, then the extra length is added to the right for the horizontal dimension and to the bottom for the vertical dimension.

`"VALID"` does not use zero padding, and hence the output image dimension would be smaller than the input image dimensions, even for a stride of 1.

TensorFlow Usage

```
def conv2d(num_filters,kernel_size=3,strides=1,padding='SAME',
activation='relu'):
    conv_layer = layers.Conv2D(num_filters,kernel_size,strides=(strides,
    strides),padding=padding,activation='relu')
    return conv_layer
```

For defining a TensorFlow convolutional layer, we use `tf.keras.layers.Conv2D` where we need to define the number of output filters for convolution, the size of each of the convolution filters, the stride size, and the padding type. Also, we add a bias for each output feature map. Finally, we use the rectified linear units (ReLUs) as activations to add nonlinearity into the system.

Pooling Layer

A pooling operation on an image generally summarizes a locality of an image, the locality being given by the size of the filter kernel, also called the receptive field. The summarization generally happens in the form of max pooling or average pooling. In max pooling, the maximum pixel intensity of a locality is taken as the representative of that locality. In average pooling, the average of the pixel intensities around a locality is taken as the representative of that locality. Pooling reduces the spatial dimensions of an image. The kernel size that determines the locality is generally chosen as 2 × 2, whereas the stride is chosen as 2. This reduces the image size to about half the size of the original image.

TensorFlow Usage

```
""" P O O L I N G L A Y E R """

def maxpool2d(ksize=2,strides=2,padding='SAME'):
    return layers.MaxPool2D(pool_size=(ksize, ksize),strides=strides,
    padding=padding)
```

The tf.keras.layers.MaxPool2D definition is used to define a max pooling layer, while tf.keras.layers.AveragePooling2D is used to define an average pooling layer. Apart from the input, we need to input the receptive field or kernel size of max pooling through the ksize parameter. Also, we need to provide the strides to be used for the max pooling. To ensure that the values in each spatial location of the output feature map from pooling are from independent neighborhoods in the input, the stride in each spatial dimension should be chosen to be equal to the kernel size in the corresponding spatial dimension.

Backpropagation Through the Convolutional Layer

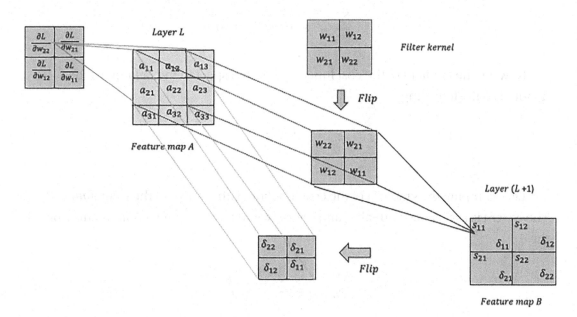

Figure 3-21. *Backpropagation through the convolutional layer*

Backpropagation through a convolution layer is much like backpropagation for a multi-layer Perceptron network. The only difference is that the weight connections are sparse since the same weights are shared by different input neighborhoods to create an output feature map. Each output feature map is the result of the convolution of an image or a feature map from the previous layer with a filter kernel whose values are the weights that we need to learn through backpropagation. The weights in the filter kernel are shared for a specific input–output feature-map combination.

In Figure 3-21, feature map A in layer L convolves with a filter kernel to produce an output feature map B in layer $(L + 1)$.

The values of the output feature map are the results of convolution and can be expressed as $s_{ij} \; \forall \; i, j \in \{1, 2\}$:

$$s_{11} = w_{22}{}^{*}a_{11} + w_{21}{}^{*}a_{12} + w_{12}{}^{*}a_{21} + w_{11}{}^{*}a_{22}$$
$$s_{12} = w_{22}{}^{*}a_{12} + w_{21}{}^{*}a_{13} + w_{12}{}^{*}a_{22} + w_{11}{}^{*}a_{23}$$
$$s_{21} = w_{22}{}^{*}a_{21} + w_{21}{}^{*}a_{22} + w_{12}{}^{*}a_{31} + w_{11}{}^{*}a_{32}$$
$$s_{22} = w_{22}{}^{*}a_{22} + w_{23}{}^{*}a_{22} + w_{12}{}^{*}a_{32} + w_{11}{}^{*}a_{33}$$

In generalized way,

$$s_{ij} = \sum_{n=1}^{2}\sum_{m=1}^{2} w_{(3-m)(3-n)} {}^{*} a_{(i-1+m)(j-1+n)}$$

Now, let the gradient of the cost function L with respect to the net input s_{ij} be denoted by the following:

$$\frac{\partial L}{\partial s_{ij}} = \delta_{ij}$$

Let's compute the gradient of the cost function with respect to the weight w_{22}. The weight is associated with all s_{ij} and hence would have gradient components from all the δ_{ij}:

$$\frac{\partial L}{\partial w_{22}} = \sum_{j=1}^{2}\sum_{i=1}^{2} \frac{\partial L}{\partial s_{ij}} \frac{\partial s_{ij}}{\partial w_{22}}$$

$$= \sum_{j=1}^{2}\sum_{i=1}^{2} \delta_{ij} \frac{\partial s_{ij}}{\partial w_{22}}$$

Also, from the preceding equations for different s_{ij}, the following can be derived:

$$\frac{\partial s_{11}}{\partial w_{22}} = a_{11}, \frac{\partial s_{12}}{\partial w_{22}} = a_{12}, \frac{\partial s_{13}}{\partial w_{22}} = a_{21}, \frac{\partial s_{14}}{\partial w_{22}} = a_{22}$$

Hence,

$$\frac{\partial L}{\partial w_{22}} = \delta_{11}{}^{*} a_{11} + \delta_{12}{}^{*} a_{12} + \delta_{21}{}^{*} a_{21} + \delta_{22}{}^{*} a_{22}$$

Similarly,

$$\frac{\partial L}{\partial w_{21}} = \sum_{j=1}^{2}\sum_{i=1}^{2}\frac{\partial L}{\partial s_{ij}}\frac{\partial s_{ij}}{\partial w_{21}}$$

$$= \sum_{j=1}^{2}\sum_{i=1}^{2}\delta_{ij}\frac{\partial s_{ij}}{\partial w_{21}}$$

Again, $\dfrac{\partial s_{11}}{\partial w_{21}} = a_{12}$, $\dfrac{\partial s_{12}}{\partial w_{21}} = a_{13}$, $\dfrac{\partial s_{21}}{\partial w_{21}} = a_{22}$, $\dfrac{\partial s_{22}}{\partial w_{21}} = a_{23}$

Hence,

$$\frac{\partial L}{\partial w_{21}} = \delta_{11}{}^{*}a_{12} + \delta_{12}{}^{*}a_{13} + \delta_{21}{}^{*}a_{22} + \delta_{22}{}^{*}a_{23}$$

Proceeding by the same approach for the other two weights, we get the following:

$$\frac{\partial L}{\partial w_{11}} = \sum_{j=1}^{2}\sum_{i=1}^{2}\frac{\partial L}{\partial s_{ij}}\frac{\partial s_{ij}}{\partial w_{11}}$$

$$= \sum_{j=1}^{2}\sum_{i=1}^{2}\delta_{ij}\frac{\partial s_{ij}}{\partial w_{11}}$$

$$\frac{\partial s_{11}}{\partial w_{11}} = a_{22}, \frac{\partial s_{12}}{\partial w_{11}} = a_{23}, \frac{\partial s_{21}}{\partial w_{11}} = a_{32}, \frac{\partial s_{22}}{\partial w_{21}} = a_{33}$$

$$\frac{\partial L}{\partial w_{11}} = \delta_{11}{}^{*}a_{22} + \delta_{12}{}^{*}a_{23} + \delta_{21}{}^{*}a_{32} + \delta_{22}{}^{*}a_{33}$$

$$\frac{\partial L}{\partial w_{12}} = \sum_{j=1}^{2}\sum_{i=1}^{2}\frac{\partial L}{\partial s_{ij}}\frac{\partial s_{ij}}{\partial w_{12}}$$

$$\frac{\partial s_{11}}{\partial w_{12}} = a_{21}, \frac{\partial s_{12}}{\partial w_{12}} = a_{22}, \frac{\partial s_{21}}{\partial w_{12}} = a_{31}, \frac{\partial s_{22}}{\partial w_{22}} = a_{32}$$

$$\frac{\partial L}{\partial w_{12}} = \delta_{11}{}^{*}a_{21} + \delta_{12}{}^{*}a_{22} + \delta_{21}{}^{*}a_{31} + \delta_{22}{}^{*}a_{32}$$

241

Based on the preceding gradients of the cost function L with respect to the four weights of the filter kernel, we get the following relationship:

$$\frac{\partial L}{\partial w_{ij}} = \sum_{n=1}^{2}\sum_{m=1}^{2} \delta_{mn}{}^{*}a_{(i-1+m)(j-1+n)}$$

When arranged in matrix form, we get the following relationship; (x) denotes the cross-correlation:

$$\begin{bmatrix} \dfrac{\partial L}{\partial w_{22}} & \dfrac{\partial L}{\partial w_{21}} \\[2mm] \dfrac{\partial L}{\partial w_{12}} & \dfrac{\partial L}{\partial w_{11}} \end{bmatrix} = \begin{bmatrix} a_{11} & a_{12} & a_{13} \\ a_{21} & a_{22} & a_{23} \\ a_{31} & a_{32} & a_{33} \end{bmatrix} (x) \begin{bmatrix} \delta_{11} & \delta_{12} \\ \delta_{21} & \delta_{22} \end{bmatrix}$$

The cross-correlation of $\begin{bmatrix} a_{11} & a_{12} & a_{13} \\ a_{21} & a_{22} & a_{23} \\ a_{31} & a_{32} & a_{33} \end{bmatrix}$ with $\begin{bmatrix} \delta_{11} & \delta_{12} \\ \delta_{21} & \delta_{22} \end{bmatrix}$ can also be thought of as

the convolution of $\begin{bmatrix} a_{11} & a_{12} & a_{13} \\ a_{21} & a_{22} & a_{23} \\ a_{31} & a_{32} & a_{33} \end{bmatrix}$ with flipped $\begin{bmatrix} \delta_{11} & \delta_{12} \\ \delta_{21} & \delta_{22} \end{bmatrix}$; i.e., $\begin{bmatrix} \delta_{22} & \delta_{21} \\ \delta_{12} & \delta_{11} \end{bmatrix}$.

Hence, the flip of the gradient matrix is the convolution of $\begin{bmatrix} a_{11} & a_{12} & a_{13} \\ a_{21} & a_{22} & a_{23} \\ a_{31} & a_{32} & a_{33} \end{bmatrix}$ with $\begin{bmatrix} \delta_{22} & \delta_{21} \\ \delta_{12} & \delta_{11} \end{bmatrix}$; i.e.,

$$\begin{bmatrix} \dfrac{\partial L}{\partial w_{22}} & \dfrac{\partial L}{\partial w_{21}} \\[2mm] \dfrac{\partial L}{\partial w_{12}} & \dfrac{\partial L}{\partial w_{11}} \end{bmatrix} = \begin{bmatrix} a_{11} & a_{12} & a_{13} \\ a_{21} & a_{22} & a_{23} \\ a_{31} & a_{32} & a_{33} \end{bmatrix} (^{*}) \begin{bmatrix} \delta_{22} & \delta_{12} \\ \delta_{21} & \delta_{11} \end{bmatrix}$$

In terms of the layers, one can say the flip of the gradient matrix turns out to be a cross-correlation of the gradient at the $(L+1)$ layer with the outputs of the feature map at layer L. Also, equivalently, the flip of the gradient matrix turns out to be a convolution of the flip of the gradient matrix at the $(L+1)$ layer with the outputs of the feature map at layer L.

Backpropagation Through the Pooling Layers

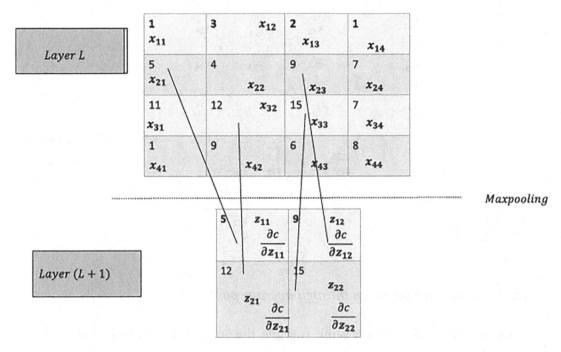

Figure 3-22. *Backpropagation through max pooling layer*

Figure 3-22 illustrates the max pooling operation. Let a feature map, after going through convolution and ReLU activations at layer L, go through the max pooling operation at layer $(L + 1)$ to produce the output feature map. The kernel or receptive field for max pooling is of size 2×2, and the stride size is 2. The output of the max pooling layer is $\frac{1}{4}$ the size of the input feature map, and its output values are represented by $z_{ij}, \forall\, i, j \in \{1, 2\}$.

We can see z_{11} gets the value of 5 since the maximum in the 2×2 block is 5. If the error derivative at z_{11} is $\frac{\partial C}{\partial z_{ij}}$, then the whole gradient is passed onto x_{21} with a value of 5, and the rest of the elements in its block—x_{11}, x_{12}, and x_{22}—receive zero gradients from z_{11}.

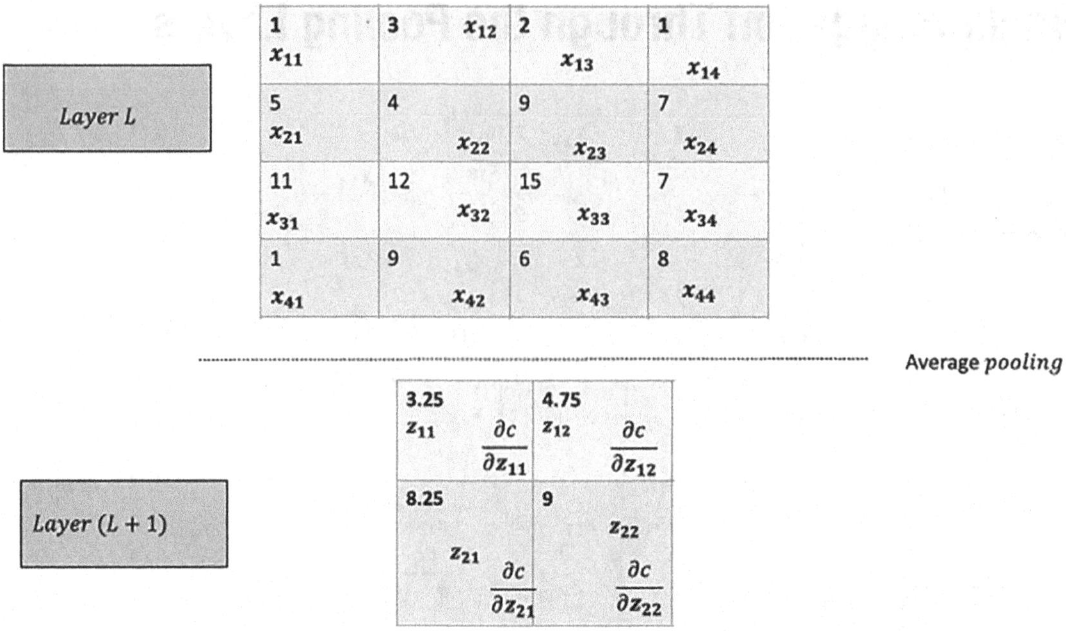

Layer L			
1 x_{11}	3 x_{12}	2 x_{13}	1 x_{14}
5 x_{21}	4 x_{22}	9 x_{23}	7 x_{24}
11 x_{31}	12 x_{32}	15 x_{33}	7 x_{34}
1 x_{41}	9 x_{42}	6 x_{43}	8 x_{44}

·· Average pooling

Layer (L + 1)	
3.25 z_{11} $\dfrac{\partial c}{\partial z_{11}}$	4.75 z_{12} $\dfrac{\partial c}{\partial z_{12}}$
8.25 z_{21} $\dfrac{\partial c}{\partial z_{21}}$	9 z_{22} $\dfrac{\partial c}{\partial z_{22}}$

Figure 3-23. *Backpropagation through average pooling layer*

To use average pooling for the same example, the output is the average of the values in the 2 × 2 block of the input. Therefore, z_{11} gets the average of the values x_{11}, x_{12}, x_{21}, and x_{22}. Here, the error gradient $\dfrac{\partial C}{\partial z_{11}}$ at z_{11} would be shared equally by x_{11}, x_{12}, x_{21}, and x_{22}. Hence,

$$\frac{\partial C}{\partial x_{11}} = \frac{\partial C}{\partial x_{12}} = \frac{\partial C}{\partial x_{21}} = \frac{\partial C}{\partial x_{22}} = \frac{1}{4}\frac{\partial C}{\partial z_{11}}$$

Weight Sharing Through Convolution and Its Advantages

Weight sharing through convolution greatly reduces the number of parameters in the convolutional neural network. Imagine we created a feature map of size $k \times k$ from an image of $n \times n$ size with full connections instead of convolutions. There would be $k^2 n^2$ weights for that one feature map alone, which is a lot of weights to learn. Instead, since in convolution the same weights are shared across locations defined by the filter-kernel size, the number of parameters to learn is reduced by a huge factor. In cases of convolution, as in this scenario, we just need to learn the weights for the specific filter kernel. Since the filter size is relatively small with respect to the image, the number of weights is reduced significantly. For any image, we generate several feature maps corresponding to different filter kernels. Each filter kernel learns to detect a different kind of feature. The feature maps created are again convolved with other filter kernels to learn even more complex features in subsequent layers.

Translation Equivariance

The convolution operation provides translational equivariance. That is, if a feature A in an input produces a specific feature B in the output, then even if feature A is translated around in the image, feature B would continue to be generated at different locations of the output.

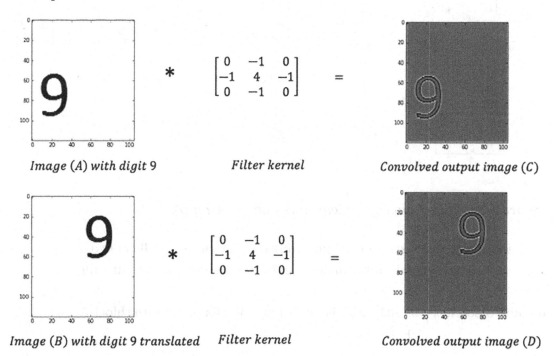

Image (A) with digit 9   Filter kernel   Convolved output image (C)

Image (B) with digit 9 translated   Filter kernel   Convolved output image (D)

Figure 3-24. *Translational equivariance illustration*

In Figure 3-24, we can see that the digit 9 has been translated in Image (B) from its position in Image (A). Both the input images (A) and (B) have been convolved with the same filter kernel, and the same feature has been detected for the digit 9 in both output images (C) and (D) at different locations based on its location in the input. Convolution still produced the same feature for the digit irrespective of the translation. This property of convolution is called *translational equivariance*. In fact, if the digit is represented by a set of pixel intensities x, and f is the translation operation on x, while g is the convolution operation with a filter kernel, then the following holds true for convolution:

$$g(f(x)) = f(g(x))$$

In our case, $f(x)$ produces the translated 9 in Image (B), and the translated 9 is convolved through g to produce the activated feature for 9, as seen in Image (D). This activated feature for 9 in Image (D) (i.e., $g(f(x))$) could also have been achieved by translating the activated 9 in Image (C) (i.e., $g(x)$) through the same translation f.

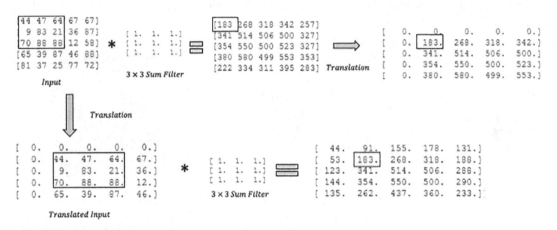

Figure 3-25. *Illustration of equivariance with an example*

It is a little easier to see equivariance with a small example, as illustrated in Figure 3-25. The part of the input image or 2D signal we are interested in is the left topmost block, i.e., $\begin{bmatrix} 44 & 47 & 64 \\ 9 & 83 & 21 \\ 70 & 88 & 88 \end{bmatrix}$. For easy reference, let us name the block A.

On convolving the input with the sum filter—i.e., $\begin{bmatrix} 1 & 1 & 1 \\ 1 & 1 & 1 \\ 1 & 1 & 1 \end{bmatrix}$—block A would correspond to an output value of 183, which could be treated as the feature detector for A.

On convolving the same sum filter with the translated image, the shifted block A would still produce an output value of 183. Also, we can see that if we were to apply the same translation to the original convoluted image output, the value of 183 would appear at the same location as that of the output of the convoluted image after translation.

Translation Invariance Due to Pooling

Pooling provides some form of translational invariance based on the receptor field kernel size of the pooling. Let us take the case of max pooling, as illustrated in Figure 3-26. The digit in image A at a specific position is detected through a convolution filter H in the form of values 100 and 80 in the output feature map P. Also, the same digit appears in another image B in a slightly translated position with respect to image A. On convolving image B with filter H, the digit 9 is detected in the form of the same values of 100 and 80 in the feature map P', but at a slightly displaced position from the one in. When these feature maps pass through the receptor field kernels of size 2×2 with stride 2 because of max pooling, the 100 and 80 values appear at the same location in both the output M and M'. In this way, max pooling provides some translational invariance to feature detection if the translation distance is not very high with respect to the size of the receptor field or kernel for max pooling.

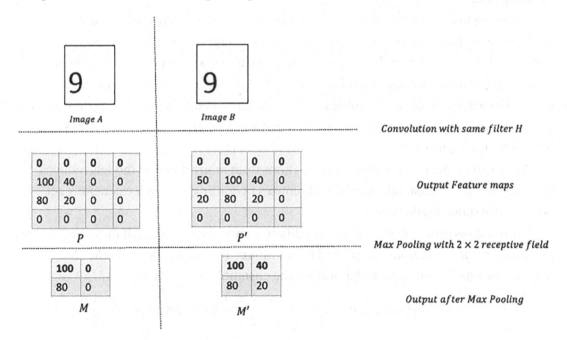

Figure 3-26. *Translational invariance through max pooling*

Similarly, average pooling takes the average of the values in a locality of a feature map based on the size of the receptor field kernel. So, if a specific feature is detected by high values in its feature map in a locality—let's say at regions of edges—then the averages would continue to be high even if the image were a little translated.

Dropout Layers and Regularization

Dropout is an activity to regularize weights in the fully connected layers of a convolutional neural network to avoid overfitting. However, it is not restricted to convolutional neural networks but rather applies to all feed-forward neural networks. A specified proportion of neural network units, both hidden and visible, is randomly dropped at training time for each training sample in a mini-batch so that the remaining neurons can learn important features all by themselves and not rely on cooperation from other neurons. When the neurons are dropped randomly, all the incoming and outgoing connections to those neurons are also dropped. Too much cooperation between neurons makes the neurons dependent on each other, and they fail to learn distinct features. This high cooperation leads to overfitting since it does well on the training dataset, while if the test dataset is somewhat different from the training dataset, the predictions on test dataset go haywire.

When the neuron units are dropped randomly, each such setting of remaining available neurons produces a different network. Let's suppose we have a network with N neural units; the number of possible neural network configuration possible is N^2. For each training sample in a mini-batch, a different set of neurons is chosen at random based on the dropout probability. So, training a neural network with dropout is equivalent to training a set of different neural networks where each network very seldom gets trained, if at all.

As we can surmise, averaging the predictions from many different models reduces the variances of the ensemble model and reduces overfitting, and so we generally get better, more stable predictions.

For two class problem, trained on two different models M_1 and M_2, if the class probabilities for a data point are p_{11} and p_{12} for Model M_1 and p_{21} and p_{22} for Model M_2, then we take the average probability for the ensemble model of M_1 and M_2. The ensemble model would have a probability of $\frac{(p_{11} + p_{21})}{2}$ for Class 1 and $\frac{(p_{12} + p_{22})}{2}$ for Class 2.

Another averaging method would be to take the geometric mean of the predictions from different models. In this case, we would need to normalize over the geometric means to get the sum of new probabilities as 1.

The new probabilities for the ensemble model for the preceding example would be

$$\frac{\sqrt{p_{11} \times p_{21}}}{\sqrt{p_{11} \times p_{21}} + \sqrt{p_{12} \times p_{22}}} \text{ and } \frac{\sqrt{p_{12} \times p_{22}}}{\sqrt{p_{11} \times p_{21}} + \sqrt{p_{12} \times p_{22}}}, \text{ respectively.}$$

At test time, it is not possible to compute the predictions from all such possible networks and then average it out. Instead, the single neural network with all the weights and connections is used—but with weight adjustments. If a neural network unit is retained with probability p during training, then the outgoing weights from that unit are scaled down by multiplying the weights by probability p. In general, this approximation for prediction on test datasets works well. It can be proven that for a model with a SoftMax output layer, the preceding arrangement is equivalent to taking predictions out of those individual models resulting from dropout and then computing their geometric mean.

In Figure 3-27, a neural network whose three units have been dropped randomly is presented. As we can see, all the input and output connections of the dropped units have also been dropped.

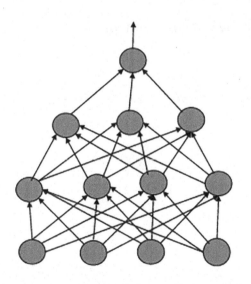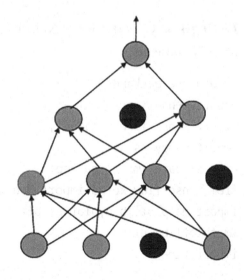

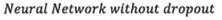

Neural Network without dropout   *Neural Network with three units dropped*

Figure 3-27. *Neural network with three units dropped randomly*

For a convolutional neural network, the units in the fully connected layers and their corresponding incoming and outgoing connections are usually dropped. Hence, the different filter-kernel weights do not require any adjustment while predicting for the test dataset.

Convolutional Neural Network for Digit Recognition on the MNIST Dataset

Now that we have gone through the basic building blocks for a convolutional neural network, let us see how good a CNN is at learning to classify the MNIST dataset. The detailed logic for a basic implementation in TensorFlow is documented in Listing 3-6. The CNN takes in images of height 28, width 28, and depth 3 corresponding to the RGB channels. The images go through the series of convolution, ReLU activations, and max pooling operations twice before being fed into the fully connected layer and finally to the output layer. The first convolution layer produces 64 feature maps, the second convolution layer provides 128 feature maps, and the fully connected layer has 1024 units. The max pooling layers have been chosen to reduce the feature map size by $\frac{1}{4}$. The feature maps can be considered 2D images.

Listing 3-6. Convolutional Neural Network for Digit Recognition on the MNIST Dataset

```
# Load the packages
import tensorflow as tf
print('tensorflow version', tf.__version__)
import numpy as np
from sklearn import datasets
from tensorflow.keras import Model, layers
import matplotlib.pyplot as plt
%matplotlib inline
import time
```

```python
# Function to Read the MNIST dataset along with the labels
def read_infile():
    (train_X, train_Y), (test_X, test_Y) = tf.keras.datasets.mnist.
    load_data()
    train_X,test_X = np.expand_dims(train_X, -1), np.expand_dims(test_X,-1)
    #print(train_X.shape,test_X.shape)
    return train_X, train_Y,test_X, test_Y

# Normalize the images
def normalize(train_X):
    return train_X/255.0

# Convolution Layer function
def conv2d(num_filters,kernel_size=3,strides=1,padding='SAME',activation=
'relu'):
    conv_layer = layers.Conv2D(num_filters,kernel_size,strides=(strides,
    strides),
                               padding=padding, activation='relu')
    return conv_layer

#Pooling Layer function
def maxpool2d(ksize=2,strides=2,padding='SAME'):
    return layers.MaxPool2D(pool_size=(ksize, ksize),strides=strides,
    padding=padding)

# Convolution Model class
class conv_model(Model):

    def __init__(self,input_size=28,filters=[64,128],fc_units=1024,kernel_
    size=3,strides=1,
                                  padding='SAME',ksize=2,n_classes=10):
        super(conv_model, self).__init__()
        self.conv1 = conv2d(num_filters=filters[0],kernel_size=kernel_
        size,strides=strides,
                                  padding=padding,activation='relu')
        self.conv2 = conv2d(num_filters=filters[1],kernel_size=kernel_
        size,strides=strides,
                                  padding=padding,activation='relu')
```

```
        self.maxpool1 = maxpool2d(ksize=ksize,strides=ksize,padding='SAME')
        self.maxpool2 = maxpool2d(ksize=ksize,strides=ksize,padding='SAME')
        self.fc = layers.Dense(fc_units,activation='relu')
        self.out = layers.Dense(n_classes)

    # Forward pass for the Model
    def call(self, x):
        x = self.conv1(x)
        x = self.maxpool1(x)
        x = self.conv2(x)
        x = self.maxpool2(x)
        x = tf.reshape(x,(tf.shape(x)[0],-1))
        x = self.fc(x)
        x = self.out(x)
        return x

# Define the loss to be Categorical CrossEntropy to work with the
Softmax logits
loss_fn = tf.keras.losses.SparseCategoricalCrossentropy(from_
logits=True,reduction=tf.keras.losses.Reduction.SUM)

#Learning rate
learning_rate = 0.01
# Define optimizer
optimizer = tf.keras.optimizers.Adam(learning_rate)

# Process the training and test data
X_train, y_train, X_test, y_test = read_infile()
X_train, X_test = normalize(X_train), normalize(X_test)
num_train_recs, num_test_recs = X_train.shape[0], X_test.shape[0]
print(X_train.shape,y_train.shape)

# build the model
model = conv_model(input_size=28,filters=[64,128],fc_units=1024,kernel_
size=3,strides=1,
padding='SAME',ksize=2,n_classes=10)
# Construct a  Tensorflow graph for the Model
model_graph = tf.function(model)
```

```
epochs = 20
batch_size = 256

loss_trace = []
accuracy_trace = []

num_train_recs,num_test_recs = X_train.shape[0], X_test.shape[0]
num_batches = num_train_recs // batch_size
order_ = np.arange(num_train_recs)

start_time = time.time()
for i in range(epochs):
    loss, accuracy = 0,0
    np.random.shuffle(order_)
    X_train,y_train = X_train[order_], y_train[order_]
    for j in range(num_batches):
        X_train_batch = tf.constant(X_train[j*batch_size:(j+1)*batch_
        size],dtype=tf.float32)
        y_train_batch = tf.constant(y_train[j*batch_size:(j+1)*batch_size])
        #print(X_train_batch,y_train_batch)
        with tf.GradientTape() as tape:
            y_pred_batch = model_graph(X_train_batch)
            loss_ = loss_fn(y_train_batch,y_pred_batch)

            # compute gradient
        gradients = tape.gradient(loss_, model.trainable_variables)
            # update the parameters
        optimizer.apply_gradients(zip(gradients, model.trainable_variables))

        accuracy += np.sum(y_train_batch.numpy() == np.argmax(y_pred_batch.
        numpy(),axis=1))
        loss += loss_.numpy()
    loss /= num_train_recs
    accuracy /= num_train_recs
    loss_trace.append(loss)
    accuracy_trace.append(accuracy)
    print(f"Epoch {i+1} : loss: {np.round(loss,4)} ,accuracy:{np.
    round(accuracy,4)}\n")
```

```
X_test, y_test = tf.constant(X_test), tf.constant(y_test)
y_pred_test = model(X_test)
loss_test = loss_fn(y_test,y_pred_test).numpy()/num_test_recs
accuracy_test = np.mean(y_test.numpy() == np.argmax(y_pred_test.
numpy(),axis=1))
print('Results on Test Dataset:','loss:',np.round(loss_
test,4),'accuracy:',np.round(accuracy_test,4))
f, a = plt.subplots(1, 10, figsize=(10, 2))
for i in range(10):
        a[i].imshow(X_test.numpy()[i,:])
        print(y_test.numpy()[i])
print(f"Total processing time: {time.time() - start_time} secs")
```

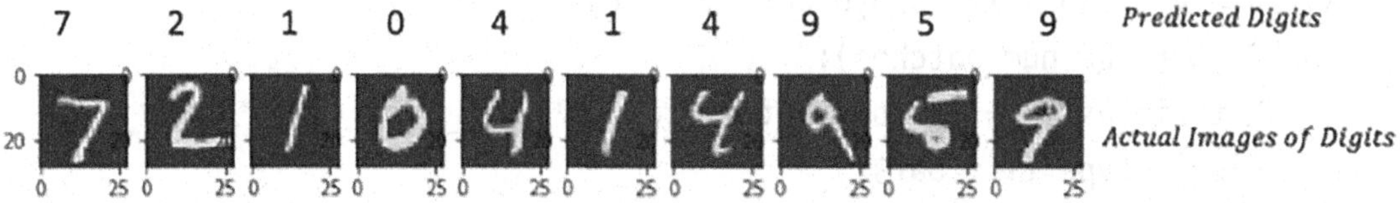

Figure 3-28.   Predicted digits vs. actual digits from CNN model

With the preceding basic convolutional neural network, which comprises two convolutional–max pooling–ReLU pairs along with a fully connected layer before the final output SoftMax unit, we can achieve a test-set accuracy of 0.9867 in just 20 epochs. As we saw previously through the multi-layer Perceptron approach in Chapter 2, with that method we were merely able to get around 90% accuracy with 1000 epochs. This is a testimony that for image-recognition problems, convolutional neural networks work best.

One more thing I want to emphasize is the importance of tuning the model with the correct set of hyperparameters and prior information. Parameters such as the learning-rate selection can be very tricky since the cost function for neural networks is generally non-convex. A large learning rate can lead to faster convergence to a local minimum but might introduce oscillations, whereas a low learning rate will lead to very slow convergence. Ideally, the learning rate should be low enough that network parameters can converge to a meaningful local minimum, and at the same time it should be high enough that the models can reach the minima faster. Generally, for the preceding neural network, a learning rate of 0.01 is a little on the higher side, but since we are only training the data on 20 epochs, it works well. A lower learning rate would not have achieved such a high accuracy with just 20 epochs. Similarly, the batch size chosen for the mini-batch version of stochastic gradient descent influences the convergence of the training process. A larger batch size might be good since the gradient estimates are less noisy; however, it may come at the cost of increased computation. One also needs to try out different filter sizes as well as experiment with different numbers of feature maps in each convolution layer. The kind of model architecture we choose works as a prior knowledge to the network.

Convolutional Neural Network for Solving Real-World Problems

We will now briefly discuss how to work on real-world image-analytics problems by going through a problem hosted in Kaggle by Intel that involved classifying different types of cervical cancer. The dataset for the problem can be located at www.kaggle. com/c/intel-mobileodt-cervical-cancer-screening. In this competition, a model needs to be built that identifies a woman's cervix type based on images. Doing so will allow for the effective treatment of patients. Images specific to three types of cancer were provided for the competition. So, the business problem boils down to being a three-class image-classification problem. A basic solution approach to the problem is provided in Listing 3-7.

Listing 3-7. Real-World Use of Convolutional Neural Network

```
# Load the relevant libraries
import glob
import cv2
import time
import os
from pathlib import Path
import tensorflow as tf
print('tensorflow version', tf.__version__)
import numpy as np
from tensorflow.keras import Model, layers
import matplotlib.pyplot as plt
```

```python
import time
from elapsedtimer import ElapsedTimer
!pip install pandas
import pandas as pd

# Create functions for different layer

""" C O N V O L U T I O N   L A Y E R """

def conv2d(num_filters, kernel_size=3, strides=1, padding='SAME',
activation='relu'):
    conv_layer = layers.Conv2D(num_filters, kernel_size, strides=(strides,
    strides), padding=padding, activation='relu')
    return conv_layer

""" P O O L I N G   L A Y E R """

def maxpool2d(ksize=2, strides=2, padding='SAME'):
    return layers.MaxPool2D(pool_size=(ksize, ksize), strides=strides,
    padding=padding)

# Convolution Model class
class conv_model(Model):
    # Set layers.
    def __init__(self, filters=[64, 128,256], fc_units=[512,512],
                 kernel_size=3, strides=1, padding='SAME',
                 ksize=2, n_classes=3,dropout=0.5):

        super(conv_model, self).__init__()
        self.conv1 = conv2d(num_filters=filters[0], kernel_size=kernel_
        size, strides=strides, padding=padding,
                        activation='relu')
        self.conv2 = conv2d(num_filters=filters[1], kernel_size=kernel_
size, strides=strides, padding=padding,
                        activation='relu')
        self.conv3 = conv2d(num_filters=filters[2], kernel_size=kernel_
        size, strides=strides, padding=padding,
                        activation='relu')
```

```python
        self.maxpool1 = maxpool2d(ksize=ksize, strides=ksize,
        padding='SAME')
        self.maxpool2 = maxpool2d(ksize=ksize, strides=ksize,
        padding='SAME')
        self.maxpool3 = maxpool2d(ksize=ksize, strides=ksize,
        padding='SAME')
        self.fc1 = layers.Dense(fc_units[0], activation='relu')
        self.fc2 = layers.Dense(fc_units[1], activation='relu')
        self.out = layers.Dense(n_classes)
        self.dropout1 = layers.Dropout(rate=dropout)
        self.dropout2 = layers.Dropout(rate=dropout)

    # Forward pass.
    def call(self, x):
        x = self.conv1(x)
        x = self.maxpool1(x)
        x = self.conv2(x)
        x = self.maxpool2(x)
        x = self.conv3(x)
        x = self.maxpool3(x)
        x = tf.reshape(x, (tf.shape(x)[0], -1))
        x = self.fc1(x)
        x = self.dropout1(x)
        x = self.fc2(x)
        x = self.dropout2(x)
        x = self.out(x)
        probs = tf.nn.softmax(x,axis=1)
        return x,probs

# Utility to read an image to numpy
def get_im_cv2(path,input_size):
    """

    :param path: Image path
    :return: np.ndarray
    """
```

```python
    img = cv2.imread(path)
    resized = cv2.resize(img, (input_size, input_size), cv2.INTER_LINEAR)
    return resized

# Utility to process all training images and store as numpy arrays
def load_train(path,input_size):
    """
    :param path: Training images path
    :return: train and test labels in np.array
    """
    assert Path(path).exists()

    X_train = []
    X_train_id = []
    y_train = []
    start_time = time.time()

    folders = ['Type_1', 'Type_2', 'Type_3']
    for fld in folders:
        index = folders.index(fld)
        path_glob = f"{Path(path)}/train/{fld}/*.jpg"
        files = glob.glob(path_glob)

        for fl in files:
            flbase = os.path.basename(fl)
            img = get_im_cv2(fl,input_size=input_size)
            X_train.append(img)
            X_train_id.append(flbase)
            y_train.append(index)

    for fld in folders:
        index = folders.index(fld)
        path_glob = f"{Path(path)}/Additional/{fld}/*.jpg"
        files = glob.glob(path_glob)

        for fl in files:
            flbase = os.path.basename(fl)
            # print fl
            img = get_im_cv2(fl,input_size=input_size)
```

```python
            X_train.append(img)
            X_train_id.append(flbase)
            y_train.append(index)
    return X_train,  y_train,  X_train_id

# Utility to process all test images and store as numpy arrays
def load_test(path,input_size):

    path_glob = os.path.join(f'{Path(path)}/test/*.jpg')
    files = sorted(glob.glob(path_glob))

    X_test = []
    X_test_id = []
    for fl in files:
        flbase = os.path.basename(fl)
        img = get_im_cv2(fl,input_size)
        X_test.append(img)
        X_test_id.append(flbase)
    path_glob = os.path.join(f'{Path(path)}/test_stg2/*.jpg')
    files = sorted(glob.glob(path_glob))
    for fl in files:
        flbase = os.path.basename(fl)
        img = get_im_cv2(fl,input_size)
        X_test.append(img)
    return X_test, X_test_id

# Data Processing pipeline for train data
def read_and_normalize_train_data(train_path,input_size):
    train_data, train_target, train_id = load_train(train_path,input_size)

    print('Convert to numpy...')
    train_data = np.array(train_data, dtype=np.uint8)
    train_target = np.array(train_target, dtype=np.uint8)

    print('Reshape...')
    train_data = train_data.transpose((0, 2, 3, 1))
    train_data = train_data.transpose((0, 1, 3, 2))
```

```python
    print('Convert to float...')
    train_data = train_data.astype('float32')
    train_data = train_data / 255
    #train_target = np_utils.to_categorical(train_target, 3)
    print('Train shape:', train_data.shape)
    print(train_data.shape[0], 'train samples')
    return train_data, train_target, train_id

# Data Processing pipeline for test data
def read_and_normalize_test_data(test_path,input_size):

    test_data, test_id = load_test(test_path,input_size)

    test_data = np.array(test_data, dtype=np.uint8)
    print("test data shape",test_data.shape)
    test_data = test_data.transpose((0, 2, 3, 1))
    test_data = test_data.transpose((0, 1, 3, 2))

    test_data = test_data.astype('float32')
    test_data = test_data / 255
    print('Test shape:', test_data.shape)
    print(test_data.shape[0], 'test samples')

    return test_data, test_id

# Shuffle the input training data to aid Stochastic gradient descent
def shuffle_train(X_train,y_train):
    num_recs_train = X_train.shape[0]
    indices = np.arange(num_recs_train)
    np.random.shuffle(indices)
    return X_train[indices],y_train[indices]

# train routine
def train(X_train, y_train, train_id,lr=0.01,epochs=200,batch_size=128, \
        n_classes=3,display_step=1,in_channels=3,filter_size=3,strides=1,
        maxpool_ksize=2,\
        filters=[64, 128,256],fc_units=[512,512],dropout=0.5):
```

```
 # Define the model
model = conv_model(filters=filters, fc_units=fc_units, kernel_
size=filter_size, \
                   strides=strides, padding='SAME', ksize=maxpool_
                   ksize, n_classes=n_classes,dropout=dropout)

model_graph = tf.function(model)

# Define the loss
loss_fn = tf.keras.losses.SparseCategoricalCrossentropy
(from_logits=True,

                                             reduction=tf.keras.
                                             losses.Reduction.SUM)
# Define the optimizer
optimizer = tf.keras.optimizers.Adam(lr)

num_train_recs, num_test_recs = X_train.shape[0], X_test.shape[0]
num_batches = num_train_recs // batch_size
loss_trace , accuracy_trace = [],[]
start_time = time.time()

for i in range(epochs):

    loss, accuracy = 0, 0
    X_train, y_train = shuffle_train(X_train,y_train)

    for j in range(num_batches):
        X_train_batch = tf.constant(X_train[j * batch_size:(j + 1) *
        batch_size], dtype=tf.float32)
        y_train_batch = tf.constant(y_train[j * batch_size:(j + 1) *
        batch_size])

        with tf.GradientTape() as tape:
            y_pred_batch,_ = model_graph(X_train_batch,training=True)
            loss_ = loss_fn(y_train_batch, y_pred_batch)

        # compute gradient
        gradients = tape.gradient(loss_, model.trainable_variables)
```

```
            # update the parameters
            optimizer.apply_gradients(zip(gradients, model.trainable_
            variables))

            accuracy += np.sum(y_train_batch.numpy() == np.argmax(y_pred_
            batch.numpy(), axis=1))
            loss += loss_.numpy()
        loss /= num_train_recs
        accuracy /= num_train_recs
        loss_trace.append(loss)
        accuracy_trace.append(accuracy)
        print(f"------------------------------------------------------------\n")
        print(f"Epoch {i + 1} : loss: {np.round(loss, 4)} ,accuracy:{np.
        round(accuracy, 4)}\n")
        print(f"------------------------------------------------------------\n")

    return model_graph

# Prediction routine
def prediction(model,X_test,test_id,batch_size=32):
    batches = X_test.shape[0]//batch_size
    probs_array = []
    for i in range(batches):
        X_batch = tf.constant(X_test[(i+1)*batch_size: i*batch_size])
        _,probs = model(X_batch,training=False)
        probs = list(probs.numpy().tolist())
        probs_array += probs
    probs_array = np.array(probs_array)

def main(train_path, test_path,input_size=64,  \
        lr=0.01,epochs=200,batch_size=128,\
        n_classes=3,display_step=1,in_channels=3,\
        filter_size=3,strides=1, maxpool_ksize=2,\
        filters=[64, 128,256],fc_units=[512,512],dropout=0.5):

    with ElapsedTimer(f'Training data processing..\n'):
        X_train, y_train, train_id = read_and_normalize_train_data(train_
        path,input_size)
```

```
with ElapsedTimer(f'Test data processing..\n'):
    X_test, test_id = read_and_normalize_test_data(test_path,
    input_size)

with ElapsedTimer("Training...\n"):
    model = train(X_train, y_train, train_id,\
        lr=lr, epochs=epochs, batch_size=batch_size,\
        n_classes=n_classes, display_step=1, in_channels=in_channels,\
        filter_size=filter_size, strides=strides, maxpool_
        ksize=maxpool_ksize,\
        filters=filters, fc_units=fc_units, dropout=dropout)

with ElapsedTimer(f"Predictions on test data.."):
    out = prediction(model,X_test,test_id,batch_size=32)
    df = pd.DataFrame(out, columns=['Type_1','Type_2','Type_3'])
    df['image_name'] = test_id
    df.to_csv('results.csv',index=False)

main(train_path='/media/santanu/9eb9b6dc-b380-486e-b4fd-c424a325b976/
Kaggle Competitions/Intel',
    test_path='/media/santanu/9eb9b6dc-b380-486e-b4fd-c424a325b976/
    Kaggle Competitions/Intel',epochs=200)
```

Readers need to update the train_path with the data location accordingly before running the training and prediction for this problem.

```
    -- output --
 Epoch 186 : loss: 1.0009 ,accuracy:0.5284
 Epoch 187 : loss: 1.0007 ,accuracy:0.5285
Epoch 188 : loss: 1.0007 ,accuracy:0.5283
Epoch 189 : loss: 1.0004 ,accuracy:0.5285
Epoch 190 : loss: 1.0006 ,accuracy:0.5286
Epoch 191 : loss: 1.0007 ,accuracy:0.5284
Epoch 192 : loss: 1.0002 ,accuracy:0.5289
Epoch 193 : loss: 1.0001 ,accuracy:0.5292
Epoch 194 : loss: 1.0003 ,accuracy:0.5285
Epoch 195 : loss: 1.0007 ,accuracy:0.5286
Epoch 196 : loss: 1.0005 ,accuracy:0.5284
```

```
Epoch 197 : loss: 1.0006 ,accuracy:0.5284
Epoch 198 : loss: 1.0006 ,accuracy:0.5286
Epoch 199 : loss: 1.0006 ,accuracy:0.5286
Epoch 200 : loss: 1.0007 ,accuracy:0.5283
```

This model achieves a log-loss of around 1.0007 in the competition leaderboard, whereas the best model for this competition achieved around 0.78 log-loss. This is because the model is a basic implementation that does not take care of other advanced concepts in image processing. We will study one such advanced technique called transfer learning later in the book that works well when the number of images provided is less. A few points about the implementation that might be of interest to the reader are as follows:

- The images have been read as a three-dimensional Numpy array and resized through OpenCV and then appended to a list. The list is then converted to Numpy array, and hence we get a four-dimensional Numpy array or tensor for both the training and testing datasets. The training and testing image tensors have been transposed to have the dimensions aligned in order of image number, location along height of image, location along width of image, and image channels.

- The images have been normalized to have values between 0 and 1 by dividing by the maximum value of pixel intensity; i.e., 255. This aids the gradient-based optimization.

- The images have been shuffled randomly so that the mini-batches have images of the three classes randomly arranged.

- The rest of the network implementation is like the MNIST classification problem but with three layers of the convolution– ReLU–max pooling combination and two layers of fully connected layers before the final SoftMax output layer.

- The code involving prediction and submission has been left out here.

Batch Normalization

Batch normalization was invented by Sergey Ioffe and Christian Szegedy and is one of the pioneering elements in the deep-learning space. The original paper for batch normalization is titled "Batch Normalization: Accelerating Deep Network Training by Reducing Internal Covariate Shift" and can be located at https://arxiv.org/abs/1502.03167.

When training a neural network through stochastic gradient descent, the distribution of the inputs to each layer changes due to the update of weights on the preceding layers. This slows down the training process and makes it difficult to train very deep neural networks. The training process for neural networks is complicated by the fact that the input to any layer is dependent on the parameters for all preceding layers, and thus even small parameter changes can have an amplified effect as the network grows. This leads to input-distribution changes in a layer.

Now, let's try to understand what might go wrong when the input distribution to the activation functions in a layer change because of weight changes in the preceding layer.

A sigmoid or tanh activation function has good linear gradients only within a specified range of its input, and the gradient drops to zero once the inputs grow large.

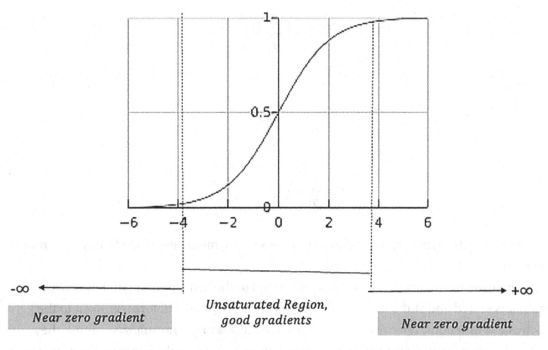

Figure 3-29. Sigmoid function with its small unsaturated region

The parameter change in the preceding layers might change the input probability distribution to a sigmoid unit layer in such a way that most of the inputs to the sigmoids belong to the saturation zone and hence produce near-zero gradients, as shown in Figure 3-29. Because of these zero or near-zero gradients, the learning becomes terribly slow or stops entirely. One way to avoid this problem is to have rectified linear units (ReLUs). The other way to avoid this problem is to keep the distribution of inputs to the sigmoid units stable within the unsaturated zone so that stochastic gradient descent does not get stuck in a saturated zone.

This phenomenon of change in the distribution of the input to the internal network units has been referred to by the inventors of the batch normalization process as *internal covariate shift.*

Batch normalization reduces the internal covariate shift by normalizing the inputs to a layer to have a zero mean and unit standard deviation. While training, the mean and standard deviation are estimated from the mini-batch samples at each layer, whereas at test-prediction time, generally the population variance and mean are used.

If a layer receives a vector $x = [x_1 x_2 ... x_n]^T \in R^{n \times 1}$ of input activations from the preceding layers, then at each mini-batch consisting of m data points, the input activations are normalized as follows:

$$\hat{x}_i = \frac{x_i - E[x_i]}{\sqrt{Var[x_i] + \epsilon}}$$

where

$$u_B = \frac{1}{m} \sum_{k=1}^{m} x_i^{(k)}$$

$$\sigma_B^2 = \frac{1}{m} \sum_{k=1}^{m} \left(x_i^{(k)} - E[x_i] \right)^2$$

Statistically, u_B and σ_B^2 are nothing but the sample mean and biased sample standard variance.

Once the normalization is done, $\hat{x}_i$ is not fed to the activation function directly but rather is scaled and shifted by introducing parameters γ and β before feeding it to the activation functions. If we restrict the input activations to the normalized values, they may change what the layer can represent. So, the idea is to apply a linear transformation on the normalized value through the following transformation so that if the network, through training, feels that the original values before any transformation are good for the network, it can recover the original values. The actual transformed input activation y_i fed to the activation function is given by the following:

$$y_i = \gamma \hat{x}_i + \beta$$

The parameters u_B, $\sigma_B{}^2$, γ, and β are to be learned by backpropagation much like the other parameters. As stated earlier, the model might learn $\gamma = Var[x_i]$ and $\beta = E[x_i]$ if it feels the original values from the network are more desirable.

A very natural question that may come up is why we are taking the mini-batch mean u_B and variance $\sigma_B{}^2$ as parameters to be learned through batch propagation and not estimating them as running averages over mini-batches for normalization purposes. This does not work, since the u_B and $\sigma_B{}^2$ are dependent on other parameters of the model through x_i, and when we directly estimate these as running averages, this dependency is not accounted for in the optimization process. To keep those dependencies intact, u_B and $\sigma_B{}^2$ should participate in the optimization process as parameters since the gradients of the u_B and $\sigma_B{}^2$ with respect to the other parameters on which x_i depends are critical to the learning process. The overall effect of this optimization modifies the model in such a way that the input $\hat{x}_i$ keeps zero mean and unit standard deviation.

During inference or testing time, the population statistics $E[x_i]$ and $Var[x_i]$ are used for normalization by keeping a running average of the mini-batch statistics.

$$E[x_i] = E[u_B]$$

$$Var[x_i] = \left(\frac{m}{m-1}\right) E[\sigma_B^2]$$

This correction factor is required to get an unbiased estimate of the population variance.

A couple of the advantages of batch normalization are as follows:

- Models can be trained faster because of the removal or reduction of internal covariate shift. A smaller number of training iterations would be required to achieve good model parameters.

- Batch normalization has some regularizing power and at times eliminates the need for dropout.

- Batch normalization works well with convolutional neural networks wherein there is one set of γ and β for each output feature map.

Different Architectures in Convolutional Neural Networks

In this section, we will go through a few widely used convolutional neural network architectures used today. These network architectures not only are used for classification but also, with minor modification, are used in segmentation, localization, and detection. Also, there are pretrained versions of each of these networks that enable the community to do transfer learning or fine-tune the models. Except LeNet, almost all the CNN models have won the ImageNet competition for classification of a thousand classes.

LeNet

The first successful convolutional neural network was developed by Yann LeCun in 1990 for classifying handwritten digits successfully for OCR-based activities such as reading ZIP codes, checks, and so on. LeNet5 is the latest offering from Yann LeCun and his colleagues. It takes in 32 × 32-size images as input and passes them through a convolutional layer to produce six feature maps of size 28x28. The six feature maps are then subsampled to produce six output images of size 14x14. Subsampling can be thought of as a pooling operation. The second convolutional layer has 16 feature maps of size 28 × 28, while the second subsampling layer reduces the feature map sizes to 14 × 14. This is followed by two fully connected layers of 120 and 84 units, respectively, followed by the output layer of ten classes corresponding to ten digits. Figure 3-30 represents the LeNet5 architectural diagrams.

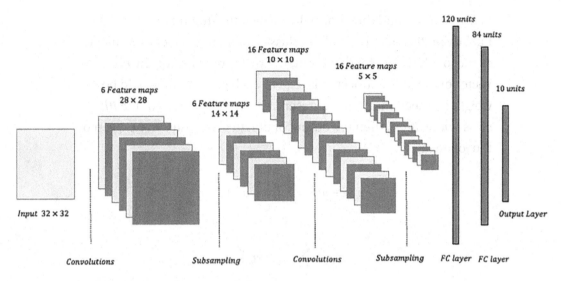

Figure 3-30. *LeNet5 architectural diagram*

Key features of the LeNet5 network are as follows:

- The pooling through subsampling takes 2 × 2 neighborhood patches and sums up the four-pixel intensity values. The sum is scaled by a trainable weight and a bias and then fed through a sigmoid activation function. This is a little different from what is done for max pooling and average pooling.

- The filter kernel used for convolution is of size 5 × 5. The output units are radial basis function (RBF) units instead of the SoftMax functions that we generally use. The 84 units of the fully connected layers had 84 connections to each of the classes and hence 84 corresponding weights. The 84 weights/classes represent each class's characteristics. If the inputs to those 84 units are very close to the weights corresponding to a class, then the inputs are more likely to belong to that class. In a SoftMax, we look at the dot product of the inputs to each of the class's weight vectors, while in RBF units we look at the Euclidean distance between the input and the output class representative's weight vectors. The greater the Euclidean distance, the smaller the chance is of the input belonging to that class. The same can be converted to probability by exponentiating the negative of the distance and then normalizing over the different classes. The Euclidean distances over all the classes for an input record would act as the loss function for that input. Let $x = [x_1 \, x_2 . . x_{84}]^T$ be the output vector of the fully connected layer. For each class, there would be 84 weight connections. If the representative weight vector for the *ith* class is $w_i \in R^{84 \times 1}$, then the output of the *ith* class unit can be given by the following:

$$x - w_{i2}^2 = \sum_{j=1}^{84}\left(x_j - w_{ij}\right)^2$$

$$y_i = \sum_{j=1}^{84}\left(x_j - w_{ij}\right)^2$$

- The representative weights for each class are fixed beforehand and are not learned weights.

AlexNet

The AlexNet CNN architecture was developed by Alex Krizhevsky, Ilya Sutskever, and Geoffrey Hinton in 2012 to win the 2012 ImageNet ILSVRC (ImageNet Large-Scale Visual Recognition Challenge). The original paper pertaining to AlexNet is titled "ImageNet Classification with Deep Convolutional Neural Networks" and can be located at https://papers.nips.cc/paper/4824-imagenet-classification-with-deep-convolutional-neural-networks.pdf.

It was the first time that a CNN architecture beat other methods by a huge margin. Their network achieved an error rate of 15.4% on its top 5 predictions as compared to the 26.2% error rate for the second-best entry. The architectural diagram of AlexNet is represented in Figure 3-31.

AlexNet consists of five convolutional layers, max pooling layers, and dropout layers and three fully connected layers in addition to the input and output layer of a thousand class units. The inputs to the network are images of size 224 × 224 × 3. The first convolutional layer produces 96 feature maps corresponding to 96 filter kernels of size 11 × 11 × 3 with strides of four-pixel units. The second convolutional layer produces 256 feature maps corresponding to filter kernels of size 5 × 5 × 48. The first two convolutional layers are followed by max pooling layers, whereas the next three convolutional layers are placed one after another without any intermediate max pooling layers. The fifth convolutional layer is followed by a max pooling layer, two fully connected layers of 4096 units, and finally a SoftMax output layer of 1000 classes. The third convolutional layer has 384 filter kernels of size 3 × 3 × 256, whereas the fourth and fifth convolutional layers have 384 and 256 filter kernels each of size 3 × 3 × 192. A dropout of 0.5 was used in the last two fully connected layers. You will notice that the depth of the filter kernels for convolutions is half the number of feature maps in the preceding layer for all but the third convolutional layer. This is because AlexNet was at that time computationally expensive and hence the training had to be split between two separate GPUs. However, if you observe carefully, for the third convolutional activity, there is cross-connectivity for convolution, and so the filter kernel is of dimension 3 × 3 × 256 and not 3 × 3 × 128. The same kind of cross-connectivity applies to the fully connected layers, and hence they behave as ordinary fully connected layers with 4096 units.

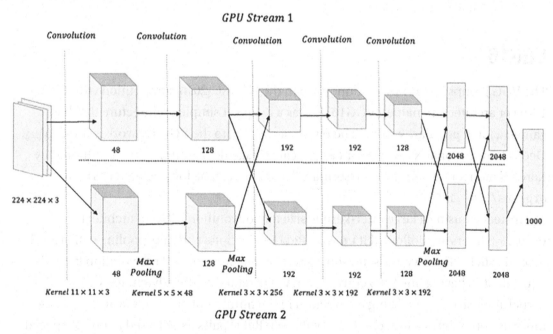

Figure 3-31. *AlexNet architecture*

The key features of AlexNet are as follows:

- ReLU activation functions were used for nonlinearity. They had a huge impact since RELUs are significantly easier to compute and have constant non-saturating gradients as opposed to sigmoid and tanh activation functions, whose gradients tend to zero for very high and low values of input.

- Dropout was used to reduce overfitting in the model.

- Overlapping pooling was used as opposed to non-overlapping pooling.

- The model was trained on two GPU GTX 580 for around five days for fast computation.

- The dataset size was increased through data-augmentation techniques, such as image translations, horizontal reflections, and patch extractions.

VGG16

The VGG groups in 2014 were runners up in the ILSVRC-2014 competition with a 16-layer architecture named VGG16. It uses a deep yet simple architecture that has gained a lot of popularity since. The paper pertaining to the VGG network is titled "Very Deep Convolutional Networks for Large-Scale Image Recognition" and is authored by Karen Simonyan and Andrew Zisserman. The paper can be located at `https://arxiv.org/abs/1409.1556`.

Instead of using a large filter-kernel size for convolution, VGG16 architecture used 3 × 3 filters and followed it up with ReLU activations and max pooling with a 2 × 2 receptive field. The inventors' reasoning was that using two 3 × 3 convolution layers is equivalent to having one 5 × 5 convolution while retaining the advantages of a smaller kernel-filter size, i.e., realizing a reduction in the number of parameters and realizing more nonlinearity because of two convolution–ReLU pairs as opposed to one. A special property of this network is that as the spatial dimensions of the input volume reduces because of convolution and max pooling, the number of feature maps increases due to the increase in the number of filters as we go deep into the network.

All Convolutions with stride 1 and pad 1
Max Pooling with stride 2

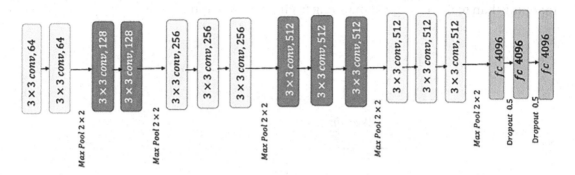

Figure 3-32. *VGG16 architecture*

Figure 3-32 represents the architecture of VGG16. The input to the network are images of size 224 × 224 × 3. The first two convolutional layers produce 64 feature maps, each followed by max pooling. The filters for convolution are of spatial size 3 × 3 with a stride of 1 and pad of 1. Max pooling is of size 2 × 2 with stride of 2 for the whole network. The third and fourth convolutional layers produce 128 feature maps, each followed by a max pooling layer. The rest of the network follows in a similar fashion, as shown in the Figure 3-32. At the end of the network, there are three fully connected layers of 4096 units, each followed by the output SoftMax layer of a thousand classes. Dropout is set at 0.5 for the fully connected layers. All the units in the network have ReLU activations.

ResNet

ResNet is a 152-layer-deep convolutional neural network from Microsoft that won the ILSVRC 2015 competition with an error rate of only 3.6%, which is perceived to be better than the human error rate of 5–10%. The paper on ResNet, authored by Kaiming He, Xiangyu Zhang, Shaoqing Ren, and Jian Sun, is titled "Deep Residual Learning for Image Recognition" and can be located at https://arxiv.org/abs/1512.03385. Apart from being deep, ResNet implements a unique idea of residual block. After each series of convolution–ReLU–convolution operations, the input to the operation is fed back to the output of the operation. In traditional methods while doing convolution and other transformations, we try to fit an underlying mapping to the original data to solve the classification task. However, with ResNet's residual block concept, we try to learn a residual mapping and not a direct mapping from the input to output. Formally, in each

273

small block of activities, we add the input to the block to the output. This is illustrated in Figure 3-33. This concept is based on the hypothesis that it is easier to fit a residual mapping than to fit the original mapping from input to output.

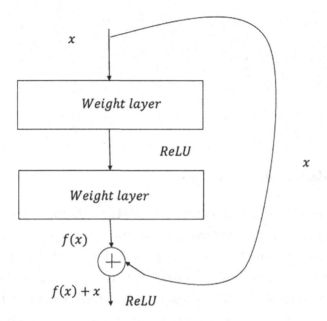

Figure 3-33. *Residual block*

Transfer Learning

Transfer learning in a broad sense refers to storing knowledge gained while solving a problem and using that knowledge for a different problem in a similar domain. Transfer learning has been hugely successful in the field of deep learning for a variety of reasons.

Deep-learning models in general have a huge number of parameters because of the nature of the hidden layers and the connectivity scheme within the different units. To train such a huge model, lots of data are required, or the model will suffer from overfitting problems. In many problems, the huge amount of data required to train the model is not available, but the nature of the problem requires a deep-learning solution in order to have a reasonable impact. For instance, in image processing for object recognition, deep-learning models are known to provide state-of-the-art solutions. In such cases, transfer learning can be used to generate generic features from a pretrained deep-learning model and then use those features to build a simple model to solve the problem. So, the only parameters for this problem are the ones used to build the simple model. The pretrained models are generally trained on a huge corpus of data and thus have reliable parameters.

When we process images through several layers of convolutions, the initial layers learn to detect very generic features such as curls and edges. As the network grows deeper, the convolutional layers in the deeper layers learn to detect more complex features relevant to the specific kind of dataset. For example, in a classification, the deeper layers would learn to detect features such as the eyes, nose, face, and so forth.

Let us assume we have a VGG16 architecture model trained on 1000 categories of the ImageNet dataset. Now, if we get a smaller dataset that has fewer categories of images like those of the VGG16 pretrained model dataset, then we can use the same VGG16 model up to the fully connected layer and then replace the output layer with the new classes. Also, we keep the weights of the network fixed till the fully connected layer and only train the model to learn the weights from the fully connected layer to the output layer. This is because the dataset's nature is the same as the smaller datasets, and thus the features learned in the pretrained model through the different parameters are good enough for the new classification problem, and we only need to learn the weights from the fully connected layer to the output layer. This is a huge reduction in the number of parameters to learn, and it will reduce the overfitting. Had we trained the small dataset using VGG16 architecture, it might have suffered from severe overfitting because of the large number of parameters to learn on a small dataset.

What do you do when the dataset's nature is very different from that of the dataset used for the pretrained model?

Well, in that case, we can use the same pretrained model but fix only the parameters for the first couple of sets of convolution–ReLU–max pooling layers and then add a couple of convolution–ReLU–max pooling layers that would learn to detect features intrinsic to the new dataset. Finally, we would have to have a fully connected layer followed by the output layer. Since we are using the weights of the initial sets of convolution–ReLU–max pooling layers from the pretrained VGG16 network, the parameters with respect to those layers need not be learned. As mentioned earlier, the early layers of convolution learn very generic features, such as edges and curves, that are applicable to all kinds of images. The rest of the network would need to be trained to learn specific features inherent to the specific problem dataset.

Guidelines for Using Transfer Learning

The following are a few guidelines as to when and how to use a pretrained model for transfer learning:

- The size of the problem dataset is large, and the dataset is like the one used for the pretrained model—this is the ideal scenario. We can retain the whole model architecture as it is except maybe the output layer when it has a different number of classes than the pretrained one. We can then train the model using the weights of the pretrained model as initial weights for the model.

- The size of the problem dataset is large, but the dataset is dissimilar to the one used for the pretrained model—in this case, since the dataset is large, we can train the model from scratch. The pretrained model will not give any gains here since the dataset's nature is very dissimilar, and since we have a large dataset, we can afford to train the whole network from scratch without overfitting related to large networks trained on small datasets.

- The size of the problem dataset is small, and the dataset is like the one used for the pretrained model—this is the case that we discussed earlier. Since the dataset content is similar, we can reuse the existing weights of most of the model and only change the output layer based on the classes in the problem dataset. Then, we train the model only for the weights in the last layer. For example, if we get images like ImageNet for only dogs and cats, we can pick up a VGG16 model pretrained on ImageNet and just modify the output layer to have two classes instead of a thousand. For the new network model, we just need to train the weights specific to the final output layer, keeping all the other weights the same as those of the pretrained VGG16 model.

- The size of the problem dataset is small, and the dataset is dissimilar to the one used in the pretrained model—this is not such a good situation to be in. As discussed earlier, we can freeze the weights of a few initial layers of a pretrained network and then train the rest of the model on the problem dataset. The output layer, as usual, needs to be changed as per the number of classes in the problem dataset. Since we do not have a large dataset, we are trying to reduce the number of parameters as much as possible by reusing weights of the initial layers of the pretrained model. Since the first few layers of CNN learn generic features inherent to any kind of image, this is possible.

Transfer Learning with Google's InceptionV3

InceptionV3 is one of the state-of-the-art convolutional neural networks from Google. It is an advanced version of GoogLeNet that won the ImageNetILSVRC-2014 competition with its out-of-the-box convolutional neural network architecture. The details of the network are documented in the paper titled "Rethinking the Inception Architecture for Computer Vision" by Christian Szegedy and his collaborators. The paper can be located at https://arxiv.org/abs/1512.00567. The core element of GoogLeNet and its modified versions is the introduction of an inception module to do the convolution and pooling. In traditional convolutional neural networks, after a convolution layer, we either perform another convolution or max pooling, whereas in the inception module, a series of convolutions and max pooling is done in parallel at each layer, and later the feature maps are merged. Also, in each layer, convolution is not done with one kernel-filter size but rather with multiple kernel-filter sizes. An inception module is presented in Figure 3-34. As we can see, there is a series of convolutions in parallel along with max pooling, and finally all the output feature maps merge in the filter concatenation block. 1×1 convolutions do a dimensionality reduction and perform an operation like average pooling. For example, let's say we have an input volume of $224 \times 224 \times 160$, with 160 being the number of feature maps. A convolution with a $1 \times 1 \times 20$ filter-kernel will create an output volume of $224 \times 224 \times 20$.

This kind of network works well since the different kernel sizes extract feature information at different granular levels based on the size of the filter's receptive field. Receptive fields of 3×3 will extract much more granular information than will a 5×5 receptive field.

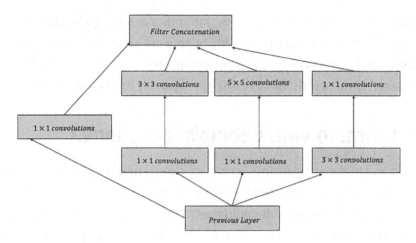

Figure 3-34. Inception module

Google's TensorFlow provides a pretrained model that is trained on the ImageNet data. It can be used for transfer learning. We use the pretrained model from Google and retrain it on a set of cat vs. dog images extracted from www.kaggle.com/c/dogs-vs-cats/ data. The train.zip dataset contains 25,000 images, with 12,500 images each for cats and dogs.

Transfer Learning with Pretrained VGG16

In this section, we will perform transfer learning by using a VGG16 network pretrained on a thousand classes of ImageNet to classify the cats vs. dogs dataset from Kaggle. The link to the dataset is https://www.kaggle.com/c/dogs-vs-cats/data. First, we would import the VGG16 model from *TensorFlow Slim* and then load the pretrained weights in the VGG16 network. The weights are from a VGG16 trained on the thousand classes of the ImageNet dataset. Since for our problem we have only two classes, we will take the output from the last fully connected layer and combine it with a new set of weights, leading to the output layer with one neuron to do a binary classification of the cats and dogs dataset from Kaggle. The idea is to use the pretrained weights to generate features, and finally we learn just one set of weights at the end, leading to the output. In this way, we learn a relatively small set of weights and can afford to train the model on a smaller amount of data. Please find the detailed implementation in Listing 3-8.

Listing 3-8. Transfer Learning with Pre-trained VGG16

```
# Load packages
import sys
from sklearn.model_selection import train_test_split
import cv2
import os
import tensorflow as tf
from pathlib import Path
from glob import glob

print('tensorflow version', tf.__version__)
import numpy as np
from tensorflow.keras import Model, layers
import matplotlib.pyplot as plt
import time
from elapsedtimer import ElapsedTimer
import pandas as pd
from tensorflow.keras.applications.vgg16 import VGG16
from tensorflow.keras.applications.vgg16 import preprocess_input
from tensorflow.keras import layers, Model

# Mean value for image normalization
MEAN_VALUE = np.array([103.939, 116.779, 123.68])

# Model class
class vgg16_customized(Model):

    def __init__(self,activation='relu', n_classes=2, input_size=224):
        super(vgg16_customized,self).__init__()
        self.base_model = VGG16(weights='imagenet', include_top=False,
        input_shape=(input_size, input_size, 3))
        self.base_model.trainable = False
        self.flatten = layers.Flatten()
        self.out = layers.Dense(n_classes-1)

    # Forward pass function
    def call(self, x):
        x = self.base_model(x)
```

```python
        x = self.flatten(x)
        #x = self.fc1(x)
        #x = self.fc2(x)
        out = self.out(x)
        probs = tf.nn.sigmoid(out)
        return out, probs

# Routine to read the Images and also do mean correction
def image_preprocess(img_path, width, height):
    img = cv2.imread(img_path)
    img = cv2.resize(img, (width, height))
    img = img - MEAN_VALUE
    return img

# Create generator for Image batches so that only the processed batch is
in memory
def data_gen_batch(images, batch_size, width, height):

    while True:
        ix = np.random.choice(np.arange(len(images)), batch_size)
        imgs = []
        labels = []
        for i in ix:

            if images[i].split('/')[-1].split('.')[0] == 'cat':
                labels.append(1)

            else:
                if images[i].split('/')[-1].split('.')[0] == 'dog':
                    labels.append(0)

            img_path = f"{Path(images[i])}"
            array_img = image_preprocess(img_path, width, height)
            imgs.append(array_img)

        imgs = np.array(imgs)
        labels = np.array(labels)
        labels = np.reshape(labels, (batch_size, 1))
        yield imgs, labels
```

```python
# Train function
def train(train_data_dir, lr=0.01, input_size=224,n_classes=2, batch_
size=32, output_dir=None,epochs=10):

    if output_dir is None:
        output_dir = os.getcwd()
    else:
        if not Path(output_dir).exists():
            os.makedirs(output_dir)

    all_images = glob(f"{train_data_dir}/*/*")
    train_images, val_images = train_test_split(all_images, train_size=0.8,
    test_size=0.2)

    print(f"Number of training images: {len(train_images)}")
    print(f"Number of validation images: {len(val_images)}")

    # Define the train and val batch generator
    train_gen = data_gen_batch(train_images, batch_size, 224, 224)
    val_gen = data_gen_batch(val_images, batch_size, 224, 224)

    # Build the model
    model = vgg16_customized(activation='relu', n_classes=n_classes,
    input_size=input_size)
    model_graph = tf.function(model)

    # Define the loss function
    loss_fn = tf.keras.losses.BinaryCrossentropy(from_logits=True,
    reduction=tf.keras.losses.Reduction.SUM)

    # Define the optimizer
    optimizer = tf.keras.optimizers.Adam(lr)

    batches = len(train_images)//batch_size
    batches_val = len(val_images) // batch_size
    loss_trace, accuracy_trace = [], []

    for epoch in range(epochs):
        loss, accuracy = 0, 0
        num_train_recs = 0
```

```
    for batch in range(batches):
        x_train_batch, y_train_batch = next(train_gen)
        x_train_batch, y_train_batch = tf.constant(x_train_batch),
        tf.constant(y_train_batch)
        num_train_recs += x_train_batch.shape[0]
        with tf.GradientTape() as tape:
            y_pred_batch,y_pred_probs = model_graph(x_train_batch,
            training=True)
            loss_ = loss_fn(y_train_batch, y_pred_batch)

        # compute gradient
        gradients = tape.gradient(loss_, model.trainable_variables)
        # update the parameters
        optimizer.apply_gradients(zip(gradients, model.trainable_
        variables))

        y_pred_probs, y_train_batch = y_pred_probs.numpy() , y_train_
        batch.numpy()
        y_pred_probs[y_pred_probs >= 0.5] = 1.0
        accuracy += np.sum(y_train_batch == y_pred_probs)
        loss += loss_.numpy()
        #print(f"Loss for Epoch {epoch} Batch {batch}: {loss_.numpy()}")
    loss /= num_train_recs
    accuracy /= num_train_recs
    loss_trace.append(loss)
    accuracy_trace.append(accuracy)
    print(f"------------------------------------------------------\n")
    print(f"Epoch {epoch} : loss: {np.round(loss, 4)} ,accuracy:{np.
    round(accuracy, 4)}\n")
    print(f"------------------------------------------------------\n")
accuracy_val, num_val_recs = 0, 0
for batch in range(batches_val):
    X_val_batch, y_val_batch = next(val_gen)
    X_val_batch = tf.constant(X_val_batch)
    num_val_recs += X_val_batch.shape[0]
    _,y_probs = model_graph(X_val_batch,training=False)
```

```
    y_probs = y_probs.numpy()
    y_probs[y_probs >= 0.5] = 1.0
    #print(y_probs.shape,y_val_batch.shape)
    accuracy_val += np.sum(y_val_batch == y_probs)
accuracy_val /= num_val_recs
print(f"Validation Accuracy at Epoch {epoch}: {accuracy_val}")
return model_graph, X_val_batch, y_val_batch, y_probs
```

```
model_graph, X_val_batch, y_val_batch, y_probs = train(train_data_dir=
'/media/santanu/9eb9b6dc-b380-486e-b4fd-c424a325b976/CatvsDog/train',
batch_size=32,epochs=1)
```

The readers should change the train_data_dir accordingly in the call to the train function based on the location of the data.

```
    --output--
Epoch 0 : loss: 5.692 ,accuracy:0.9669

------------------------------------------------------------------------------

Validation Accuracy at Epoch 0: 0.9769
```

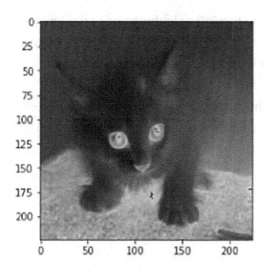 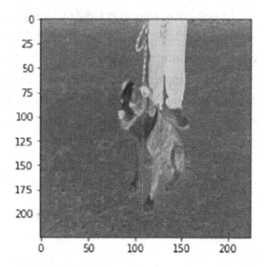

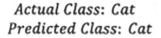

Actual Class: Cat *Actual Class: Dog*
Predicted Class: Cat *Predicted Class: Dog*

Figure 3-35. *Validation-set images and their actual vs. predicted classes*

We see that the validation accuracy is 98% after training the model for random batches of size 32 for approximately one epoch. Here the epoch does not necessarily mean a complete pass over the training records since the batches are generated randomly. Hence an epoch in this content can have several images sampled more than once and several samples which did not get sampled even once. In Figure 3-35, a couple of validation-set images have been plotted along with their actual and predicted classes to illustrate the correctness of the predictions. Hence, proper utilization of transfer learning helps us to reuse feature detectors learned for one problem in solving a new problem. Transfer learning greatly reduces the number of parameters that need to be learned and hence reduces the computational burden on the network. Also, the training data-size constraints are reduced since fewer parameters require less data to be trained.

Dilated Convolution

Dilated convolution, as an alternative to regular convolution, was first introduced in the paper "Semantic Image Segmentation with Deep Convolutional Nets and Fully Connected CRFs" (https://arxiv.org/abs/1412.7062). The central idea behind dilated convolution is to apply the filter kernel weights to pixels that might not be direct neighbors of each other.

In general convolution, a 3x3 filter gets applied to a 3x3 neighborhood of the image, whereas in dilated convolution, a 3x3 filter would get applied to a 5x5 neighborhood in which every alternate pixel is skipped. The latter is called a one-dilation convolution as we are skipping every alternate pixel. Using dilated convolution one can increase the receptive field of the kernel without increasing the parameters. For instance, a 3x3 filter kernel in a general convolution has a receptive field of 3x3 while the same kernel for a one-dilation has a receptive field of 5x5.

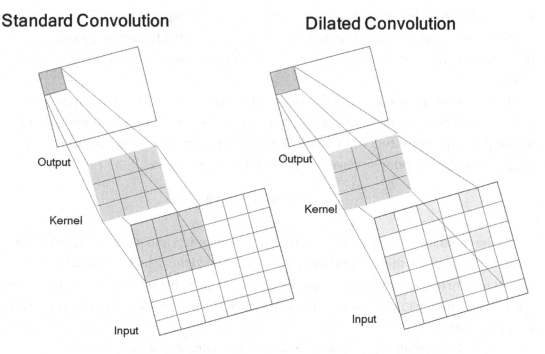

Standard Convolution

Dilated Convolution

Figure 3-36. *Standard convolution vs. dilated convolution illustration*

The difference between standard convolution and dilated convolution has been illustrated in Figure 3-36. We can see the 3x3 filter-kernel in standard convolution acts on a 3x3 image patch to produce a 1x1 output. In the dilated convolution, the 3x3 filter kernel works on a 5x5 patch to produce a 1x1 output.

Depthwise Separable Convolution

Depthwise separable convolution is a two-step convolution process which reduces the number of parameters drastically in a convolution layer. In the first step, depthwise convolution is performed in which each input channel is convolved with one filter. So, if there are m input channels, then there would be m convolution filters in the depthwise convolution step which would yield m output feature maps. In the second step, pointwise convolution is applied to create a linear combination of the output feature maps from depthwise convolution for each output channel. Pointwise convolution here essentially means taking the weighted average of the output feature maps' activation at each given location. If there are n output channels each output channel would the linear sum of the m output feature maps from Depthwise Convolution step. The linear sum would be based on m parameters corresponding to each output channel. The pointwise convolution step going from m input feature maps to n output feature maps can be thought of as standard convolution with 1×1 filters with number of parameters equal to $m \times n \times 1 \times 1$.

If we think of the width and height of each filter kernel to be w and h, respectively, then for depthwise convolution, we would have $m \times w \times h$ number of parameters, and for the pointwise convolution, we would have $m \times n$ parameters. The total number of parameters N_d for depthwise separable convolution is as follows:

$$N_c = m \times w \times h + m \times n$$

In the standard convolution layer with m input channels and n output channels for each output channel, there are m filters corresponding to the m input channels. The output feature maps from the m convolutions are averaged to produce the output feature map corresponding to each output channel. Hence, if there are n output channels, there are $m \times n$ filters.

Hence, the number of parameters N_s in standard convolution is as follows:

$$N_s = m \times n \times w \times h$$

In general, $N_s \gg N_d$, and hence depthwise separable convolution helps in reducing the number of parameters in each convolution layer. In general, the performance of models using depthwise separable convolution is comparable to the standard convolution.

The steps of depthwise separable convolution for a convolution layer with three input channels and four output channels are illustrated in Figure 3-37.

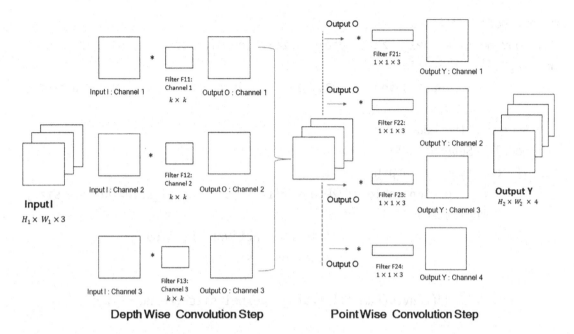

Depth Wise Convolution Step   **Point Wise Convolution Step**

Figure 3-37. *Depthwise separable convolution*

To illustrate the fact that the model using depthwise separable convolution has comparable performance to model with standard convolution, we train and validate the MNIST-based model using both standard convolution and depthwise separable convolution layers. Listing 3-9a through 3-9c illustrate the same. We have used all the Keras features within TensorFlow 2 for this exercise.

Listing 3-9a. Model Definition and Training Routine for Depthwise Separable Convolution-Based and Standard Convolution-Based Models

```
# Load Packages
import tensorflow
from tensorflow.keras.models import Sequential
from tensorflow.keras.datasets import mnist
from tensorflow.keras.layers import Dense, Dropout, Flatten
from tensorflow.keras.layers import Conv2D, SeparableConv2D, MaxPooling2D
```

```python
# Conv Model definition
def conv_model(kernel_size=3, pool_ksize=2, num_filters=[32, 64], hidden_
units=[256], activation='relu',
               input_shape=(28, 28, 1), n_classes=10, conv_type='standard',
               dropout=0.25):
    # Create the model
    model = Sequential()
    if conv_type == 'depth_wise':
        model.add(SeparableConv2D(num_filters[0], kernel_size=(kernel_size,
        kernel_size),
                                  activation=activation, input_
                                  shape=input_shape))
    else:
        model.add(Conv2D(num_filters[0], kernel_size=(kernel_size,
        kernel_size),
                         activation=activation, input_
                         shape=input_shape))

    model.add(MaxPooling2D(pool_size=(pool_ksize, pool_ksize)))
    model.add(Dropout(dropout))

    if conv_type == 'depth_wise':
        model.add(SeparableConv2D(num_filters[1], kernel_size=(kernel_size,
        kernel_size), activation=activation))
    else:
        model.add(Conv2D(num_filters[1], kernel_size=(kernel_size,
        kernel_size), activation=activation))

    model.add(MaxPooling2D(pool_size=(2, 2)))
    model.add(Dropout(dropout))
    model.add(Flatten())
    model.add(Dense(hidden_units[0], activation=activation))
    model.add(Dense(n_classes, activation='softmax'))
    return model
```

```python
# Train Routine
def train(input_size=28, input_channels=1, epochs=10, n_classes=10,
        val_ratio=0.3, kernel_size=3, pool_ksize=2, num_filters=[32, 64],
        hidden_units=[256], activation='relu',
        dropout=0.25, lr=0.01, batch_size=128, conv_type='standard'):

    # Load MNIST dataset and format
    (input_train, target_train), (input_test, target_test) = mnist.
    load_data()

    # Reshape the data
    input_train = input_train.reshape(input_train.shape[0], input_size,
    input_size, 1)
    input_test = input_test.reshape(input_test.shape[0], input_size,
    input_size, 1)
    input_shape = (input_size, input_size, input_channels)

    input_train = input_train.astype('float32')
    input_test = input_test.astype('float32')

    print(f"Normalize the input..\n")
    input_train = input_train / 255
    input_test = input_test / 255

    # Convert target vectors to categorical targets
    target_train = tensorflow.keras.utils.to_categorical(target_train,
    n_classes)
    target_test = tensorflow.keras.utils.to_categorical(target_test,
    n_classes)

    # Build Model
    model = conv_model(kernel_size=kernel_size, pool_ksize=pool_ksize, num_
    filters=num_filters,
                    hidden_units=hidden_units, activation=activation,
                    input_shape=input_shape,
                    n_classes=n_classes, conv_type=conv_type,
                    dropout=dropout)
```

```
# Compile the model
model.compile(loss=tensorflow.keras.losses.categorical_crossentropy,
              optimizer=tensorflow.keras.optimizers.Adam(),
              metrics=['accuracy'])

# Train Model
model.fit(input_train, target_train,
          batch_size=batch_size,
          epochs=epochs,
          verbose=1,
          validation_split=val_ratio)

print(f"Evaluation test data..\n")
score = model.evaluate(input_test, target_test, verbose=0)
print(f'Test loss: {score[0]} / Test accuracy: {score[1]}')
```

Listing 3-9b. Run Standard Convolution

```
train(conv_type='standard')
—output—
Test loss: 0.028370419517159462 / Test accuracy: 0.9904999732971191
```

Listing 3-9c. Run Depthwise Separable Convolution

```
train(conv_type='depth_wise')
—output—
Test loss: 0.04192721098661423 / Test accuracy: 0.9858999848365784
```

We can see from the output log for Listing 3-9b and 3-9c that the test accuracy of both the models on MNIST dataset is around ~99%.

Summary

In this chapter, we learned about the convolution operation and how it is used to construct a convolutional neural network. Also, we learned about the various key components of CNN and the backpropagation method of training the convolutional layers and the pooling layers. We discussed two critical concepts of CNN responsible for its success in image processing—the equivariance property provided by convolution and the translation invariance provided by the pooling operation. Further, we discussed several established CNN architectures and how to perform transfer learning using the pretrained versions of these CNNs. In the next chapter, we will discuss recurrent neural networks and their variants in the realm of natural language processing.

## CHAPTER 4

Natural Language Processing

In the modern age of information and analytics, natural language processing (NLP) is one of the most important technologies out there. Making sense of complex structures in language and deriving insights and actions from it are crucial from an artificial intelligence perspective. In several domains, the importance of natural language processing is paramount and ever growing, as digital information in the form of language is ubiquitous. Applications of natural language processing include language translation, sentiment analysis, web search applications, customer service automation, text classification, topic detection from text, language modeling, and so forth. Traditional methods of natural language processing relied on the bag of word models, the vector space of word model, and on-hand coded knowledge bases and ontologies. One of the key areas for natural language processing is the syntactic and semantic analysis of language. Syntactic analysis refers to how words are grouped and connected in a sentence. The main tasks in syntactic analysis are tagging parts of speech, detecting syntactic classes (such as verbs, nouns, noun phrases, etc.), and assembling sentences by constructing syntax trees. Semantic analysis refers to complex tasks such as finding synonyms, performing word-verb disambiguation, and so on. As part of this chapter, we will begin with a brief introduction of the traditional natural language processing techniques and then move on to deep learning-based methods involving recurrent neural networks, attention, and transformers.

Vector Space Model (VSM)

In NLP information-retrieval systems, a document is generally represented as simply a vector of the count of the words it contains. For retrieving documents similar to a specific document, the cosine of the angle or the dot product between the document and other

© Santanu Pattanayak 2023
S. Pattanayak, *Pro Deep Learning with TensorFlow 2.0*, https://doi.org/10.1007/978-1-4842-8931-0_4

documents in the corpus is computed. The cosine of the angle between two vectors gives a similarity measure based on the similarity between their vector compositions. To illustrate this fact, let us look at two vectors $x, y \in R^{2 \times 1}$ as shown here:

$$x = \begin{bmatrix} 2 & 3 \end{bmatrix}^T$$
$$y = \begin{bmatrix} 4 & 6 \end{bmatrix}^T$$

Although vectors x and y are different, their cosine similarity is the maximum possible, i.e., 1. This is because the two vectors are identical in their component compositions. The ratio of the first component to the second component for both vectors is $\frac{2}{3}$, and hence content-composition-wise, they are treated as being similar. Hence, documents with high cosine similarity are generally considered similar in nature.

Let's say we have two sentences:

$$Doc1 = \begin{bmatrix} The \ dog \ chased \ the \ cat \end{bmatrix}$$

$$Doc2 = \begin{bmatrix} The \ cat \ was \ chased \ down \ by \ the \ dog \end{bmatrix}$$

The number of distinct words in the two sentences would be the vector space dimension for this problem. The distinct words are *The, dog, chased, the, cat, down, by,* and *was,* and hence we can represent each document as an eight-dimensional vector of word counts.

$$'The' \ dog' \ chased' \ the' \ cat' \ down' \ by' \ was'$$

$$Doc1 = \begin{bmatrix} 1 1 1 1 1 0 0 0 \end{bmatrix}$$

$$Doc2 = \begin{bmatrix} 1 1 1 1 1 1 1 1 \end{bmatrix}$$

If we represent *Doc*1 by v_1 and *Doc*2 by v_2, then the cosine similarity can be expressed as follows:

$$\cos(v_1, v_2) = \frac{\left(v_1^T v_2\right)}{v_1 v_2} = \frac{1 \times 1 + 1 \times 1 + 1 \times 1 + 1 \times 1 + 1 \times 1}{\sqrt{5}\sqrt{8}} = \frac{5}{\sqrt{40}}$$

where $\|v_1\|$ is the magnitude or the l^2 norm of the vector v_1.

As stated earlier, cosine similarity gives a measure of the similarity based on the component composition of each vector. If the components of the document vectors are in somewhat similar proportion, the cosine distance would be high. It does not take the magnitude of the vector into consideration.

In certain cases, when the documents are of highly varying lengths, the dot product between the document vectors is taken instead of the cosine similarity. This is done when, along with the content of the document, the size of the document is also compared. For instance, we can have a tweet in which the words *global* and *economics* might have word counts of 1 and 2, respectively, while a newspaper article might have word counts of 50 and 100, respectively, for the same words. Assuming the other words in both documents have insignificant counts, the cosine similarity between the tweet and the newspaper article would be close to 1. Since the tweet sizes are significantly smaller, the word counts proportion of 1:2 for *global* and *economics* does not really compare to the proportion of 1:2 for these words in the newspaper article. Hence, it does not really make sense to assign such a high similarity measure to these documents for several applications. In that case, taking the dot product as a similarity measure rather than the cosine similarity helps since it scales up the cosine similarity by the magnitude of the word vectors for the two documents. For comparable cosine similarities, documents with higher magnitudes would have higher dot product similarity since they have enough text to justify their word composition. The word composition for small texts might just be by chance and not be the true representation of its intended representation. For most applications where the documents are of comparable lengths, cosine similarity is a fair enough measure.

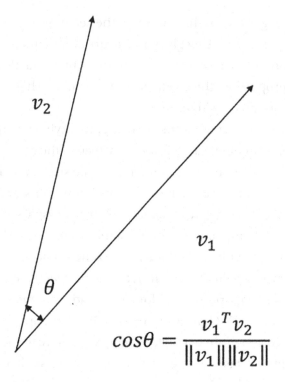

Figure 4-1. *Cosine similarity between two word vectors*

Figure 4-1 illustrates two vectors v_1 and v_2 with cosine similarity as the cosine of the angle θ between them.

At times, it makes sense to work with the distance counterpart of the cosine similarity. The cosine distance is defined as the square of the Euclidean distance between the unit vectors in the direction of the original vectors for which the distance needs to be computed. For two vectors v_1 and v_2 at an angle of θ between them, the cosine distance is given by $2(1 - \cos \theta)$.

This can be easily deduced by taking the square of the Euclidean distance between the unit vectors $u_1 = \dfrac{v_1}{\|v_1\|}$ and $u_2 = \dfrac{v_2}{\|v_2\|}$ as shown here:

$$
\begin{aligned}
u_1 - u_2{}^2 &= \left(u_1 - u_2\right)^T \left(u_1 - u_2\right) \\
&= u_1{}^T u_1 + u_2{}^T u_2 - 2u_1{}^T u_2 \\
&= \|u_1\|^2 + \|u_2\|^2 - 2\|u_1\| \|u_2\| \cos\theta
\end{aligned}
$$

Now, u_1 and u_2 being unit vectors, their magnitudes $\|u_1\|$ and $\|u_2\|$, respectively, are both equal to 1 and hence

$$\|u_1\|^2 + \|u_2\|^2 - 2\|u_1\|\,\|u_2\|cos\theta = 1 + 1 - 2cos\theta = 2(1 - cos\theta)$$

Generally, when working with document term-frequency vectors, the raw term/word counts are not taken and instead are normalized by how frequently the word is used in the corpus. For example, the term *the* is a frequently occurring word in any corpus, and it is likely that the term has a high count in two documents. This high count for *the* is likely to increase the cosine similarity, whereas we know that the term is a frequently occurring word in any corpus and should contribute very little to document similarity. The count of such words in the document term vector is penalized by a factor called *inverse document frequency*.

For a term word t occurring n times in a document d and occurring in N documents out of M documents in the corpus, the normalized count after applying inverse document frequency is as follows:

$$Normalized\ count = (Term\ freqency) \times (Inverse\ document\ frequency)$$

$$= nlog\left(\frac{M}{N}\right)$$

As we can see, as N increases with respect to M, the $log\left(\frac{M}{N}\right)$ component diminishes until it's zero for $M = N$. So, if a word is highly popular in the corpus, then it would not contribute much to the individual document term vector. A word that has high frequency in the document but is less frequent across the corpus would contribute more to the document term vector. This normalizing scheme is popularly known as *tf – idf*, a short-form representation for term-frequency inverse document frequency. Generally, for practical purposes, the $(N + 1)$ is taken as a denominator to avoid zeros that make the *log* function undefined. Hence, the inverse document frequency can be rephrased as $log\left(\frac{M}{N+1}\right)$.

Normalizing schemes are even applied to the term-frequency n to make it nonlinear. A popular such normalizing scheme is BM25 where the document frequency contribution is linear for small values of n and then the contribution is made to saturate as n increases. The term frequency is normalized as follows in BM25:

$$BM25(n) = \frac{(k+1)n}{k+n}$$

where k is a parameter that provides different shapes for different values of k and one needs to optimize k based on the corpus.

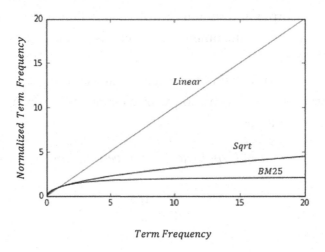

Figure 4-2. *Normalized term frequency vs. term frequency for different methods*

In Figure 4-2, the normalized term frequency for different normalizing schemes has been plotted against the term frequency. Square-root transformation makes the dependency sublinear, whereas the BM25 plot for $k = 1.2$ is very aggressive and the curve saturates beyond a term frequency of 5. As stated earlier, the k can be optimized through cross-validation or other methods based on the problem needs.

Vector Representation of Words

Just as the documents are expressed as vectors of different word counts, a word in a corpus can also be expressed as a vector, with the components being word counts in each document.

Other ways of expressing words as vectors would be to have the component specific to a document set to 1 if the word is present in the document or zero if the word does not exist in the document.

$$\begin{array}{c}
\textit{'The' 'dog' 'chased' 'the' 'cat' 'down' 'by' 'was'} \\
Doc11 = \begin{bmatrix} 1 & 1 & 1 & 1 & 1 & 0 & 0 & 0 \end{bmatrix} \in \mathbb{R}^{8\times1} \\
Doc12 = \begin{bmatrix} 1 & 1 & 1 & 1 & 1 & 1 & 1 & 1 \end{bmatrix} \in \mathbb{R}^{8\times1}
\end{array}$$

Reusing the same example, the word *The* can be expressed as a two-dimensional vector $[1\ 1]^T$ in the corpus of two documents. In a huge corpus of documents, the dimensionality of the word vector would be large as well. Like document similarity, word similarity can be computed through either cosine similarity or dot product.

Another way to represent words in a corpus is to one-hot encode them. In that case, the dimensionality of each word would be the number of unique words in the corpus. Each word would correspond to an index that would be set to 1 for the word, and all other remaining entries would be set to 0. So, each would be extremely sparse. Even similar words would have entries set to 1 for different indexes, and hence any kind of similarity measure would not work.

To represent word vectors better so that the similarity of the words can be captured more meaningfully, and to render less dimensionality to word vectors, Word2Vec was introduced.

Word2Vec

Word2Vec is an intelligent way of expressing a word as a vector by training the word with words in its neighborhood as context. Words that are contextually similar the given word would produce high cosine similarity or dot product when their Word2Vec representations are considered.

Generally, the words in the corpus are trained with respect to the words in their neighborhood to derive the set of the Word2Vec representations. The two most popular methods of extracting Word2Vec representations are the CBOW (continuous bag of words) method and Skip-gram method. The core idea behind CBOW is expressed in Figure 4-3.

Continuous Bag of Words (CBOW)

The CBOW method tries to predict the center word from the context of the neighboring words in a specific window length. Let's look at the following sentence and consider a window of 5 as a neighborhood.

"The cat jumped over the fence and crossed the road"

In the first instance, we will try to predict the word **jumped** from its neighborhood **The cat over the**. In the second instance, as we slide the window by one position, we will try to predict the word **over** from the neighboring words **cat jumped the fence**. This process would be repeated for the entire corpus.

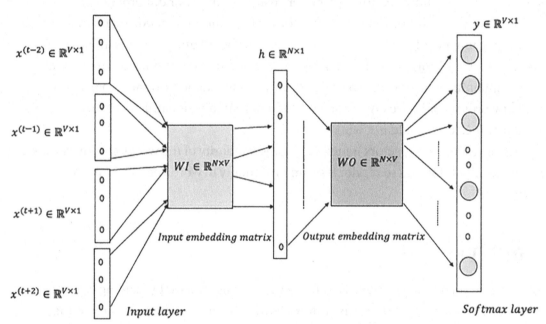

Figure 4-3. *Continuous bag of word model for word embeddings*

As shown in Figure 4-3, the continuous bag of word model (CBOW) is trained on the context words as input and the center word as the output. The words in the input layer are expressed as one-hot encoded vectors where the component for the specific word is set to 1 and all other components are set to 0. The number of unique words V in the corpus determines the dimensionality of these one-hot encoded vectors, and hence $x^{(t)} \in R^{V \times 1}$. Each one-hot encoded vector $x^{(t)}$ is multiplied by the input embedding matrix $WI \in R^{N \times V}$ to extract the word-embedding vector $u^{(k)} \in R^{N \times 1}$ specific to that word. The index k in $u^{(k)}$ signifies that $u^{(k)}$ is the word embedded for the kth word in the vocabulary. The hidden-layer vector h is the average of the input embedding vectors for all the context words in the window, and hence $h \in R^{N \times 1}$ has the same dimension as that of the word-embedding vectors.

$$h = \frac{1}{l-1} \sum_{\substack{k=(t-2) \\ k \neq t}}^{(t+2)} (WI) x^{(k)}$$

where l is the length of the window size.

Just for clarity, let's say we have a six-word vocabulary—i.e., $V = 6$—with the words being *cat*, *rat*, *chased*, *garden*, *the*, and *was*.

Let their one-hot encodings occupy the indexes in order so they can be represented as follows:

$$x_{cat} = \begin{bmatrix} 1 \\ 0 \\ 0 \\ 0 \\ 0 \\ 0 \end{bmatrix} \quad x_{rat} = \begin{bmatrix} 0 \\ 1 \\ 0 \\ 0 \\ 0 \\ 0 \end{bmatrix} \quad x_{chased} = \begin{bmatrix} 0 \\ 0 \\ 1 \\ 0 \\ 0 \\ 0 \end{bmatrix} \quad x_{garden} = \begin{bmatrix} 0 \\ 0 \\ 0 \\ 1 \\ 0 \\ 0 \end{bmatrix} \quad x_{the} = \begin{bmatrix} 0 \\ 0 \\ 0 \\ 0 \\ 1 \\ 0 \end{bmatrix} \quad x_{was} = \begin{bmatrix} 0 \\ 0 \\ 0 \\ 0 \\ 0 \\ 1 \end{bmatrix}$$

Let the input embedding where the embedding vector for each word is of dimensionality 5 be expressed as follows:

$$WI = \begin{matrix} cat & rat & chased & garden & the & was \\ \begin{bmatrix} 0.5 & 0.3 & 0.1 & 0.01 & 0.2 & 0.2 \\ 0.7 & 0.2 & 0.1 & 0.02 & 0.3 & 0.3 \\ 0.9 & 0.7 & 0.3 & 0.4 & 0.4 & 0.33 \\ 0.8 & 0.6 & 0.3 & 0.53 & 0.91 & 0.4 \\ 0.6 & 0.5 & 0.2 & 0.76 & 0.6 & 0.5 \end{bmatrix} \end{matrix}$$

Once we multiply the word-embedding matrix by the one-hot encoded vector for a word, we get the word-embedding vector for that word. Hence, by multiplying the one-hot vector for *cat* (i.e., x_{cat}) by the input embedding matrix *WI*, one would get the first column of the *WI* matrix that corresponds to the cat, as shown here:

$$[WI][x_{cat}]$$

$$= \begin{bmatrix} 0.5 & 0.3 & 0.1 & 0.01 & 0.2 & 0.2 \\ 0.7 & 0.2 & 0.1 & 0.02 & 0.3 & 0.3 \\ 0.9 & 0.7 & 0.3 & 0.4 & 0.4 & 0.33 \\ 0.8 & 0.6 & 0.3 & 0.53 & 0.91 & 0.4 \\ 0.6 & 0.5 & 0.2 & 0.76 & 0.6 & 0.4 \end{bmatrix} \begin{bmatrix} 1 \\ 0 \\ 0 \\ 0 \\ 0 \\ 0 \end{bmatrix} = \begin{bmatrix} 0.5 \\ 0.7 \\ 0.9 \\ 0.8 \\ 0.6 \end{bmatrix}$$

$$\begin{bmatrix} 0.5 \\ 0.7 \\ 0.9 \\ 0.8 \\ 0.6 \end{bmatrix}$$ is the word-embedding vector for the word *cat*.

Similarly, all the word-embedding vectors for the input words are extracted, and their average is the output of the hidden layer.

The output of the hidden layer h is supposed to represent the embedding of the target word.

All the words in the vocabulary have another set of word embedding housed in the output embedding matrix $WO \in R^{V \times N}$. Let the word embeddings in *WO* be represented by $v^{(j)} \in R^{N \times 1}$, where the index j denotes the *jth* word in the vocabulary in order as maintained in both the one-hot encoding scheme and the input embedding matrix.

$$WO = \begin{bmatrix} v^{(1)T} \rightarrow \\ v^{(2)T} \rightarrow \\ \cdot \\ v^{(j)T} \rightarrow \\ \cdot \\ v^{(V)T} \rightarrow \end{bmatrix}$$

The dot product of the hidden-layer embedding h is computed with each of the $v^{(j)}$ by multiplying the matrix WO by h. The dot product, as we know, would give a similarity measure for each of the output word embedding $v^{(j)}$ $\forall j \in \{1, 2, ..., N\}$ and the hidden-layer computed embedding h. The dot products are normalized to probability through a SoftMax, and, based on the target word $w^{(t)}$, the categorical cross-entropy loss is computed and backpropagated through gradient descent to update the matrices' weights for both the input and output embedding matrices.

Input to the SoftMax layer can be expressed as follows:

$$[WO][h] = \begin{bmatrix} v^{(1)T} \rightarrow \\ v^{(2)T} \rightarrow \\ \cdot \\ v^{(j)T} \rightarrow \\ \cdot \\ v^{(V)T} \rightarrow \end{bmatrix} [h] = \begin{bmatrix} v^{(1)T}h & v^{(2)T}h & \ldots v^{(j)T}h & v^{(V)T}h \end{bmatrix}$$

The SoftMax output probability for the *jth* word of the vocabulary $w^{(j)}$ given the context words is given by the following:

$$P\left(w = w^{(j)} / h\right) = p^{(j)} = \frac{e^{v^{(j)T}h}}{\sum\limits_{k=1}^{V} e^{v^{(k)T}h}}$$

If the actual output is represented by a one-hot encoded vector $y = [y_1 y_2 .... y_j . y_v]^T \in R^{V \times 1}$, where only one of the y_j is 1 (i.e., $\sum\limits_{j=1}^{V} y_j = 1$), then the loss function for the particular combination of target word and its context words can be given by the following:

$$C = -\sum\limits_{j=1}^{V} y_j \log\left(p^{(j)}\right)$$

The different $p^{(j)}$s are dependent on the input and output embeddings, which are parameters to the cost function C. The cost function can be minimized with respect to these embedding parameters through backpropagation gradient-descent techniques.

To make this more intuitive, let us say our target variable is cat. If the hidden-layer vector h gives the maximum dot product with the outer matrix word-embedding vector for cat while the dot product with the other outer word embedding is low, then the embedding vectors are more or less correct, and very little error or log-loss will be backpropagated to correct the embedding matrices. However, let us say the dot product of h with cat is less and that of the other outer embedding vectors is more; the loss of the SoftMax is going to be significantly high, and thus more errors/log-loss are going to be backpropagated to reduce the error.

Continuous Bag of Words Implementation in TensorFlow

The continuous bag of words TensorFlow implementation has been illustrated in this section. The neighboring words within a distance of two from either side are used to predict the middle word. The output layer is a big SoftMax over the entire vocabulary. The word-embedding vectors are chosen to be of size 128. The detailed implementation is outlined in Listing 4-1a. See also Figure 4-4.

Listing 4-1a. Continuous Bag of Words Implementation in TensorFlow

```
import tensorflow as tf
print(tf.__version__)

import numpy as np
from tensorflow.keras import Model, layers
import time
from sklearn.manifold import TSNE
import matplotlib.pyplot as plt
%matplotlib inline

emb_dims = 128

#----------------------------------------------------
# to one hot the words
#----------------------------------------------------
def one_hot(ind,vocab_size):
    rec = np.zeros(vocab_size)
    rec[ind] = 1
    return rec
```

```
#-------------------------------------------------------
# Create training data
#-------------------------------------------------------
def create_training_data(corpus_raw,WINDOW_SIZE = 2):
    words_list = []

    for sent in corpus_raw.split('.'):
        for w in sent.split():
            if w != '.':
                words_list.append(w.split('.')[0])   # Remove if delimiter
                                                     #   is tied to the end
                                                     #   of a word

    words_list = set(words_list)     # Remove the duplicates for each word

    word2ind = {}     # Define the dictionary for converting a word to index
    ind2word = {}     # Define dictionary for retrieving a word from
                      #     its index

    vocab_size = len(words_list)    # Count of unique words in the
                                    #     vocabulary

    for i,w in enumerate(words_list):    # Build the dictionaries
        word2ind[w] = i
        ind2word[i] = w

    #print(word2ind)
    sentences_list = corpus_raw.split('.')
    sentences = []

    for sent in sentences_list:
        sent_array = sent.split()
        sent_array = [s.split('.')[0] for s in sent_array]
        sentences.append(sent_array)     # finally sentences would hold
                                         #     arrays of word array for
                                         #     sentences

    data_recs = []    # Holder for the input output record
```

```
    for sent in sentences:
        for ind,w in enumerate(sent):
            rec = []
            for nb_w in sent[max(ind - WINDOW_SIZE, 0) : min(ind + WINDOW_
            SIZE, len(sent)) + 1] :
                if nb_w != w:
                    rec.append(nb_w)
                data_recs.append([rec,w])

    x_train,y_train = [],[]

    for rec in data_recs:
        input_ = np.zeros(vocab_size)
        for i in range(len(rec[0])):
            input_ += one_hot(word2ind[ rec[0][i] ], vocab_size)
        input_ = input_/len(rec[0])
        x_train.append(input_)
        y_train.append(one_hot(word2ind[ rec[1] ], vocab_size))

    return x_train,y_train,word2ind,ind2word,vocab_size

class CBOW(Model):

    def __init__(self,vocab_size,embedding_size):
        super(CBOW,self).__init__()
        self.vocab_size = vocab_size
        self.embedding_size = embedding_size

        self.embedding_layer = layers.Dense(self.embedding_size)
        # input is vocab_size: vocab_size x embedding_size
        self.output_layer = layers.Dense(self.vocab_size)
        # embedding_size x vocab size

    def call(self,x):
        x = self.embedding_layer(x)
        x = self.output_layer(x)
        return x
```

```python
def shuffle_train(X_train,y_train):
    num_recs_train = X_train.shape[0]
    indices = np.arange(num_recs_train)
    np.random.shuffle(indices)
    return X_train[indices],y_train[indices]

def train_embeddings(training_corpus,epochs=100,lr=0.01,
batch_size=32,embedding_size=32):

    training_corpus = (training_corpus).lower()
    #-------------------------------------------------------------------
    # Invoke the training data generation the corpus data
    #-------------------------------------------------------------------
    X_train,y_train,word2ind,ind2word,vocab_size= create_training_
    data(training_corpus,2)

    print(f"Vocab size: {vocab_size}")
    X_train = np.array(X_train)
    y_train = np.array(y_train)

    model = CBOW(vocab_size=vocab_size,embedding_size=embedding_size)
    model_graph = tf.function(model)
    loss_fn = tf.keras.losses.CategoricalCrossentropy(
    from_logits=True,reduction=tf.keras.losses.Reduction.SUM)
    optimizer = tf.keras.optimizers.Adam(lr)

    num_train_recs = X_train.shape[0]
    num_batches = num_train_recs // batch_size
    loss_trace , accuracy_trace = [],[]
    start_time = time.time()

    for i in range(epochs):

        loss = 0
        X_train, y_train = shuffle_train(X_train,y_train)
```

```
    for j in range(num_batches):
        X_train_batch = tf.constant(X_train[j * batch_size:(j + 1) *
        batch_size], dtype=tf.float32)
        y_train_batch = tf.constant(y_train[j * batch_size:(j + 1) *
        batch_size])

        with tf.GradientTape() as tape:
            y_pred_batch = model_graph(X_train_batch,training=True)
            loss_ = loss_fn(y_train_batch, y_pred_batch)

        # compute gradient
        gradients = tape.gradient(loss_, model.trainable_variables)
        # update the parameters
        optimizer.apply_gradients(zip(gradients, model.trainable_
        variables))

        loss += loss_.numpy()
    loss /= num_train_recs
    loss_trace.append(loss)
    if i % 5 == 0:
        print(f"Epoch {i} : loss: {np.round(loss, 4)}  \n")

embeddings =  model_graph.embedding_layer.get_weights()[0]
print(f"Emeddings shape : {embeddings.shape} ")
return embeddings,ind2word
```

training_corpus = "Deep learning has evolved from artificial neural networks which has been there since the 1940s. Neural networks are interconnected networks of processing units called artificial neurons that loosely mimic axons in a biological brain. In a biological neuron, the dendrites receive input signals from various neighboring neurons, typically greater than 1000. These modified signals are then passed on to the cell body or soma of the neuron where these signals are summed together and then passed on to the axon of the neuron. If the received input signal is more than a specified threshold, the axon will release a signal which again will pass on to neighboring dendrites of other neurons. Figure 2-1 depicts the structure of a biological neuron for reference. The artificial neuron units are inspired from the biological neurons with some modifications as per convenience. Much like the dendrites, the input connections to the neuron carry the attenuated or amplified input signals from other neighboring neurons.

The signals are passed onto the neuron where the input signals are summed up, and then a decision is taken what to output based on the total input received. For instance, for a binary threshold neuron output value of 1 is provided when the total input exceeds a predefined threshold; otherwise, the output stays at 0. Several other types of neurons are used in artificial neural network, and their implementation only differs with respect to the activation function on the total input to produce the neuron output. In Figure 2-2, the different biological equivalents are tagged in the artificial neuron for easy analogy and interpretation."

```
embeddings,ind2word =  train_embeddings(training_corpus)
W_embedded = TSNE(n_components=2).fit_transform(embeddings)
plt.figure(figsize=(9,9))
for i in range(len(W_embedded)):
    plt.text(W_embedded[i,0],W_embedded[i,1],ind2word[i])

plt.xlim(-9,9)
plt.ylim(-9,9)
print("TSNE plot of the CBOW based Word Vector Embeddings")

--output--
2.9.1
Vocab size: 128
Epoch 95 : loss: 0.0105
```

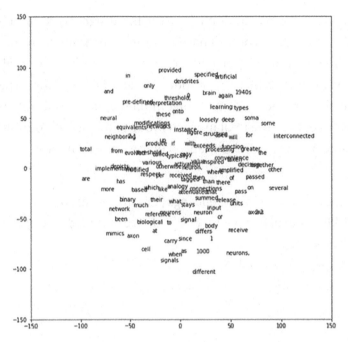

Figure 4-4. *TSNE plot for the word-embedding vectors learned from CBOW*

The word embeddings learned have been projected to a 2D plane through the TSNE plot. The TSNE plot gives a rough idea of the neighborhood of a given word. We can see that the word-embedding vectors learned are reasonable. For instance, the words *deep* and *learning* are very close to each other. Similarly, the words *biological* and *reference* are also very close to each other.

Skip-Gram Model for Word Embedding

Skip-gram models work the other way around. Instead of trying to predict the current word from the context words, as in continuous bag of words, in Skip-gram models, the context words are predicted based on the current word. Generally, given a current word, context words are taken in its neighborhood in each window. For a given window of five words, there would be four context words that one needs to predict based on the current word. Figure 4-5 shows the high-level design of a Skip-gram model. Much like continuous bag of words, in the Skip-gram model, one needs to learn two sets of word embedding: one for the input words and one for the output context words. A Skip-gram model can be seen as a reversed continuous bag of word model.

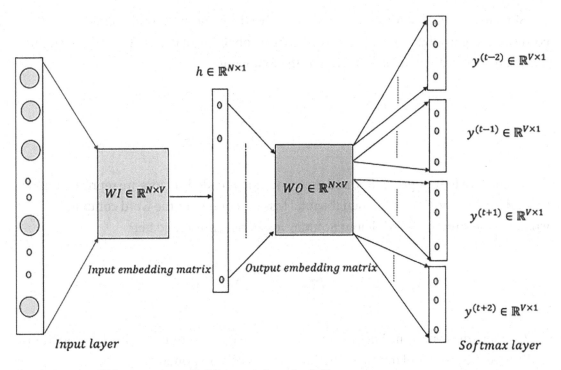

Figure 4-5. *Skip-gram model for word embeddings*

In the CBOW model, the input to the model is a one-hot encoded vector $x^{(t)} \in R^{V \times 1}$ for the current word, where V is the size of the vocabulary of the corpus. However, unlike CBOW, here the input is the current word and not the context words. The input $x^{(t)}$, when multiplied by the input word-embedding matrix WI, produces the word-embedding vector $u^{(k)} \in R^{N \times 1}$ given that $x^{(t)}$ represents the kth word in the vocabulary list. N, as before, represents the word-embedding dimensionality. The hidden-layer output h is nothing but $u^{(k)}$.

The dot product of the hidden-layer output h is computed with every word vector $v^{(j)}$ of the outer embedding matrix $WO \in R^{V \times N}$ by computing $[WO][h]$ just as in CBOW. However, instead of one SoftMax output layer, there are multiple SoftMax layers based on the number of context words that we are going to predict. For example, in Figure 4-5, there are four SoftMax output layers corresponding to the four context words. The input to each of these SoftMax layers is the same set of dot products in $[WO][h]$ representing how similar the input word is to each word in the vocabulary.

$$[WO][h] = \left[v^{(1)^T} h \quad v^{(2)^T} h \ldots v^{(j)^T} h .. v^{(V)^T} h \right]$$

Similarly, all the SoftMax layers would receive the same set of probabilities corresponding to all the vocabulary words. The probability of the *jth* word $w^{(j)}$ given the current or the center word $w^{(k)}$ is given by the following:

$$P\left(w=w^{(j)}\,/\,w=w^{(k)}\right)=p^{(j)}=\frac{e^{v^{(j)T}h}}{\sum\limits_{k=1}^{V}e^{v^{(k)T}h}}=\frac{e^{v^{(j)T}w^{(k)}}}{\sum\limits_{k=1}^{V}e^{v^{(k)T}w^{(k)}}}$$

If there are four target words, and their one-hot encoded vectors are represented by $y^{(t-2)}$, $y^{(t-1)}$, $y^{(t+1)}$, $y^{(t+2)} \in R^{V \times 1}$, then the total loss function C for the word combination would be the summation of all four SoftMax losses as represented here:

$$C=-\sum_{\substack{m=t-2 \\ m \neq t}}^{t+2}\sum_{j=1}^{V}y_j^{(m)}\log\left(p^{(j)}\right)$$

Gradient descent using backpropagation can be used to minimize the cost function and derive the input and output embedding matrices' components.

Here are a few salient features about the Skip-gram and CBOW models:

- For Skip-gram models, the window size is not generally fixed. Given a maximum window size, the window size at each current word is randomly chosen so that smaller windows are chosen more frequently than larger ones. With Skip-gram, one can generate a lot of training samples from a limited amount of text, and infrequent words and phrases are also very well represented.

- CBOW is much faster to train than Skip-gram and has slightly better accuracy for frequent words.

- Both Skip-gram and CBOW look at local windows for word co-occurrences and then try to predict either the context words from the center word (as with Skip-gram) or the center word from the context words (as with CBOW). So, basically, we observe in Skip-gram that locally within each window the probability of the co-occurrence of the context word w_c and the current word w_t, given by $P(w_c/w_t)$, is assumed to be proportional to the exponential of the dot product of their word-embedding vectors. For example,

$$P\left(w_c \,/\, w_t\right) \propto e^{u^T v}$$

where u and v are the input and output word-embedding vectors for the current and context words, respectively. Since the co-occurrence is measured locally, these models miss utilizing the global co-occurrence statistics for word pairs within certain window lengths. Next, we are going to explore a basic method to look at the global co-occurrence statistics over a corpus and then use SVD (singular value decomposition) to generate word vectors.

Skip-Gram Implementation in TensorFlow

In this section, we will illustrate the Skip-gram model for learning word-vector embeddings with a TensorFlow implementation. The model is trained on a small dataset for easy representation. However, the model can be used to train large corpuses as desired. As illustrated in the Skip-gram section, the model is trained as a classification network. However, we are more interested in the word-embedding matrix than in the actual classification of words. The size of the word embeddings has been chosen to be 128. The detailed code is represented in Listing 4-1b. Once the word-embedding vectors are learned, they are projected via TSNE on a two-dimensional surface for visual interpretation.

Listing 4-1b. Skip-Gram Implementation in TensorFlow

```
import numpy as np
import tensorflow as tf
from sklearn.manifold import TSNE
import matplotlib.pyplot as plt
%matplotlib inline
from sklearn.manifold import TSNE
import matplotlib.pyplot as plt
%matplotlib inline
```

```
#----------------------------------------------------------------
# Function to one hot encode the words
#----------------------------------------------------------------
def one_hot(ind,vocab_size):
    rec = np.zeros(vocab_size)
    rec[ind] = 1
    return rec

#--------------------------------------------------------
# Function to create the training data from the corpus
#--------------------------------------------------------
def create_training_data(corpus_raw,WINDOW_SIZE = 2):
    words_list = []

    for sent in corpus_raw.split('.'):
        for w in sent.split():
            if w != '.':
                words_list.append(w.split('.')[0])   # Remove if delimiter
                                                      #   is tied to the end
                                                      #   of a word

    words_list = set(words_list)   # Remove the duplicates for each word

    word2ind = {}        # Define the dictionary for converting a word
                         #   to index
    ind2word = {}        # Define dictionary for retrieving a word from
                         #   its index

    vocab_size = len(words_list)    # Count of unique words in the
                                    #   vocabulary

    for i,w in enumerate(words_list):  # Build the dictionaries
        word2ind[w] = i
        ind2word[i] = w

    sentences_list = corpus_raw.split('.')
    sentences = []
```

```
    for sent in sentences_list:
        sent_array = sent.split()
        sent_array = [s.split('.')[0] for s in sent_array]
        sentences.append(sent_array)    # finally sentences would hold
                                        arrays of word array for
                                        sentences

    data_recs = []      # Holder for the input output record

    for sent in sentences:
        for ind,w in enumerate(sent):
            for nb_w in sent[max(ind - WINDOW_SIZE, 0) : min(ind + WINDOW_
            SIZE, len(sent)) + 1] :
                if nb_w != w:
                    data_recs.append([w,nb_w])

    x_train,y_train = [],[]

    #print(data_recs)

    for rec in data_recs:
        x_train.append(one_hot(word2ind[ rec[0] ], vocab_size))
        y_train.append(one_hot(word2ind[ rec[1] ], vocab_size))

    return x_train,y_train,word2ind,ind2word,vocab_size

class SkipGram(Model):

    def __init__(self,vocab_size,embedding_size):
        super(SkipGram,self).__init__()
        self.vocab_size = vocab_size
        self.embedding_size = embedding_size

        self.embedding_layer = layers.Dense(self.embedding_size) # input is
        vocab_size: vocab_size x embedding_size
        self.output_layer = layers.Dense(self.vocab_size) # embedding_size
        x vocab size
```

```python
    def call(self,x):
        x = self.embedding_layer(x)
        x = self.output_layer(x)
        return x

def shuffle_train(X_train,y_train):
    num_recs_train = X_train.shape[0]
    indices = np.arange(num_recs_train)
    np.random.shuffle(indices)
    return X_train[indices],y_train[indices]

def train_embeddings(training_corpus,epochs=100,lr=0.01,batch_
size=32,embedding_size=32):

    training_corpus = (training_corpus).lower()
    #-------------------------------------------------------------------------
    # Invoke the training data generation the corpus data
    #-------------------------------------------------------------------------
    X_train,y_train,word2ind,ind2word,vocab_size= create_training_data
    (training_corpus,2)

    print(f"Vocab size: {vocab_size}")
    X_train = np.array(X_train)
    y_train = np.array(y_train)

    model = SkipGram(vocab_size=vocab_size,embedding_size=embedding_size)
    model_graph = tf.function(model)
    loss_fn = tf.keras.losses.CategoricalCrossentropy(
    from_logits=True,reduction=tf.keras.losses.Reduction.SUM)

    #loss_fn = tf.nn.sampled_softmax_loss(num_sampled=4,
    #          num_classes=vocab_size,
    #          num_true=1)

    optimizer = tf.keras.optimizers.Adam(lr)

    num_train_recs = X_train.shape[0]
    num_batches = num_train_recs // batch_size
    loss_trace , accuracy_trace = [],[]
```

```
start_time = time.time()

for i in range(epochs):

    loss = 0
    X_train, y_train = shuffle_train(X_train,y_train)

    for j in range(num_batches):
        X_train_batch = tf.constant(X_train[j * batch_size:(j + 1) *
        batch_size], dtype=tf.float32)
        y_train_batch = tf.constant(y_train[j * batch_size:(j + 1) *
        batch_size])

        with tf.GradientTape() as tape:
            y_pred_batch = model_graph(X_train_batch,training=True)
            loss_ = loss_fn(y_train_batch, y_pred_batch)

        # compute gradient
        gradients = tape.gradient(loss_, model.trainable_variables)
        # update the parameters
        optimizer.apply_gradients(zip(gradients, model.trainable_
        variables))

        loss += loss_.numpy()
    loss /= num_train_recs
    loss_trace.append(loss)
    if i % 5 == 0:
        print(f"Epoch {i} : loss: {np.round(loss, 4)}  \n")

embeddings =  model_graph.embedding_layer.get_weights()[0]
print(f"Emeddings shape : {embeddings.shape} ")
return embeddings, ind2word
```

training_corpus = "Deep Learning has evolved from Artificial Neural Networks which has been there since the 1940s. Neural Networks are interconnected networks of processing units called artificial neurons, that loosely mimics axons in a biological brain. In a biological neuron, the Dendrites receive input signals from various neighboring neurons, typically greater than 1000. These modified signals are then passed on to the cell body or soma of the neuron where these signals are summed together and then passed on to the Axon of the neuron. If the received input signal is more than a specified threshold, the axon will release a signal which again will pass on to neighboring dendrites of other neurons. Figure 2-1 depicts the structure of a biological neuron for reference.The artificial neuron units are inspired from the biological neurons with some modifications as per convenience. Much like the dendrites the input connections to the neuron carry the attenuated or amplified input signals from other neighboring neurons. The signals are passed onto the neuron where the input signals are summed up and then a decision is taken what to output based on the total input received. For instance, for a binary threshold neuron output value of 1 is provided when the total input exceeds a pre-defined threshold, otherwise the output stays at 0. Several other types of neurons are used in artificial neural network and their implementation only differs with respect to the activation function on the total input to produce the neuron output. In Figure 2-2 the different biological equivalents are tagged in the artificial neuron for easy analogy and interpretation."

```
embeddings, ind2word = train_embeddings(training_corpus)
W_embedded = TSNE(n_components=2).fit_transform(embeddings)
plt.figure(figsize=(10,10))
for i in range(len(W_embedded)):
    plt.text(W_embedded[i,0],W_embedded[i,1],ind2word[i])

plt.xlim(-14,14)
plt.ylim(-14,14)
print("TSNE plot of the SkipGram based Word Vector Embeddings")

--output--
Epoch 95 : loss: 2.3412
```

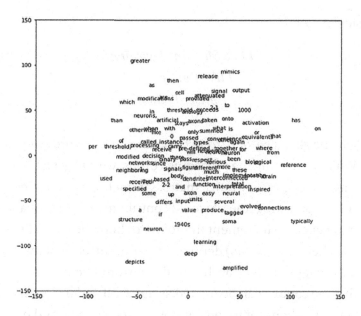

Figure 4-6. *TSNE plot of word-embedding vectors learned from Skip-gram model*

Much like the word-embedding vectors from the continuous bag of words method, the embedding vectors learned from Skip-gram method are reasonable. For instance, the words *deep* and *learning* are very close to each other in Skip-grams too, as we can see from Figure 4-6. Also, we see other interesting patterns, such as the word *attenuated* being very close to the word *signal*.

Global Co-occurrence Statistics-Based Word Vectors

The global co-occurrence methods, where the global counts of co-occurrences of words in each window over the whole corpus are collected, can be used to derive meaningful word vectors. Initially, we will look at a method that does matrix factorization through SVD (singular value decomposition) on the global co-occurrence matrix to derive a meaningful lower-dimensional representation of words. Later, we will look at the GloVe technique for word-vector representation, which combines the best of global co-occurrence statistics and the prediction methods of CBOW and/or Skip-gram to represent word vectors.

Let us consider a corpus:

'I like Machine Learning.'
'I like Tensor Flow'
'I prefer Python'.

We first collect the global co-occurrence counts for each word combination within a window of 1. While processing the preceding corpus, we will get a co-occurrence matrix. Also, we make the co-occurrence matrix symmetric by assuming that whenever two words w_1 and w_2 appear together it would contribute to both probabilities $P(w_1/w_2)$ and $P(w_2/w_1)$, and hence we increment the count for both the count buckets $c(w_1/w_2)$ and $c(w_2/w_1)$ by 1. The term $c(w_1/w_2)$ denotes the co-occurrence of the words w_1 and w_2, where w_2 acts as the context and w_1 as the word. For word-occurrence pairs, the roles can be reversed so that the context can be treated as the word and the word as the context. For this precise reason, whenever we encounter a co-occurring word pair (w_1, w_2), both count buckets $c(w_1/w_2)$ and $c(w_2/w_1)$ are incremented.

Coming to the incremental count, we need not always increment by 1 for a co-occurrence of two words. If we are looking at a window of K for populating the co-occurrence matrix, we can define a differential weighting scheme to provide more weight for words co-occurring at less distance from the context and penalize them as the distance increases. One such weighing scheme would be to increment the co-occurrence counter by $\left(\dfrac{1}{k}\right)$, where k is the offset between the word and the context. When the word and the context are next to each other, then the offset is 1 and the co-occurrence counter can be incremented by 1, while when the offset is at maximum for a window of K, the counter increment is at minimum at $\left(\dfrac{1}{k}\right)$.

In the SVD method of generating the word-vector embedding, the assumption is that the global co-occurrence count $c(w_i/w_j)$ between a word w_i and context w_j can be expressed as the dot product of the word-vector embeddings for the word w_i and for the context w_j. Generally, two sets of word embeddings are considered, one for the words and the other for the contexts. If $u_i \in R^{D \times 1}$ and $v_i \in R^{D \times 1}$ denote the word vector and the context vector for the *ith* word w_i in the corpus, respectively, then the co-occurrence count can be expressed as follows:

$$c\left(w_i / w_j\right) = u_i^T v_j$$

Let's look at a three-word corpus and represent the co-occurrence matrix $X \in R^{3 \times 3}$ in terms of the dot products of the words and the context vectors. Further, let the words be w_i, $\forall\, i = \{1, 2, 3\}$ and their corresponding word vectors and context vectors be u_i, $\forall\, i = \{1, 2, 3\}$ and v_i, $\forall\, i = \{1, 2, 3\}$, respectively.

As we can see, the co-occurrence matrix turns out to be the product of two matrices, which are nothing but the word-vector embedding matrices for the words and contexts, respectively. The word-vector embedding matrix $W \in R^{3 \times D}$ and the context-word-embedding matrix $C \in R^{D \times 3}$, where D is the dimension of the word and context embedding vectors.

Now that we know that the word co-occurrence matrix is a product of the word-vector embedding matrix and the context embedding matrix, we can factorize the co-occurrence matrix by any applicable matrix factorization technique. Singular value decomposition (SVD) is a well-adopted method because it works, even if the matrices are not square or symmetric.

As we know from SVD, any rectangular matrix X can be decomposed into three matrices U, Σ, and V such that

$$X = [U][\Sigma]\left[V^T\right]$$

The matrix U is generally chosen as the word-vector embedding matrix W, while ΣV^T is chosen as the context-vector embedding matrix C, but there is no such restriction, and whichever works well on the given corpus can be chosen. One can very well choose W as $U\Sigma^{1/2}$ and C as $\Sigma^{1/2}V^T$. Generally, fewer dimensions in the data based on the significant singular values are chosen to reduce the size of U, Σ, and V. If $X \in R^{m \times n}$, then $U \in R^{m \times m}$. However, with truncated SVD, we take only a few significant directions along which the data has maximum variability and ignore the rest as insignificant or as noise. If we chose D dimensions, the new word-vector embedding matrix $U' \in R^{m \times D}$, where D is the dimension of every word-vector embedding.

The co-occurrence matrix $X \in R^{m \times n}$ is generally obtained in a more generalized setting by making a pass through the entire corpus once. However, since the corpus might get new documents or contents over time, those new documents or contents can be processed incrementally. Figure 4-7 illustrates the derivation of the word vectors or word embeddings in a three-step process.

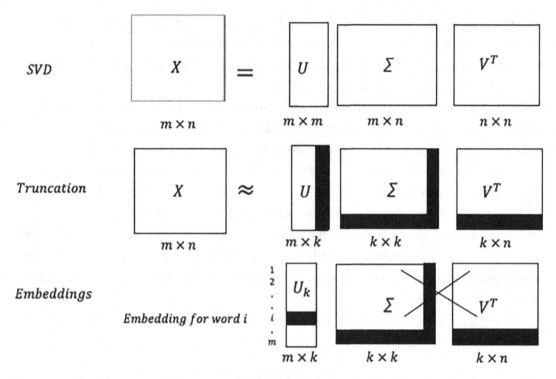

Figure 4-7. *Extraction of word embeddings through SVD of word co-occurrence matrix*

- In the first step, singular value decomposition (SVD) is performed on the co-occurrence matrix $X \in R^{m \times n}$ to produce $U \in R^{m \times m}$, which contains the left singular vectors; $\Sigma \in R^{m \times n}$, which contains the singular values; and $V \in R^{n \times n}$, which contains the right singular vectors.

$$\left[X \right]_{m \times n} = \left[U \right]_{m \times m} \left[\Sigma \right]_{m \times n} \left[V^T \right]_{n \times n}$$

- Generally, for word-to-word co-occurrence matrices, the dimensions m and n should be equal. However, sometimes instead of expressing words by words, words are expressed by contexts, and hence for generalization we have taken separate m and n.

- In the second step, the co-occurrence matrix is approximated by taking only k significant singular values from Σ that explain the maximum variance in data and by also choosing the corresponding k left singular and right singular vectors in U and V.

If we start with

$$U = \left[u_1 u_2 \ u_3 \ldots u_m \right], \Sigma = \begin{bmatrix} \sigma_1 & \cdots & 0 \\ \vdots & \ddots & \vdots \\ 0 & \cdots & \sigma_m \end{bmatrix}, V^T = \begin{bmatrix} v_1^T \rightarrow \\ v_2^T \rightarrow \\ \cdots \\ v_n^T \rightarrow \end{bmatrix}$$

after truncation, we would have the following:

$$U' = \left[u_1 u_2 \ u_3 \ldots u_k \right], \Sigma' = \begin{bmatrix} \sigma_1 & \cdots & 0 \\ \vdots & \ddots & \vdots \\ 0 & \cdots & \sigma_k \end{bmatrix}, \quad V'^T = \begin{bmatrix} v_1^T \rightarrow \\ v_2^T \rightarrow \\ \cdots \\ v_k^T \rightarrow \end{bmatrix}$$

- In the third step, Σ' and V'^T are discarded and the matrix $U' \in R^{m \times 1}$ is taken to the word-embedding vector matrix. The word vectors have k dense dimensions corresponding to the chosen k singular values. So, from a sparse co-occurrence matrix, we get a dense representation of the word-vector embeddings. There would be m word embeddings corresponding to each word of the processed corpus.

Mentioned in Listing 4-1c is the logic for building word vectors from a given corpus by factorizing the co-occurrence matrix of different words through SVD. Accompanying the listing is the plot of the derived word-vector embeddings in Figure 4-8.

Listing 4-1c.

```
import numpy as np
import matplotlib.pyplot as plt

corpus = ['I like Machine Learning.','I like TensorFlow.','I prefer Python.']
corpus_words_unique = set()
corpus_processed_docs = []
```

```python
# Process the documents in the corpus to create the Co-occrence count
for doc in corpus:
    corpus_words_ = []
    corpus_words = doc.split()
    print(corpus_words)
    for x in corpus_words:
        if len(x.split('.')) == 2:
            corpus_words_ += [x.split('.')[0]] + ['.']
        else:
            corpus_words_ += x.split('.')

    corpus_processed_docs.append(corpus_words_)
    corpus_words_unique.update(corpus_words_)
corpus_words_unique = np.array(list(corpus_words_unique))
co_occurence_matrix = np.zeros((len(corpus_words_unique),
len(corpus_words_unique)))

for corpus_words_ in corpus_processed_docs:
    for i in range(1,len(corpus_words_)) :
        index_1 = np.argwhere(corpus_words_unique == corpus_words_[i])
        index_2 = np.argwhere(corpus_words_unique == corpus_words_[i-1])

        co_occurence_matrix[index_1,index_2] += 1
        co_occurence_matrix[index_2,index_1] += 1

U,S,V = np.linalg.svd(co_occurence_matrix,full_matrices=False)
print(f'co_occurence_matrix follows:')
print(co_occurence_matrix)

for i in range(len(corpus_words_unique)):
    plt.text(U[i,0],U[i,1],corpus_words_unique[i])

plt.xlim((-0.75,0.75))
plt.ylim((-0.75,0.75))
plt.show()
```

```
--output--
co_occurence_matrix follows:
[[ 0.  2.  0.  0.  1.  0.  0.  1.]
 [ 2.  0.  1.  0.  0.  0.  0.  0.]
 [ 0.  1.  0.  0.  0.  1.  0.  0.]
 [ 0.  0.  0.  0.  0.  1.  1.  1.]
 [ 1.  0.  0.  0.  0.  0.  1.  0.]
 [ 0.  0.  1.  1.  0.  0.  0.  0.]
 [ 0.  0.  0.  1.  1.  0.  0.  0.]
 [ 1.  0.  0.  1.  0.  0.  0.  0.]]
Word-Embeddings Plot
```

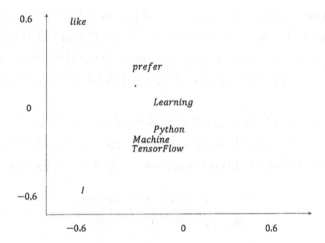

Figure 4-8. *Word-embedding plot*

We can see a clear pattern in the plot of the word-vector embedding in Figure 4-8, even with this small corpus. Here are a few of the findings:

- Common words like *I* and *like* are far away from others.

- *Machine, Learning, Python*, and *TensorFlow*, being associated with different areas of learning, are clustered close to each other.

Next, we move on to global vectors, commonly known as GloVe, for generating word-vector embeddings.

GloVe

GloVe is a pretrained, readily available word-embedding vectors library from Stanford University. The training method for GloVe is significantly different from those for CBOW and Skip-gram. Instead of basing predictions on local-running windows for words, GloVe uses global word-to-word co-occurrence statistics from a corpus to train the model and derive the GloVe vectors. GloVe stands for *global vectors*. Pretrained GloVe word embeddings are available at `https://nlp.stanford.edu/projects/glove/`. Jeffrey Pennington, Richard Socher, and Christopher D. Manning are the inventors of GloVe vectors, and they have documented GloVe vectors in their paper titled "GloVe: Global Vectors for Word Representation." The paper can be located at `https://nlp.stanford.edu/pubs/glove.pdf`.

Like SVD methods, GloVe looks at the global co-occurrence statistics, but the relation of the word and context vectors with respect to the co-occurrence count is a little different. If there are two words w_i and w_j and a context word w_k, then the ratio of the probabilities $P(w_k/w_i)$ and $P(w_k/w_j)$ provides more information than the probabilities themselves.

Let's consider two words, w_i = "*garden*" and w_j = "*market*", and a couple of context words, $w_k \in$ {"*plants*","*shops*"}. The individual co-occurrence probability might be low; however, if we take the ratio of the co-occurrence probabilities, for instance,

$$\frac{P\left(w_k = "plants"/w_i = "garden"\right)}{P\left(w_k = "plants"/w_j = "market"\right)}$$

The preceding ratio is going to be much greater than one indicating that *plants* are much more likely to be associated with *garden* than with *market*.

Similarly, let us consider w_k = "*shops*" and look at the ratio of the following co-occurrence probability:

$$\frac{P\left(w_k = "shops"/w_i = "garden"\right)}{P\left(w_k = "shops"/w_j = "market"\right)}$$

In this case, the ratio is going to be very small, signifying that the word *shop* is much more likely to be associated with the word *market* than with *garden*.

Hence, we can see that the ratio of the co-occurrence probabilities provides much more discrimination between words. Since we are trying to learn word vectors, this discrimination should be encoded by the difference of the word vectors, as in a linear vector space that's the best way to represent the discrimination between vectors. Similarly, the most convenient way to represent the similarity between vectors in a linear vector space is to consider their dot product, and hence the co-occurrence probability would be well represented by some function of the dot product between the word and the context vector. Taking all this into consideration helps one derive the logic for the GloVe vector derivation on a high level.

If u_i, u_j are the word-vector embeddings for the words w_i and w_j, and v_k is the context vector for word w_k, then the ratio of the two co-occurrence probabilities can be expressed as some function of the difference of the word vectors (i.e., $(u_i - u_j)$) and the context vector v_k. A logical function should work on the dot product of the difference of the word vector and the context vector, primarily because it preserves the linear structure between the vectors before it is manipulated by the function. Had we not taken the dot product, the function could have worked on the vectors in a way that would have disrupted the linear structure. Based on the preceding explanation, the ratio of the two co-occurrence probabilities can be expressed as follows:

$$\frac{P(w_k / w_i)}{P(w_k / w_j)} = f\left(\left(u_i - u_j\right)^T v_k\right) \tag{4.1.1}$$

In the above expression, f is a given function that we seek to find out.

Also, as discussed, the co-occurrence probability $P(w_k/w_i)$ should be encoded by some form of similarity between vectors in a linear vector space, and the best operation to do so is to represent the co-occurrence probability by some function g of the dot product between the word vector w_i and context vector w_k, as expressed here:

$$P(w_i|w_k) = g\left(u_i^T v_k\right) \tag{4.1.2}$$

Combining (4.1.1) and (4.1.2), we have the following:

$$\frac{g\left(u_i^T v_k\right)}{g\left(u_j^T v_k\right)} = f\left(\left(u_i - u_j\right)^T v_k\right) \tag{4.1.3}$$

Now the task is to determine meaningful functions f and g for the preceding equation to make sense. If we choose f and g to be the exponential function, it enables the ratio of the probabilities to encode the difference of the word vectors and at the same time keeps the co-occurrence probability dependent on the dot product. The dot product and difference of vectors keep the notion of similarity and discrimination of vectors in a linear space. Had the functions f and g been some kind of kernel functions, then the measure of similarity and dissimilarity wouldn't have been restricted to a linear vector space and would have made the interpretability of word vectors very difficult.

Replacing f and g with the exponential function in (4.1.3), we get the following:

$$\frac{e^{\left(u_i^T v_k\right)}}{e^{u_j^T v_k}} = e^{\left(u_i - u_j\right)^T v_k}$$

which gives us the following:

$$P\left(w_k / w_i\right) = e^{u_i^T v_k} \Rightarrow \log P\left(w_k / w_i\right) = u_i^T v_k \qquad (4.1.4)$$

Interested readers with some knowledge of group theory in abstract algebra can see that the function $f(x) = e^x$ has been chosen so as to define a group homomorphism between the groups $(R, +)$ and $(R > 0, \times)$.

The co-occurrence probability of the word w_i and the context word w_k can be denoted as follows:

$$P\left(w_k / w_i\right) = \frac{c\left(w_i, w_k\right)}{c\left(w_i\right)} \qquad (4.1.5)$$

where $c(w_i, w_k)$ denotes the co-occurrence count of word w_i with the context word w_k and $c(w_i)$ denotes the total occurrences of the word w_i. The total count of any word can be computed by summing up its co-occurrence count with all other words in the corpus, as shown here:

$$c\left(w_i\right) = \sum_k c\left(w_i, w_k\right)$$

Combining (4.1.4) and (4.1.5), we get the following:

$$\log c\left(w_i, w_k\right) - \log c\left(w_i\right) = u_i^T v_k$$

$\log c(w_i)$ can be expressed as a bias b_i for the word w_i, and an additional bias $\tilde{b}_k$ for the word w_k is also introduced to make the equation symmetric. So, the final relation can be expressed as follows:

$$\log c(w_i, w_k) = u_i^T v_k + \tilde{b}_k + b_i$$

Just as we have two sets of word-vector embeddings, similarly we see two sets of bias—one for context words given by $\tilde{b}_i$ and the other for words given by b_i, where i indicates the *ith* word of the corpus.

The final aim is to minimize a sum of the squared error cost function between the actual $\log c(w_i, w_k)$ and the predicted $u_i^T v_k + \tilde{b}_k + b_i$ for all word-context pairs, as follows:

$$C(U,V,\tilde{B},B) = \sum_{i,j=1}^{V} \left(u_i^T v_k + \tilde{b}_k + b_i - \log c(w_i, w_k) \right)^2$$

U and *V* are the set of parameters for the word-vector embeddings and context-vector embeddings. Similarly, $\tilde{B}$ and *B* are the parameters for the biases corresponding to the words and the contexts. The cost function $C(U,V,\tilde{B},B)$ must be minimized with respect to these parameters in $U,V,\tilde{B},B$.

One issue with this scheme of least square method is that it weights all co-occurrences equally in the cost function. This prevents the model from achieving good results since the rare co-occurrences carry very little information. One of the ways to handle this issue is to assign more weight to co-occurrences that have a higher count. The cost function can be modified to have a weight component for each co-occurrence pair that is a function of the co-occurrence count. The modified cost function can be expressed as follows:

$$C(U,V,\tilde{B},B) = \sum_{i,j=1}^{V} h\big(c(w_i, w_k)\big)\left(u_i^T v_k + \tilde{b}_k + b_i - \log c(w_i, w_k) \right)^2$$

where h is the newly introduced function.

The function $h(x)$ (see Figure 4-9) can be chosen as follows:

$$h(x) = \left(\frac{x}{x_{max}}\right)^{\alpha} \; if \; x < x_{max}$$

$$= 1, elsewhere$$

One can experiment with different values of α, which acts as a hyperparameter to the model.

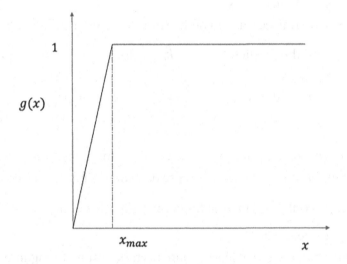

Figure 4-9. *Weightage function for the co-occurrence counts*

Word Analogy with Word Vectors

The good thing about word-vector embedding lies in its abilities to linearize analogies. We look at some analogies using the pretrained GloVe vectors in Listing 4-2a, Listing 4-2b, and Listing 4-2c.

Listing 4-2a.

```
import numpy as np
import scipy
from sklearn.manifold import TSNE
import matplotlib.pyplot as plt
%matplotlib inline

#########################
# Loading glove vector
#########################

EMBEDDING_FILE = '/media/santanu/9eb9b6dc-b380-486e-b4fd-c424a325b976/
glove.6B.300d.txt'
print('Indexing word vectors')
embeddings_index = {}
f = open(EMBEDDING_FILE)

count = 0

for line in f:
    if count == 0:
        count = 1
        continue
    values = line.split()
    word = values[0]

    coefs = np.asarray(values[1:], dtype='float32')
    embeddings_index[word] = coefs
f.close()

print('Found %d word vectors of glove.' % len(embeddings_index))
```

Listing 4-2b.

```
# queen - woman +man ~ king
king_wordvec = embeddings_index['king']
queen_wordvec = embeddings_index['queen']
man_wordvec = embeddings_index['man']
woman_wordvec = embeddings_index['woman']
pseudo_king = queen_wordvec - woman_wordvec + man_wordvec
cosine_simi = np.dot(pseudo_king/np.linalg.norm(pseudo_king),king_wordvec/
np.linalg.norm(king_wordvec))
print(f"Cosine Similarity: {cosine_simi}")

--output --
Cosine Similarity 0.663537
```

Listing 4-2c.

```
tsne = TSNE(n_components=2)
words_array = []
word_list = ['king','queen','man','woman']

for w in word_list:
    words_array.append(embeddings_index[w])

index1 = list(embeddings_index.keys())[0:100]

for i in range(100):
    words_array.append(embeddings_index[index1[i]])
words_array = np.array(words_array)
words_tsne = tsne.fit_transform(words_array)

ax = plt.subplot(111)

for i in range(4):
    plt.text(words_tsne[i, 0], words_tsne[i, 1],word_list[i])

plt.xlim((50,125))
plt.ylim((0,80))
plt.show()

--output--
```

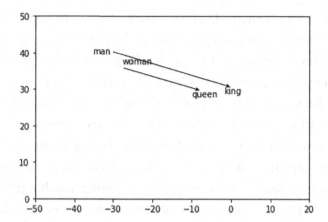

Figure 4-10. *2D TSNE vector plot for pretrained GloVe vectors*

The embedding file glove.6B.300d.txt is available at `https://nlp.stanford.edu/ projects/glove/`.

In Listing 4-2a, the pretrained GloVe vectors of dimensionality 300 are loaded and stored in a dictionary. We play around with the GloVe vectors for the words *king*, *queen*, *man*, and *woman* to find an analogy. Taking the word vectors for *queen*, *man*, and *woman*, a word vector *pseudo_king* is created as follows:

$$pseudo\_king = queen - woman + man$$

The idea is to see whether the preceding created vector somewhat represents the concept of *king* or not. The cosine of the angle between the word vectors *pseudo_king* and *king* is high at around 0.67, which is an indication that (*queen* – *woman* + *man*) very well represents the concept of *king*.

Next, in Listing 4-2c, we try to represent an analogy, and for that purpose, through TSNE, we represent the GloVe vectors of dimensionality 300 in a two-dimensional space. The results have been plotted in Figure 4-10. We can see that the word vectors for *king* and *queen* are close to each other and clustered together, and the word vectors for *man* and *woman* are clustered close to each other as well. Also, we see that the vector differences between *king* and *man* and those between *queen* and *woman* are almost parallelly aligned and of comparable lengths.

Before we move on to recurrent neural networks, one thing I want to mention is the importance of word embeddings for recurrent neural networks in the context of natural language processing. A recurrent neural network doesn't understand text, and hence each word in the text needs to have some form of number representation.

Word-embedding vectors are a great choice since words can be represented by multiple concepts given by the components of the word-embedding vector. Recurrent neural networks can be made to work both ways, either by providing the word-embedding vectors as input or by letting the network learn those embedding vectors by itself. In the latter case, the word-embedding vectors would be aligned more toward the ultimate problems being solved through the recurrent neural network. However, at times the recurrent neural network might have a lot of other parameters to learn, or the network might have very little data to train on. In such cases, having to learn the word-embedding vectors as parameters might lead to overfitting or suboptimal results. Using the pretrained word-vector embeddings might be a wiser option in such scenarios.

Introduction to Recurrent Neural Networks

Recurrent neural networks (RNNs) are designed to utilize and learn from sequential information. The RNN architecture is supposed to perform the same task for every element of a sequence, and hence the term *recurrent* in its nomenclature. RNNs have been of great use in the task of natural language processing because of the sequential dependency of words in any language. For example, in the task of predicting the next word in a sentence, the prior sequence of words that came before it is of paramount importance. Generally, at any time step of a sequence, RNNs compute some memory based on its computations thus far, i.e., prior memory and the current input. This computed memory is used to make predictions for the current time step and is passed on to the next step as an input. The basic architectural principal of a recurrent neural network is illustrated in Figure 4-11.

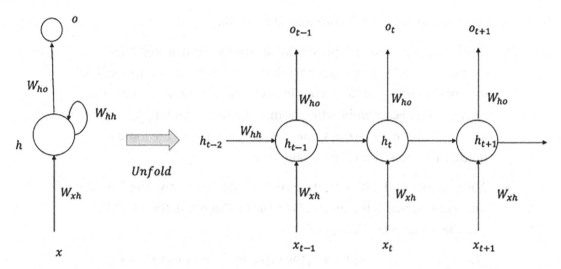

Figure 4-11. *Folded and unfolded structure of an RNN*

In Figure 4-11, the RNN architecture has been unrolled in time to depict the complete sequence. If we wish to process seven-word sequences of sentences, then the unfolded RNN architecture would represent a seven-layer feed-forward neural network, with the only difference being that the weights at each layer are common shared weights. This significantly reduces the number of parameters to learn in a recurrent neural network.

Just to get us familiar with the notations used, x_t, h_t, and o_t represent the input, computed memory or hidden states, and output, respectively, at time step t. W_{hh} represents the weight matrix from the memory states h_t at time t to the memory states h_{t+1} at time $(t + 1)$. W_{xh} represents the weight matrix from the input x_t to the hidden states h_t, whereas W_{ho} represents the weight matrix from the memory states h_t to o_t. The weight matrix W_{xh} acts as some sort of word-vector embedding matrix when the inputs are presented in a one-hot encoded form. Alternately, in cases of one-hot encoded inputs, one may choose to have a learnable separate embedding matrix so that when the one-hot encoded input vector passes through the embeddings layer, its desired embedding vector is presented as an output.

Now, let us drill down to each component in detail:

- The input x_t is a vector representing the input word at step t. For example, it can be a one-hot encoded vector with the component set to 1 for the corresponding word index in the vocabulary. It can also be the word-vector embedding from some pretrained repository such as GloVe. In general, we assume $x_t \in R^{D \times 1}$. Also, if we are looking to predict V classes, then the output $y_t \in R^{V \times 1}$.

- The memory or the hidden state vector h_t can have any length as per the user's choice. If the number of states chosen is n, then $h_t \in R^{n \times 1}$ and the weight matrix $W_{hh} \in R^{n \times n}$.

- The weight matrix connecting the input to the memory states $W_{xh} \in R^{n \times D}$, and the weight matrix connecting the memory states to the output $W_{ho} \in R^{n \times V}$.

- The memory h_t at step t is computed as follows:

  $h_t = f(W_{hh}h_{t-1} + W_{xh} x_t)$, where f is a chosen nonlinear activation function.

  The dimension of $(W_{hh}h_{t-1} + W_{xh} x_t)$ is n, i.e., $(W_{hh}h_{t-1} + W_{xh} x_t) \in R^{n \times 1}$.

  The function f works element-wise on $(W_{hh}h_{t-1} + W_{xh} x_t)$ to produce h_t, and hence $(W_{hh}h_{t-1} + W_{xh} x_t)$ and h_t have the same dimension.

  If $W_{hh}h_{t-1} + W_{xh} \ x_t = \begin{bmatrix} s_{1t} \\ s_{2t} \\ . \\ . \\ s_{nt} \end{bmatrix}$, then the following holds true for h_t:

  $$h_t = \begin{bmatrix} f(s_{1t}) \\ f(s_{2t}) \\ . \\ . \\ f(s_{nt}) \end{bmatrix}$$

- The connections from the memory states to the output are just like the connections from a fully connected layer to the output layer. When a multi-class classification problem is involved, such as predicting the next word, then the output layer would be a huge SoftMax of the number of words in the vocabulary. In such cases, the predicted output vector $o_t \in R^{V \times 1}$ can be expressed as $SoftMax(W_{ho}h_t)$. Just to keep the notations simple, the biases have not been mentioned. In every unit, we can add bias to the input before it is acted upon by the different functions. So, o_t can be represented as follows:

$$o_t = SoftMax\left(W_{ho}h_t + b_o\right)$$

  where $b_o \in R^{n \times 1}$ is the bias vector for the output units.

- Similarly, bias can be introduced in the memory units, and hence h_t can be expressed as follows:

$$h_t = f\left(W_{hh}h_{t-1} + W_{xh}\ x_t + b_h\right)$$

  where $b_h \in R^{n \times 1}$ is the bias vector at the memory units.

- For a classification problem predicting the next word for a text sequence of T time steps, the output at each time step is a big SoftMax of V classes, where V is the size of the vocabulary. So, the corresponding loss at each time step is the negative log-loss over all the vocabulary size V. The loss at each time step can be expressed as follows:

$$C_t = -\sum_{j=1}^{V} y_t^{(j)} \log o_t^{(j)}$$

- To get the overall loss over all the time steps T, all such C_t needs to be summed up or averaged out. Generally, the average of all C_t is more convenient for stochastic gradient descent to get a fair comparison when the sequence length varies. The overall cost function for all the time steps T can thus be represented by the following:

$$C = -\sum_{t=1}^{T}\sum_{j=1}^{V} y_t^{(j)} \log o_t^{(j)}$$

Language Modeling

In language modeling, the probability of a sequence of words is computed through the product rule of intersection of events. The probability of a sequence of words $w_1 w_2 w_3 \ldots w_n$ of length n is given as follows:

$$P(w_1 w_2 w_3 \ldots w_n) = P(w_1)P(w_2/w_1)P(w_3/w_1w_2)\ldots P(w_3/w_1w_2)\ldots P(w_n/w_1w_2 \ldots w_{n-1})$$

$$= P(w_1)\prod_{k=2}^{n} P(w_k/w_1 w_2 \ldots w_{k-1})$$

In traditional approaches, the probability of a word at time step k is generally not conditioned on the whole sequence of length $(k-1)$ prior to that but rather on a small window L prior to t. Hence, the probability is generally approximated as follows:

$$P(w_1 w_2 w_3 \ldots w_n) \cong P(w_1)\prod_{k=2}^{n} P(w_k/w_{k-L} \ldots w_{k-2}w_{k-1})$$

This method of conditioning a state based on L recent states is called the Markov assumption of chain rule probability. Although it is an approximation, it is a necessary one for traditional language models since the words cannot be conditioned on a large sequence of words because of memory constraints.

Language models are generally used for varied tasks related to natural language processing, such as sentence completion by predicting the next words, machine translation, speech recognition, and others. In machine translation, the words from another language might be translated to English but may not be correct syntactically. For example, a sentence in the Hindi language has been machine translated to produce the English sentence *beautiful very is the sky*. If one computes the probability of the machine-translated sequence (P("*beautiful very is the sky*")), it would be very much less than that of the arranged counterpart, (P("*The sky is very beautiful*")). Through language modeling, such comparisons of the probability of text sequences can be carried out.

Predicting the Next Word in a Sentence Through RNN Versus Traditional Methods

In traditional language models, the probability of a word appearing next is generally conditioned on a window of a specified number of previous words, as discussed earlier. To estimate probabilities, different n-gram counts are usually computed. From bigram and trigram counts, the conditional probabilities can be computed as follows:

$$P\left(w=w_2 \,/\, w=w_1\right)=\frac{count\left(w_1,w_2\right)}{count\left(w_1\right)}$$

$$P\left(w=w_3 \,/\, w=w_1,w=w_2\right)=\frac{count\left(w_1,w_2,w_3\right)}{count\left(w_1,w_2\right)}$$

In a similar manner, we can condition the word probabilities on longer sequences of words by keeping count of larger n-grams. Generally, when a match is not found on a higher n-gram—let's say four-gram—then a lower n-gram such as a three-gram is tried. This method is called back off and gives some performance gain over a fixed n-gram approach.

Since word prediction is conditioned on only a few previous words based on the chosen window size, this traditional method of computing probabilities through n-gram counts doesn't do as well as those models in which the whole sequence of words is taken into consideration for the next-word prediction.

In RNNs, the output at each step is conditioned on all previous words, and hence RNNs do a better job than the n-gram models at language-model tasks. To understand this, let's look at the working principals of a generative recurrent neural network while considering a sequence $(x_1x_2x_3 \ldots x_n)$ of length n.

The RNN updates its hidden state h_t recursively as $h_t = f(h_{t-1}, x_t)$. The hidden state h_{t-1} has information accumulated for sequence of words $(x_1x_2\ldots x_{t-1})$, and when the new word in the sequence x_t arrives, the updated sequence information for $(x_1x_2x_3...x_t)$ is encoded in h_t through the recursive update.

Now, if we must predict the next word based on the word sequence seen so far, i.e., $(x_1x_2x_3...x_t)$, the conditional probability distribution one needs to look at is as follows:

$$P\left(x_{n+1} = o_i / x_1x_2x_3.....x_t\right)$$

where o_i represents any generalized word in the vocabulary.

For neural networks, this probability distribution is governed by the computed hidden state h_t based on the sequence seen so far, i.e., $x_1x_2x_3 ..... x_t$, and the model parameter V, which converts the hidden states to probabilities corresponding to every word in the vocabulary.

So,

$$P\left(x_{n+1} = o_i / x_1x_2x_3.....x_t\right)$$

$$=P(x_{n+1} = o_i/h_t) \text{ or } P(x_{n+1} = o_i/x_t; h_{t-1})$$

The vector of probabilities for $P(x_{n+1} = o_i/h_t)$ corresponding to all indices i over the vocabulary is given by $Softmax(W_{ho}h_t)$.

Backpropagation Through Time (BPTT)

Backpropagation for recurrent neural networks is same as that for feed-forward neural networks, with the only difference being that the gradient is the sum of the gradient with respect to the log-loss at each step.

First, the RNN is unfolded in time, and then the forward step is executed to get the internal activations and the final predictions for output. Based on the predicted output and the actual output labels, the loss and the corresponding error at each time step are computed. The error at each time step is backpropagated to update the weights. So, any weight update is proportional to the sum of the gradients' contribution from errors at all the T time steps.

Let's look at a sequence of length T and at the weight updates through BPTT. We take the number of memory states as n (i.e., $h_t \in R^{n \times 1}$) and choose the input vector length as

D (i.e., $x_t \in R^{D \times 1}$). At each sequence step t, we predict the next word from a vocabulary of V words through a SoftMax function.

The total cost function for a sequence of length T is given as follows:

$$C = -\sum_{t=1}^{T}\sum_{j=1}^{V} y_t^{(j)} \log o_t^{(j)}$$

Let's compute the gradient of the cost function with respect to the weights connecting the hidden memory states to the output states layers—i.e., the weights belonging to the matrix W_{ho}. The weight w_{ij} denotes the weight connecting the hidden state i to the output unit j.

The gradient of the cost function C_t with respect to w_{ij} can be broken down by the chain rule of partial derivatives as the product of the partial derivative of the cost function with respect to the output of the *jth* unit (i.e., $\dfrac{\partial C_t}{\partial o_t^{(j)}}$), the partial derivative of the output of the *jth* unit with respect to the net input $s_t^{(j)}$ at the *jth* unit (i.e., $\dfrac{\partial o_t^{(j)}}{\partial s_t^{(j)}}$), and finally the partial derivative of the net input to the *jth* unit with respect to the concerned weight from the *ith* memory unit to the *jth* hidden layer (i.e., $\dfrac{\partial s_t^{(j)}}{\partial w_{ij}}$).

$$\frac{\partial C_t}{\partial w_{ij}} = \frac{\partial C_t}{\partial o_t^{(j)}}\frac{\partial o_t^{(j)}}{\partial s_t^{(j)}}\frac{\partial s_t^{(j)}}{\partial w_{ij}} \tag{4.2.1}$$

$$\frac{\partial s_t^{(j)}}{\partial w_{ij}} = h_t^{(i)} \tag{4.2.2}$$

Considering the SoftMax over vocabulary V and the actual output at time t as $y_t = \left[y_t^{(1)} y_t^{(2)} \ldots y_t^{(V)} \right]^T$,

$$\frac{\partial C_t}{\partial o_t^{(j)}} = -\frac{y_t^{(j)}}{o_t^{(j)}} \tag{4.2.3}$$

$$\frac{\partial o_t^{(j)}}{\partial s_t^{(j)}} = o_t^{(j)}\left(1 - o_t^{(j)}\right) \tag{4.2.4}$$

Substituting the expressions for individual gradients from (4.2.2), (4.2.3), and (4.2.4) into (4.2.1), we get the following:

$$\frac{\partial C_t}{\partial w_{ij}} = -\frac{y_t^{(j)}}{o_t^{(j)}} o_t^{(j)} \left(1 - o_t^{(j)}\right) = -y_t^{(j)} \left(1 - o_t^{(j)}\right) h_t^{(i)} \tag{4.2.5}$$

To get the expression of the gradient of the total cost function C with respect to w_{ij}, one needs to sum up the gradients at each sequence step. Hence, the expression for the gradient $\dfrac{\partial C}{\partial w_{ij}}$ is as follows:

$$\frac{\partial C}{\partial w_{ij}} = \sum_{t=1}^{T} \frac{\partial C_t}{\partial w_{ij}} = \sum_{t=1}^{T} -y_t^{(j)} \left(1 - o_t^{(j)}\right) h_t^{(i)} \tag{4.2.6}$$

So, we can see that determining the weights of the connections from the memory state to the output layers is pretty much the same as doing so for the fully connected layers of feed-forward neural networks, the only difference being that the effect of each sequential step is summed up to get the final gradient.

Now, let's look at the gradient of the cost function with respect to the weights connecting the memory states in one step to the memory states in the next step—i.e., the weights of the matrix W_{hh}. We take the generalized weight $u_{ki} \in W_{hh}$, where k and i are indexes of the memory units in consecutive memory units.

This will get a little more involved because of the recurrent nature of the memory units' connections. To appreciate this fact, let's look at the output of the memory unit indexed by i at step t—i.e., $h_t^{(i)}$:

$$h_t^{(i)} = g\left(\sum_{l=1}^{N} u_{li} h_{t-1}^{(l)} + \sum_{m=1}^{D} v_{mi} x_t^{(m)} + b_{hi} \right) \tag{4.2.7}$$

Now, let's look at the gradient of the cost function at step t with respect to the weight u_{ki}:

$$\frac{\partial C_t}{\partial u_{ki}} = \frac{\partial C_t}{\partial h_t^{(i)}} \frac{\partial h_t^{(i)}}{\partial u_{ki}} \tag{4.2.8}$$

We are only interested in expressing $h_t^{(i)}$ as a function of u_{ki}, and hence we rearrange (4.2.7) as follows:

$$h_t^{(i)} = g\left(u_{ki}h_{t-1}^{(k)} + u_{ii}h_{t-1}^{(i)} + \sum_{\substack{l=1\\l\neq k,i}}^{N} u_{li}h_{t-1}^{(l)} + \sum_{m=1}^{D} v_{mi}x_t^{(m)} \right)$$

$$= g\left(u_{ki}h_{t-1}^{(k)} + u_{ii}h_{t-1}^{(i)} + c_t^{(i)} \right)$$

(4.2.9)

where

$$c_t^{(i)} = \sum_{\substack{l=1\\l\neq k,i}}^{N} u_{li}h_{t-1}^{(l)} + \sum_{m=1}^{D} v_{mi}x_t^{(m)}$$

We have rearranged $h_t^{(i)}$ to be a function that contains the required weight u_{ki} and kept $h_{t-1}^{(i)}$ since it can be expressed through recurrence as $g\left(u_{ki}h_{t-2}^{(k)} + u_{ii}h_{t-2}^{(i)} + c_{t-1}^{(i)} \right)$.

This nature of recurrence at every time step t would continue up to the first step, and hence one needs to consider the summation effect of all associated gradients from $t = t$ to $t = 1$. If the gradient of $h_t^{(i)}$ with respect to the weight u_{ki} follows the recurrence, and if we take the derivative of $h_t^{(i)}$ as expressed in (4.2.9) with respect to u_{ki}, then the following will hold true:

$$\frac{\partial h_t^{(i)}}{\partial u_{ki}} = \sum_{t'=1}^{t} \frac{\partial h_t^{(i)}}{\partial h_{t'}^{(i)}} \frac{\overline{\partial} h_{t'}^{(i)}}{\partial u_{ki}}$$

(4.2.10)

Please note the bar for the expression $\dfrac{\overline{\partial} h_{t'}^{(i)}}{\partial u_{ki}}$. It denotes the local gradient of $h_t^{(i)}$ with respect to u_{ki}, holding $h_{t'-1}^{(i)}$ constant.

Replacing (4.2.9) with (4.2.8), we get the following:

$$\frac{\partial C_t}{\partial u_{ki}} = \sum_{t'=1}^{t} \frac{\partial C_t}{\partial h_t^{(i)}} \frac{\partial h_t^{(i)}}{\partial h_{t'}^{(i)}} \frac{\overline{\partial} h_{t'}^{(i)}}{\partial u_{ki}}$$

(4.2.11)

Equation (4.2.11) gives us the generalized equation of the gradient of the cost function at time t. Hence, to get the total gradient, we need to sum up the gradients with respect to the cost at each time step. Therefore, the total gradient can be represented by the following:

$$\frac{\partial C}{\partial u_{ki}} = \sum_{t=1}^{T} \sum_{t'=1}^{t} \frac{\partial C_t}{\partial h_t^{(i)}} \frac{\partial h_t^{(i)}}{\partial h_{t'}^{(i)}} \frac{\overline{\partial} h_{t'}^{(i)}}{\partial u_{ki}} \tag{4.2.12}$$

The expression $\dfrac{\partial h_t^{(i)}}{\partial h_{t'}^{(i)}}$ follows a product recurrence and hence can be formulated as follows:

$$\frac{\partial h_t^{(i)}}{\partial h_{t'}^{(i)}} = \prod_{g=t'+1}^{t} \frac{\partial h_g^{(i)}}{\partial h_{g-1}^{(i)}} \tag{4.2.13}$$

Combining (4.2.12) and (4.2.13), we get the following:

$$\frac{\partial C}{\partial u_{ki}} = \sum_{t=1}^{T} \sum_{t'=1}^{t} \frac{\partial C_t}{\partial h_t^{(i)}} \left(\prod_{g=t'+1}^{t} \frac{\partial h_g^{(i)}}{\partial h_{g-1}^{(i)}} \right) \frac{\overline{\partial} h_{t'}^{(i)}}{\partial u_{ki}} \tag{4.2.14}$$

The computation of the gradient of the cost function C for the weights of the matrix W_{xh} can be computed in a manner similar to that for the weights corresponding to the memory states.

Vanishing- and Exploding-Gradient Problem in RNN

The aim of recurrent neural networks (RNNs) is to learn long dependencies so that the interrelations between words that are far apart are captured. For example, the actual meaning that a sentence is trying to convey may be captured well by words that are not near each other. Recurrent neural networks should be able to learn those dependencies. However, RNNs suffer from an inherent problem: they fail to capture long-distance dependencies between words. This is because the gradients in instances of long sequences have a high chance of either going to zero or going to infinity very quickly. When the gradients drop to zero very quickly, the model is unable to learn the associations or correlations between events that are temporally far apart. The equations derived for the gradient of the cost function with respect to the weights of the hidden memory layers will help us understand why this vanishing-gradient problem might take place.

The gradient of the cost function C_t at step t for a generalized weight $u_{ki} \in W_{hh}$ is given by the following:

$$\frac{\partial C_t}{\partial u_{ki}} = \sum_{t'=1}^{t} \frac{\partial C_t}{\partial h_t^{(i)}} \frac{\partial h_t^{(i)}}{\partial h_{t'}^{(i)}} \frac{\partial h_{t'}^{(i)}}{\partial u_{ki}}$$

where the notations comply with their original interpretations as mentioned in the "Backpropagation Through Time (BPTT)" section.

The components summed to form $\dfrac{\partial C_t}{\partial u_{ki}}$ are called its temporal components. Each of those components measures how the weight u_{ki} at step t' influences the loss at step t. The component $\dfrac{\partial h_t^{(i)}}{\partial h_{t'}^{(i)}}$ backpropagates the error at step t back to step t'.

Also,

$$\frac{\partial h_t^{(i)}}{\partial h_{t'}^{(i)}} = \prod_{g=t'+1}^{t} \frac{\partial h_g^{(i)}}{\partial h_{g-1}^{(i)}}$$

Combining the preceding two equations, we get the following:

$$\frac{\partial C_t}{\partial u_{ki}} = \sum_{t'=1}^{t} \frac{\partial C_t}{\partial h_t^{(i)}} \left(\prod_{g=t'+1}^{t} \frac{\partial h_g^{(i)}}{\partial h_{g-1}^{(i)}} \right) \frac{\bar{\partial} h_t^{(i)}}{\partial h_{ki}}$$

Let us take the net input at a memory unit i at time step g to be $z_g^{(i)}$. So, if we take the activation at the memory units to be sigmoid, then

$$h_g^{(i)} = \sigma\left(z_g^{(i)} \right)$$

where σ is the sigmoid function.

Now,

$$\frac{\partial h_g^{(i)}}{\partial h_{g-1}^{(i)}} = \sigma'\left(z_g^{(i)} \right) \frac{\partial z_g^{(i)}}{\partial h_{g-1}^{(i)}} = \sigma'\left(z_g^{(i)} \right) u_{ii}$$

345

where $\sigma'(z_g^{(i)})$ denotes the gradient of $\sigma(z_g^{(i)})$ with respect to $z_g^{(i)}$.

If we have a long sequence, i.e., $t = t$ and $t' = k$, then the following quantity would have many derivatives of the sigmoid function, as shown here:

$$\frac{\partial h_t^{(i)}}{\partial h_k^{(i)}} = \prod_{g=k+1}^{T} \frac{\partial h_g^{(i)}}{\partial h_{g-1}^{(i)}} = \frac{\partial h_{k+1}^{(i)}}{\partial h_k^{(i)}} \frac{\partial h_{k+2}^{(i)}}{\partial h_{k+1}^{(i)}} \cdots \frac{\partial h_t^{(i)}}{\partial h_{(t-1)}^{(i)}} = \sigma'\left(z_{k+1}^{(i)}\right) u_{ii} \sigma'\left(z_{k+2}^{(i)}\right) u_{ii} .. \sigma'\left(z_t^{(i)}\right) u_{ii}$$

Combining the gradient expressions in product-notation form, this important equation can be rewritten as follows:

$$\frac{\partial h_t^{(i)}}{\partial h_k^{(i)}} = \left(u_{ii}\right)^{t-k} \prod_{g=k+1}^{t} \sigma'\left(z_g^{(i)}\right)(1)$$

Sigmoid functions have good gradients only within a small range of values and saturate very quickly. Also, the gradients of sigmoid activation functions are less than 1. So, when the error from a long-distance step at $t = T$ passes to a step at $t = 1$ there would be $(T-1)$ sigmoid activation function gradients the error must pass through, and the multiplicative effect of $(T-1)$ gradients of values less than 1 may make the gradient component $\frac{\partial h_T^{(i)}}{\partial h_1^{(i)}}$ vanish exponentially fast. As discussed, $\frac{\partial h_T^{(i)}}{\partial h_1^{(i)}}$ backpropagates the error at step $T = t$ back to step $t = 1$ so that the long-distance correlation between the words at steps $t = 1$ and $t = T$ is learned. However, because of the vanishing-gradient problem, $\frac{\partial h_T^{(i)}}{\partial h_1^{(i)}}$ may not receive enough gradient and may be close to zero, and hence it will not be possible to learn the correlation or dependencies between the long-distance words in a sentence.

RNNs can suffer from exploding gradients too. We see in the expression for $\frac{\partial h_T^{(i)}}{\partial h_1^{(i)}}$ the weight u_{ii} has been repeatedly multiplied $(T-1)$ times. If $u_{ii} > 1$, and for simplicity we assume $u_{ii} = 2$, then after 50 steps of backpropagation from the sequence step T to sequence step $(T-50)$, the gradient would magnify approximately 2^{50} times and hence would drive the model training into instability.

Solution to Vanishing- and Exploding-Gradient Problem in RNNs

There are several methods adopted by the deep-learning community to combat the vanishing-gradient problem. In this section, we will discuss those methods before moving on to a variant of RNN called long short-term memory (LSTM) recurrent neural networks. LSTMs are much more robust regarding vanishing and exploding gradients.

Gradient Clipping

Exploding gradients can be tackled by a simple technique called gradient clipping. If the magnitude of the gradient vector exceeds a specified threshold, then the magnitude of the gradient vector is set to the threshold while keeping the direction of the gradient vector intact. So, while doing backpropagation on a neural network at time t, if the gradient of the cost function C with respect to the weight vector w exceeds a threshold k, then the gradient g used for backpropagation is updated as follows:

- **Step 1:** Update $g \leftarrow \nabla C(w = w^{(t)})$

- **Step 2:** If $g > k$, then update $g \leftarrow \dfrac{k}{g} g$

Smart Initialization of the Memory-to-Memory Weight Connection Matrix and ReLU Units

Instead of randomly initializing the weights in the weight matrix W_{hh}, initializing it as an identity matrix helps prevent the vanishing-gradient problem. One of the main reasons for vanishing-gradient problem is the gradient of the hidden unit i at time t with respect to the hidden unit i at time t' where $t' << t$ is given by the following:

$$\frac{\partial h_t^{(i)}}{\partial h_{t'}^{(i)}} = \prod_{g=t'+1}^{t} \frac{\partial h_g^{(i)}}{\partial h_{g-1}^{(i)}}$$

In the case of the sigmoid activation function, each of the terms $\dfrac{\partial h_g^{(i)}}{\partial h_{g-1}^{(i)}}$ can be expanded as follows:

$$\frac{\partial h_g^{(i)}}{\partial h_{g-1}^{(i)}} = \sigma'\left(z_g^{(t)}\right)u_{ii}$$

347

where $\sigma(.)$ denotes the sigmoid function and $z_g^{(t)}$ denotes the net input to hidden unit i at step t. The parameter u_{ii} is the weight connecting the *ith* hidden memory state at any sequence step $t-1$ to the *ith* hidden unit memory at sequence step t.

The greater the distance is between sequence steps t' and t, the more sigmoid derivatives the error would go through in passing from t to t'. Since the sigmoid activation function derivatives are always less than 1 and saturate quickly, the chance of the gradient going to zero in cases of long dependencies is high.

If, however, we choose ReLU units, then the sigmoid gradients would be replaced by ReLU gradients, which have a constant value of 1 for positive net input. This would reduce the vanishing-gradient problem since when the gradient is 1, the gradients would flow unattenuated. Also, when the weight matrix is chosen to be identity, then the weight connection u_{ii} would be 1, and hence the quantity $\dfrac{\partial h_t^{(i)}}{\partial h_{t'}^{(i)}}$ would be 1 irrespective of the distance between sequence step t and sequence step t'. This means that the error that would be propagated from hidden memory state $h_t^{(i)}$ at time t to the hidden memory state $h_{t'}^{.(i)}$ at any prior time step t' would be constant irrespective of the distance from step t to t'. This will enable the RNN to learn associated correlations or dependencies between the long-distance words in a sentence.

Long Short-Term Memory (LSTM)

Long short-term memory recurrent neural networks, popularly known as LSTMs, are special versions of RNNs that can learn distant dependencies between words in a sentence. Basic RNNs are incapable of learning such correlations between distant words, as discussed earlier. The architecture of long short-term memory (LSTM) recurrent neural networks is quite a bit different than that of traditional RNNs. Represented in Figure 4-12 is a high-level representation of an LSTM.

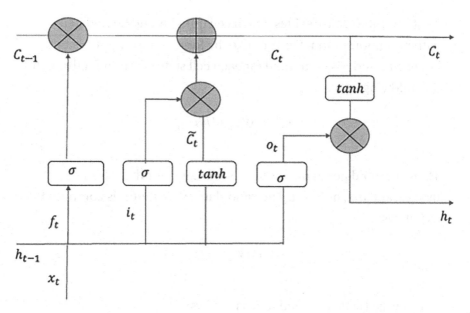

Figure 4-12. *LSTM architecture*

The basic building blocks of an LSTM and their functions are as follows:

- The new element in LSTMs is the introduction of the cell state C_t, which is regulated by three gates. The gates are composed of sigmoid functions so that they output values between 0 and 1.

At sequence step t, the input x_t and the previous step's hidden states h_{t-1} decide what information to forget from cell state C_{t-1} through the forget-gate layer. The forget gate looks at x_t and h_{t-1} and assigns a number between 0 and 1 for each element in the cell state vector C_{t-1}. An output of 0 means totally forget the state, while an output of 1 means keep the state completely. The forget-gate output is computed as follows:

$$f_t = \sigma\left(W_f x_t + U_f h_{t-1}\right)$$

- Next, the input gate decides which cell units should be updated with new information. For this, like the forget-gate output, a value between 0 and 1 is computed for each cell state component through the following logic:

$$i_t = \sigma\left(W_i x_t + U_i h_{t-1}\right)$$

Then, a candidate set of new values is created for the cell state using x_t and h_{t-1} as input. The candidate cell state $\widetilde{C}t$ is computed as follows:

$$\tilde{C}t = \tanh\left(W_c x_t + U_c h_{t-1}\right)$$

The new cell state C_t is updated as follows:

$$C_t = f_t^* C_{t-1} + i_t^* \tilde{C}t$$

- The next job is to determine which cell states to output since the cell state contains a lot of information. For this, x_t and h_{t-1} are passed through an output gate, which outputs a value between 0 and 1 for each component in the cell state vector C_t. The output of this gate is computed as follows:

$$o_t = \sigma\left(W_o x_t + U_o h_{t-1}\right)$$

- The updated hidden state h_t is computed from the cell state C_t by passing each of its elements through a *tanh* function and then doing an element-wise product with the output-gate values:

$$h_t = o_t^* \tanh\left(C_t\right)$$

Please note that the symbol * in the preceding equations denotes element-wise multiplication. This is done so that, based on the gate outputs, we can assign weights to each element of the vector it is operated on. Also, note that whatever gate output values are obtained, they are multiplied as is. The gate outputs are not converted to discrete values of 0 and 1. Continuous values between 0 and 1 through sigmoid provide smooth gradients for backpropagation.

The forget gate plays a crucial role in the LSTM; when the forget-gate units output zero, then the recurrent gradients become zero and the corresponding old cell state units are discarded. This way, the LSTM throws away information that it thinks is not going to be useful in the future. Also, when the forget-gate units output 1, the error flows through the cell units unattenuated, and the model can learn long-distance correlations between temporally distant words. We will discuss more about this in the next section.

Another important feature of the LSTM is the introduction of the output gates. The output-gate unit ensures that not all information the cell state C_t units have is exposed to the rest of the network, and hence only the relevant information is revealed in the form of h_t. This ensures that the rest of the network is not impacted by the unnecessary data while that data in the cell state is still held back in the cell state to help drive future decisions.

LSTM in Reducing Exploding- and Vanishing-Gradient Problems

LSTMs don't suffer much from vanishing- or exploding-gradient problems. The main reason for this is the introduction of the forget gate f_t and the way the current cell state is dependent on it through the following equation:

$$C_t = f_t^* C_{t-1} + i_t^* \tilde{C}t$$

This equation can be broken down on a cell state unit level that has a general index i as follows:

$$C_t^{(i)} = f_t^{(i)} C_{t-1}^{(i)} + i_t^{(i)} \tilde{C}t^{(i)}$$

It is important to note here that $C_t^{(i)}$ is linearly dependent on $C_{t-1}^{(i)}$, and hence the activation function is the identity function with a gradient of 1.

The notorious component in recurrent neural network backpropagation that may lead to vanishing or exploding gradients is the component $\dfrac{\partial h_t^{(i)}}{\partial h_k^{(i)}}$ when $(t - k)$ is large.

This component backpropagates the error at sequence step t to sequence step k so that the model learns long-distance dependencies or correlations. The expression for $\dfrac{\partial h_t^{(i)}}{\partial h_k^{(i)}}$, as we have seen in the vanishing- and exploding-gradient section, is given by the following:

$$\frac{\partial h_t^{(i)}}{\partial h_k^{(i)}} = \left(u_{ii}\right)^{t-k} \prod_{g=k+1}^{t} \sigma'\left(z_g^{(i)}\right)$$

A vanishing-gradient condition will arise when the gradients and/or weights are less than 1 since the product of $(t-k)$ in them would force the overall product to near zero. Since sigmoid gradients and tanh gradients are most of the time less than 1 and saturate fast where they have near-zero gradients, this problem of vanishing gradients would be more severe with those. Similarly, exploding gradients can happen when the weight connection u_{ii} between the *ith* hidden to the *ith* hidden unit is greater than 1 since the product of $(t-k)$ in them would make the term $(u_{ii})^{t-k}$ exponentially large.

The equivalent of $\dfrac{\partial h_t^{(i)}}{\partial h_k^{(i)}}$ in LSTM is the component $\dfrac{\partial C_t^{(i)}}{\partial C_k^{(i)}}$, which can also be expressed in the product format as follows:

$$\frac{\partial C_t^{(i)}}{\partial C_k^{(i)}} = \prod_{g=t'+1}^{t} \frac{\partial C_g^{(i)}}{\partial C_{g-1}^{(i)}} \tag{4.3.1}$$

On taking the partial derivative on both sides of the cell state update equation $C_t^{(i)} = f_t^{(i)} C_{t-1}^{(i)} + i_t^{(i)} \tilde{C}t^{(i)}$, one gets the important expression as follows:

$$\frac{\partial C_t^{(i)}}{\partial C_{t-1}^{(i)}} = f_t^{(i)} \tag{4.3.2}$$

Combining (1) and (2), one gets the following:

$$\frac{\partial C_t^{(i)}}{\partial C_k^{(i)}} = \left(f_t^{(i)}\right)^{t-k} \tag{4.3.3}$$

Equation (4.3.3) says that if the forget-gate value is held near 1, LSTM will not suffer from vanishing- or exploding-gradient problems.

MNIST Digit Identification in TensorFlow Using Recurrent Neural Networks

We see an implementation of RNN in classifying the images in the MNIST dataset through LSTM. The images of MNIST dataset are 28 × 28 in dimension. Each image would be treated as having 28 sequence steps, and each sequence step would consist of a row in an image. There would not be outputs after each sequence step but rather only one output at the end of the 28 steps for each image. The output is one of the ten classes corresponding to the ten digits from 0 to 9. So, the output layer would be a SoftMax of ten classes. The hidden cell state of the final sequence step h_{28} would be fed through weights into the output layer. So, only the final sequence step would contribute to the cost function; there would be no cost function associated with the intermediate sequence steps. If you remember, backpropagation was with respect to the cost function C_t in each individual sequence step t when there was output involved at each sequence step, and finally the gradients with respect to each of those cost functions were summed together. Here, everything else remains the same, and backpropagation would be done only with respect to the cost at sequence step 28, C_{28}. Also, as stated earlier, each row of an image would form data at a sequence step t and hence would be the input vector x_t. Finally, we would be processing images in a mini-batch, and thus for each image in the batch, a similar processing procedure would be followed, minimizing the average cost over the batch. One more important thing to note is that TensorFlow demands that there be separate tensors for each step within the mini-batch input tensor. The input tensor structure has been depicted in Figure 4-13 for ease of understanding.

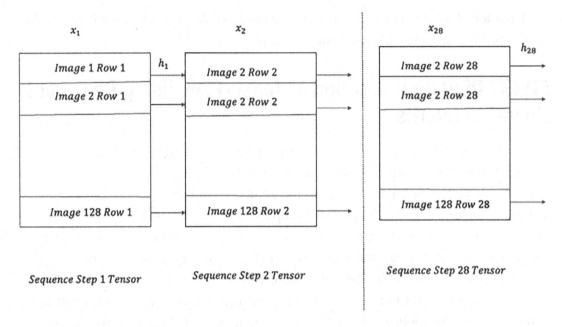

Figure 4-13. *Input tensor shape for RNN LSTM network in TensorFlow*

Now that we have some clarity about the problem and approach, we will proceed with the implementation in TensorFlow. The detailed code for training the model and validating it on the test dataset is illustrated in Listing 4-3.

Listing 4-3. TensorFlow Implementation of Recurrent Neural Network Using LSTM for Classification

```
###################################
#Import the Required Libraries
###################################

import tensorflow as tf
import numpy as np
from tensorflow.keras import Model, layers

def read_infile():
    (train_X, train_Y), (test_X, test_Y) = tf.keras.datasets.mnist.
    load_data()
    train_X,test_X = np.expand_dims(train_X, -1), np.expand_dims(test_X,-1)
```

```
    #print(train_X.shape,test_X.shape)
    return train_X, train_Y,test_X, test_Y

def normalize(train_X):
    return train_X/255.0

class RNN_network(Model):

    def __init__(self,hidden_dim,num_classes):
        super(RNN_network,self).__init__()
        self.hidden_dim = hidden_dim
        self.num_classes = num_classes

        self.lstm = layers.LSTM(hidden_dim, return_sequences=False,
        return_state=False)
        self.out = layers.Dense(num_classes)

    def call(self,x):
        x = self.lstm(x)
        x = self.out(x)
        return x

def train(epochs=10,lr=0.01,hidden_dim=64,batch_size=32,n_steps=28,
n_inputs=28):

    X_train, y_train, X_test, y_test = read_infile()
    X_train, X_test = normalize(X_train), normalize(X_test)
    num_train_recs, num_test_recs = X_train.shape[0], X_test.shape[0]

    model = RNN_network(hidden_dim=hidden_dim,num_classes=10)
    #model_graph = tf.function(model)

    loss_fn = tf.keras.losses.SparseCategoricalCrossentropy
    (from_logits=True,reduction=tf.keras.losses.Reduction.SUM)

    optimizer = tf.keras.optimizers.Adam(lr)

    num_train_recs,num_test_recs = X_train.shape[0], X_test.shape[0]
    num_batches = num_train_recs // batch_size
    num_batches_test = num_test_recs //batch_size
    order_ = np.arange(num_train_recs)
```

```
loss_trace,accuracy_trace = [], []
start_time = time.time()
for i in range(epochs):
    loss, accuracy = 0,0
    np.random.shuffle(order_)
    X_train,y_train = X_train[order_], y_train[order_]
    for j in range(num_batches):
        X_train_batch = tf.constant(X_train[j*batch_size:(j+1)*batch_
        size].reshape((batch_size, n_steps, n_inputs)),dtype=tf.
        float32)
        y_train_batch = tf.constant(y_train[j*batch_
        size:(j+1)*batch_size])
        #print(X_train_batch, y_train_batch)

        with tf.GradientTape() as tape:
            y_pred_batch = model(X_train_batch)
            loss_ = loss_fn(y_train_batch,y_pred_batch)

        # compute gradient
        gradients = tape.gradient(loss_, model.trainable_variables)
            # update the parameters
        optimizer.apply_gradients(zip(gradients,
        model.trainable_variables))

        accuracy += np.sum(y_train_batch.numpy() ==
        np.argmax(y_pred_batch.numpy(),axis=1))
        loss += loss_.numpy()
    loss /= num_train_recs
    accuracy /= num_train_recs
    loss_trace.append(loss)
    accuracy_trace.append(accuracy)
    print(f"Epoch {i+1} : loss: {np.round(loss,4)} ,accuracy:{np.
    round(accuracy,4)}\n")
```

```
    test_accuracy = 0
    for j in range(num_batches_test):
        X_test_batch = tf.constant(X_test[j*batch_size:(j+1)*batch_size].
        reshape((batch_size, n_steps, n_inputs)),dtype=tf.float32)
        y_test_batch = tf.constant(y_test[j*batch_size:(j+1)*batch_size])
        y_pred_batch = model(X_test_batch)
        test_accuracy += np.sum(y_test_batch.numpy() == np.argmax(y_pred_
        batch.numpy(),axis=1))
    print(f"Test accuracy : {test_accuracy/num_test_recs}")

train()

---output--
Epoch 1 : loss: 0.2713 ,accuracy:0.9144

Epoch 2 : loss: 0.1151 ,accuracy:0.9658

Epoch 3 : loss: 0.0947 ,accuracy:0.9716

Epoch 4 : loss: 0.0866 ,accuracy:0.9745

Epoch 5 : loss: 0.0875 ,accuracy:0.9739

Epoch 6 : loss: 0.092 ,accuracy:0.9728

Epoch 7 : loss: 0.0936 ,accuracy:0.9722

Epoch 8 : loss: 0.0899 ,accuracy:0.973

Epoch 9 : loss: 0.083 ,accuracy:0.9751

Epoch 10 : loss: 0.0813 ,accuracy:0.9759

Test accuracy : 0.9672
```

As we can see from the Listing 4-3 output, just by running two epochs, an accuracy of 95% is achieved on the test dataset.

Next-Word Prediction and Sentence Completion in TensorFlow Using Recurrent Neural Networks

We train a model on a small passage from *Alice in Wonderland* to predict the next word from the given vocabulary using LSTM. Sequences of three words have been taken as input, and the subsequent word has been taken as output for this problem. Also, a two-layered LSTM model has been chosen instead of one. The sets of inputs and outputs are chosen randomly from the corpus and fed as a mini-batch of size 1. We see the model achieves good accuracy and can learn the passage well. Later, once the model is trained, we input a three-word sentence and let the model predict the next 28 words. Each time it predicts a new word, it appends it to the updated sentence. For predicting the next word, the previous three words from the updated sentence are taken as input. The detailed implementation of the problem has been outlined in Listing 4-4.

Listing 4-4. Next-Word Prediction and Sentence Completion in TensorFlow Using Recurrent Neural Networks

```
import numpy as np
import tensorflow as tf
import random
import collections
import time

def read_data(fname):
    with open(fname) as f:
        data = f.readlines()
    data = [x.strip() for x in data]
    data = [data[i].lower().split() for i in range(len(data))]
    data = np.array(data)
    data = np.reshape(data, [-1, ])
    return data

train_file = '/home/santanu/Downloads/alice_in_wonderland.txt'
train_data = read_data(train_file)
print(train_data)
```

```python
def build_dataset(train_data):
    #print(train_data)
    count = collections.Counter(train_data).most_common()
    dictionary = dict()
    for word, _ in count:
        dictionary[word] = len(dictionary)
    reverse_dictionary = dict(zip(dictionary.values(), dictionary.keys()))
    return dictionary, reverse_dictionary

def input_one_hot(num,vocab_size):
    x = np.zeros(vocab_size)
    x[num] = 1
    return x.tolist()

class RNN_network(Model):

    def __init__(self,hidden_dim,num_classes):
        super(RNN_network,self).__init__()
        self.hidden_dim = hidden_dim
        self.num_classes = num_classes

        self.lstm = layers.LSTM(hidden_dim, return_sequences=False,
        return_state=False)
        self.out = layers.Dense(num_classes)

    def call(self,x):
        x = self.lstm(x)
        x = self.out(x)
        return x

def train(n_input=3,n_hidden=512,lr=0.001,training_iters = 50000,
display_step = 500):
    dictionary, reverse_dictionary = build_dataset(train_data)
    vocab_size = len(dictionary)
    model = RNN_network(hidden_dim=n_hidden,num_classes=vocab_size)
    print(model)
    # Launch the graph
    step = 0
```

```
offset = random.randint(0,n_input+1)
end_offset = n_input + 1
acc_total = 0
loss_total = 0
loss_fn = tf.keras.losses.CategoricalCrossentropy(
from_logits=True,reduction=tf.keras.losses.Reduction.SUM)
optimizer = tf.keras.optimizers.Adam(lr)

while step < training_iters:
    if offset > (len(train_data)-end_offset):
        offset = random.randint(0, n_input+1)

    symbols_in_keys = [ input_one_hot(dictionary[ str(train_data[i])],
    vocab_size=vocab_size) for i in range(offset, offset+n_input) ]
    symbols_in_keys = tf.constant(np.reshape(np.array(symbols_in_keys),
    [-1, n_input,vocab_size]))
    symbols_out_onehot = np.zeros([vocab_size], dtype=float)
    symbols_out_onehot[dictionary[str(train_data[offset+
    n_input])]] = 1.0
    symbols_out_onehot = tf.constant(np.reshape(symbols_out_onehot,
    [1,-1]))
    with tf.GradientTape() as tape:
        onehot_pred = model(symbols_in_keys)
        loss_ = loss_fn(symbols_out_onehot,onehot_pred)

    # compute gradient
    gradients = tape.gradient(loss_, model.trainable_variables)
    # update the parameters
    optimizer.apply_gradients(zip(gradients, model.trainable_
    variables))

    acc_total += np.sum(np.argmax(symbols_out_onehot.numpy(),axis=1) ==
    np.argmax(onehot_pred.numpy(),axis=1))
    loss_total += loss_.numpy()
```

```python
        if step % display_step == 0:
            print("Iter= " + str(step) + ", Average Loss= " + \
                    "{:.6f}".format(loss_total/display_step) + ", \
                    Average Accuracy= " + \
                    "{:.2f}%".format(100*acc_total/display_step))
            acc_total = 0
            loss_total = 0
            symbols_in = [train_data[i] for i in range(offset, offset +
            n_input)]
            symbols_out = train_data[offset + n_input]
            symbols_out_pred = reverse_dictionary[int(tf.argmax
            (onehot_pred, 1))]
            print("%s - Actual word:[%s] vs Predicted word:[%s]" %
            (symbols_in,symbols_out,symbols_out_pred))
        step += 1
        offset += (n_input+1)

    sentence = 'i only wish'
    words = sentence.split(' ')
try:
    symbols_in_keys = [ input_one_hot(dictionary[ str(train_data[i])])
    for i in range(offset, offset+n_input) ]
    for i in range(28):
        keys = np.reshape(np.array(symbols_in_keys), [-1, n_input,
        vocab_size])
        onehot_pred = model(keys)
        onehot_pred_index = int(tf.argmax(onehot_pred, 1))
        sentence = "%s %s" % (sentence,reverse_dictionary
        [onehot_pred_index])
        symbols_in_keys = symbols_in_keys[1:]
        symbols_in_keys.append(input_one_hot(onehot_pred_index))
        print(sentence)
except:
    print("Word not in dictionary")
print("Optimization Finished!")
train()
```

---output --

Iter= 30500, Average Loss= 0.073997, Average Accuracy= 99.40%
['only', 'you', 'can'] - Actual word:[find] vs Predicted word:[find]
Iter= 31000, Average Loss= 0.004558, Average Accuracy= 99.80%
['very', 'hopeful', 'tone'] - Actual word:[though] vs Predicted
word:[though]
Iter= 31500, Average Loss= 0.083401, Average Accuracy= 99.20%
['tut', ',', 'tut'] - Actual word:[,] vs Predicted word:[,]
Iter= 32000, Average Loss= 0.116754, Average Accuracy= 99.00%
['when', 'they', 'met'] - Actual word:[in] vs Predicted word:[in]
Iter= 32500, Average Loss= 0.060253, Average Accuracy= 99.20%
['it', 'in', 'a'] - Actual word:[bit] vs Predicted word:[bit]
Iter= 33000, Average Loss= 0.081280, Average Accuracy= 99.00%
['perhaps', 'it', 'was'] - Actual word:[only] vs Predicted word:[only]
Iter= 33500, Average Loss= 0.043646, Average Accuracy= 99.40%
['you', 'forget', 'to'] - Actual word:[talk] vs Predicted word:[talk]
Iter= 34000, Average Loss= 0.088316, Average Accuracy= 98.80%
[',', 'and', 'they'] - Actual word:[walked] vs Predicted word:[walked]
Iter= 34500, Average Loss= 0.154543, Average Accuracy= 97.60%
['a', 'little', 'startled'] - Actual word:[when] vs Predicted word:[when]
Iter= 35000, Average Loss= 0.105387, Average Accuracy= 98.40%
['you', 'again', ','] - Actual word:[you] vs Predicted word:[you]
Iter= 35500, Average Loss= 0.038441, Average Accuracy= 99.40%
['so', 'stingy', 'about'] - Actual word:[it] vs Predicted word:[it]
Iter= 36000, Average Loss= 0.108765, Average Accuracy= 99.00%
['like', 'to', 'be'] - Actual word:[rude] vs Predicted word:[rude]
Iter= 36500, Average Loss= 0.114396, Average Accuracy= 98.00%
['make', 'children', 'sweet-tempered'] - Actual word:[.] vs Predicted
word:[.]
Iter= 37000, Average Loss= 0.062745, Average Accuracy= 98.00%
['chin', 'upon', "alice's"] - Actual word:[shoulder] vs Predicted
word:[shoulder]
Iter= 37500, Average Loss= 0.050380, Average Accuracy= 99.20%
['sour', '\xe2\x80\x94', 'and'] - Actual word:[camomile] vs Predicted
word:[camomile]

```
Iter= 38000, Average Loss= 0.137896, Average Accuracy= 99.00%
['very', 'ugly', ';'] - Actual word:[and] vs Predicted word:[and]
Iter= 38500, Average Loss= 0.101443, Average Accuracy= 98.20%
["'", 'she', 'went'] - Actual word:[on] vs Predicted word:[on]
Iter= 39000, Average Loss= 0.064076, Average Accuracy= 99.20%
['closer', 'to', "alice's"] - Actual word:[side] vs Predicted word:[side]
Iter= 39500, Average Loss= 0.032137, Average Accuracy= 99.60%
['in', 'my', 'kitchen'] - Actual word:[at] vs Predicted word:[at]
Iter= 40000, Average Loss= 0.110244, Average Accuracy= 98.60%
[',', 'tut', ','] - Actual word:[child] vs Predicted word:[child]
Iter= 40500, Average Loss= 0.088653, Average Accuracy= 98.60%
["i'm", 'a', 'duchess'] - Actual word:[,] vs Predicted word:[,]
Iter= 41000, Average Loss= 0.122520, Average Accuracy= 98.20%
["'", "'", 'perhaps'] - Actual word:[it] vs Predicted word:[it]
Iter= 41500, Average Loss= 0.011063, Average Accuracy= 99.60%
['it', 'was', 'only'] - Actual word:[the] vs Predicted word:[the]
Iter= 42000, Average Loss= 0.057289, Average Accuracy= 99.40%
['you', 'forget', 'to'] - Actual word:[talk] vs Predicted word:[talk]
Iter= 42500, Average Loss= 0.089094, Average Accuracy= 98.60%
['and', 'they', 'walked'] - Actual word:[off] vs Predicted word:[off]
Iter= 43000, Average Loss= 0.023430, Average Accuracy= 99.20%
['heard', 'her', 'voice'] - Actual word:[close] vs Predicted word:[close]
Iter= 43500, Average Loss= 0.022014, Average Accuracy= 99.60%
['i', 'am', 'to'] - Actual word:[see] vs Predicted word:[see]
Iter= 44000, Average Loss= 0.000067, Average Accuracy= 100.00%
["wouldn't", 'be', 'so'] - Actual word:[stingy] vs Predicted word:[stingy]
Iter= 44500, Average Loss= 0.131948, Average Accuracy= 98.60%
['did', 'not', 'like'] - Actual word:[to] vs Predicted word:[to]
Iter= 45000, Average Loss= 0.074768, Average Accuracy= 99.00%
['that', 'makes', 'them'] - Actual word:[bitter] vs Predicted word:[bitter]
Iter= 45500, Average Loss= 0.001024, Average Accuracy= 100.00%
[',', 'because', 'she'] - Actual word:[was] vs Predicted word:[was]
Iter= 46000, Average Loss= 0.085342, Average Accuracy= 98.40%
['new', 'kind', 'of'] - Actual word:[rule] vs Predicted word:[rule]
Iter= 46500, Average Loss= 0.105341, Average Accuracy= 98.40%
```

```
['alice', 'did', 'not'] - Actual word:[much] vs Predicted word:[much]
Iter= 47000, Average Loss= 0.081714, Average Accuracy= 98.40%
['soup', 'does', 'very'] - Actual word:[well] vs Predicted word:[well]
Iter= 47500, Average Loss= 0.076034, Average Accuracy= 98.40%
['.', "'", "everything's"] - Actual word:[got] vs Predicted word:[got]
Iter= 48000, Average Loss= 0.099089, Average Accuracy= 98.20%
[',', "'", 'she'] - Actual word:[said] vs Predicted word:[said]
Iter= 48500, Average Loss= 0.082119, Average Accuracy= 98.60%
['.', "'", "'"] - Actual word:[perhaps] vs Predicted word:[perhaps]
Iter= 49000, Average Loss= 0.055227, Average Accuracy= 98.80%
[',', 'and', 'thought'] - Actual word:[to] vs Predicted word:[to]
Iter= 49500, Average Loss= 0.068357, Average Accuracy= 98.60%
['dear', ',', 'and'] - Actual word:[that] vs Predicted word:[that]
Iter= 50000, Average Loss= 0.043755, Average Accuracy= 99.40%
['affectionately', 'into', "alice's"] - Actual word:[,] vs Predicted
word:[,]
Training Completed!

i only wish off together . alice was very glad to find her in such a
pleasant temper, and thought to herself that perhaps it was only the
pepper that
```

We can see from the output of Listing 4-4 that the model is able to predict the actual words while training quite nicely. In the sentence-completion task, although the prediction did not start off well with the first two predictions, it did an excellent job for the rest of the 28 characters. The generated sentence is exceptionally rich in grammar and punctuation. The model accuracy can be increased by increasing the sequence length and by introducing predictions after every word in the sequence. Also, the training corpus was small. Word-prediction and sentence-completion quality will be further enhanced if the model is trained on a larger corpus of data. Listing 4-5 shows the passage from *Alice in Wonderland* that was used to train the model.

Listing 4-5.

' You can't think how glad I am to see you again, you dear old thing ! '
said the Duchess, as she tucked her arm affectionately into Alice's, and
they walked off together . Alice was very glad to find her in such a
pleasant temper, and thought to herself that perhaps it was only the pepper
that had made her so savage when they met in the kitchen . ' When I'm a
Duchess, ' she said to herself, (not in a very hopeful tone though),
' I won't have any pepper in my kitchen at all . Soup does very well
without — Maybe it's always pepper that makes people hot-tempered, ' she
went on, very much pleased at having found out a new kind of rule, ' and
vinegar that makes them sour — and camomile that makes them bitter — and —
and barley-sugar and such things that make children sweet-tempered . I
only wish people knew that : then they wouldn't be so stingy about it, you
know — 'She had quite forgotten the Duchess by this time, and was a little
startled when she heard her voice close to her ear . ' You're thinking about
something, my dear, and that makes you forget to talk . I can't tell you
just now what the moral of that is, but I shall remember it in a bit . ' '
Perhaps it hasn't one, ' Alice ventured to remark . ' Tut, tut, child ! '
said the Duchess . ' Everything's got a moral, if only you can find it .
' And she squeezed herself up closer to Alice's side as she spoke . Alice
did not much like keeping so close to her : first, because the Duchess was
very ugly ; and secondly, because she was exactly the right height to rest
her chin upon Alice's shoulder, and it was an uncomfortably sharp chin .
However, she did not like to be rude, so she bore it as well as she could .

Gated Recurrent Unit (GRU)

Much like LSTM, the gated recurrent units, popularly known as GRU, have gating units that control the flow of information inside. However, unlike LSTM, they don't have separate memory cells. The hidden memory state h_t at any time step t is a linear interpolation between previous hidden memory states h_{t-1} and candidate new hidden state $\tilde{h}_t$. The architectural diagram of a GRU is presented in Figure 4-14.

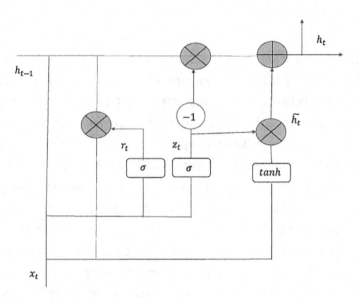

Figure 4-14. *GRU architecture*

The following are the high-level details as to how the GRU works:

- Based on the hidden memory state h_{t-1} and current input x_t, the reset gate r_t and the update gate z_t are computed as follows:

$$r_t = \sigma\left(W_r h_{t-1} + U_r x_t\right)$$

$$z_t = \sigma\left(W_z h_{t-1} + U_z x_t\right)$$

   The reset gate determines how important h_{t-1} is in determining the candidate new hidden state. The update gate determines how much the new candidate state should influence the new hidden state.

- The candidate new hidden state $\tilde{h}_t$ is computed as follows:

$$\tilde{h}_t = \tanh\left(r_t^* U h_{t-1} + W x_t\right) \tag{4.4.1}$$

- Based on the candidate state and the previous hidden state, the new hidden memory state is updated as follows:

$$h_t = \left(1 - z_t\right)^* h_{t-1} + z_t^* \tilde{h}_t \qquad (4.4.2)$$

The key point to stress in a GRU is the role of the gating functions, as discussed here:

- When the reset gate output units from r_t are close to zero, the previous hidden states for those units are ignored in the computation of the candidate new hidden state, as is evident from the equations in (4.4.1). This allows the model to drop information that would not be useful in the future.

- When the update-gate output units from z_t are close to zero, then the previous step states for those units are copied over to the current step. As we have seen before, the notorious component in recurrent neural network backpropagation that may lead to vanishing or exploding gradients is the component $\dfrac{\partial h_t^{(i)}}{\partial h_k^{(i)}}$, which backpropagates the error at sequence step t to sequence step k so that the model learns long-distance dependencies or correlations. The expression for $\dfrac{\partial h_t^{(i)}}{\partial h_k^{(i)}}$, as we saw in the vanishing- and exploding-gradient section, is given by the following:

$$\frac{\partial h_t^{(i)}}{\partial h_k^{(i)}} = \left(u_{ii}\right)^{t-k} \prod_{g=k+!}^{t} \sigma'\left(z_g^{(i)}\right)$$

When $(t - k)$ is large, a vanishing-gradient condition will arise when the gradients of the activations in the hidden state units and/or the weights are less than 1 since the product of $(t - k)$ in them would force the overall product to near zero. Sigmoid and tanh gradients are often less than 1 and saturate fast where they have near-zero gradients, thus making the problem of vanishing gradient more severe. Similarly, exploding gradients can happen when the weight connection u_{ii} between the *ith* hidden to the *ith* hidden unit is greater than 1 since the product of $(t - k)$ in them would make the term $(u_{ii})^{t-k}$ exponentially large for large values of $(t - k)$.

- Now, coming back to the GRU, when the update-gate output units in z_t are close to 0, then from Equation (4.4.2),

$$h_t^{(i)} \approx h_{t-1}^{(i)} \quad \forall i \in K \tag{4.4.3}$$

where K is the set of all hidden units for which $z_t^{(i)} \approx 0$.

On taking the partial derivative of $h_t^{(i)}$ with respect to $h_{t-1}^{(i)}$ in (4.4.3), we get the following:

$$\frac{\partial h_t^{(i)}}{\partial h_{t-1}^{(i)}} \approx 1$$

This will ensure that the notorious term $\dfrac{\partial h_t^{(i)}}{\partial h_k^{(i)}}$ is also close to 1 since it can be expressed as follows:

$$\frac{\partial h_t^{(i)}}{\partial h_k^{(i)}} = \prod_{g=t'+1}^{t} \frac{\partial h_g^{(i)}}{\partial h_{g-1}^{(i)}}$$

$$= 1.1.1....1 \ (t-k) times = 1$$

This allows the hidden states to be copied over many sequence steps without alteration, and hence the chances of a vanishing gradient diminish, and the model is able to learn temporally long-distance association or correlation between words.

Bidirectional RNN

In a standard recurrent neural network, we make predictions that take the past sequence states into account. For example, for predicting the next word in a sequence, we take the words that appear before it into consideration. However, for certain tasks in natural language processing, such as parts of speech, tagging both past words and future words with respect to a given word is crucial in determining the given word's part of speech tag. Also, for parts of speech–tagging applications, the whole sentence would be available for tagging, and hence for each given word—barring the ones at the start and end of a sentence—its past and future words would be present to make use of.

Bidirectional RNNs are a special type of RNN that makes use of both the past and future states to predict the output label at the current state. A bidirectional RNN combines two RNNs, one of which runs forward from left to right and the other of which runs backward from right to left. A high-level architecture diagram of a bidirectional RNN is depicted in Figure 4-15.

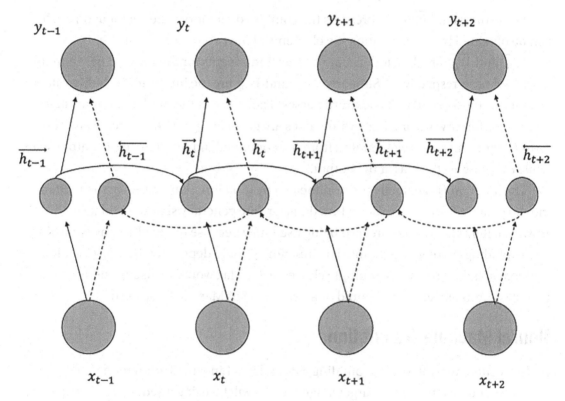

***Figure 4-15.** Bidirectional RNN architectural diagram*

For a bidirectional RNN, there are two hidden memory states for any sequence step t. The hidden memory states corresponding to the forward flow of information, as in standard RNN, can be represented as $\overrightarrow{h_t}$, and the ones corresponding to the backward flow of information can be denoted by $\overleftarrow{h_t}$. The output at any sequence step t depends on both the memory states $\overrightarrow{h_t}$ and $\overleftarrow{h_t}$. The following are the governing equations for a bidirectional RNN.

$$\vec{h_t} = f\left(\overrightarrow{W_{xh}}x_t + \overrightarrow{W_{hh}}\,\vec{h_{t-1}} + \vec{b}\right)$$

$$\overleftarrow{h_t} = f\left(\overleftarrow{W_{xh}}x_t + \overleftarrow{W_{hh}}\,\overleftarrow{h_{t-1}} + \overleftarrow{b}\right)$$

$$y_t = g\left(U\left[\vec{h_t};\overleftarrow{h_t}\right] + c\right)$$

The expression $\left[\vec{h_t};\overleftarrow{h_t}\right]$ represents the combined memory state vector at time t. It can be obtained by concatenating the elements of the two vectors $\vec{h_t}$ and $\overleftarrow{h_t}$.

$\overrightarrow{W_{hh}}$ and $\overleftarrow{W_{hh}}$ are the hidden state connection weights for the forward pass and the backward pass, respectively. Similarly, $\overrightarrow{W_{xh}}$ and $\overleftarrow{W_{xh}}$ are the inputs to the hidden state weights for the forward and backward passes. The biases at the hidden memory state activations for forward and backward passes are given by $\vec{b}$ and $\overleftarrow{b}$, respectively. The term U represents the weight matrix from the combined hidden state to the output state, while c represents the bias at the output.

The function f is generally the nonlinear activation function chosen at the hidden memory states. Chosen activation functions for f are generally sigmoid and tanh. However, ReLU activations are also being used now because they reduce the vanishing- and exploding-gradient problems. The function g would depend on the classification problem at hand. In cases of multiple classes, a SoftMax would be used, whereas for a two-class problem, either a sigmoid or a two-class SoftMax could be used.

Neural Machine Translation

In this section, we will work on building a neural machine translation model that translates source language to target language. We will be using a sequence-to-sequence architecture with LSTMs as the encoder and the decoder of the neural machine translation model. Unlike rule-based machine translation models, the neural machine translation model can be trained end to end.

Architecture of the Neural Machine Translation Model Using Seq2Seq

Illustrated in Figure 4-16 is the high-level architecture of a neural machine translation model.

The encoder is an LSTM which processes the source language words sequentially and at the end of the sentence or paragraph comes up with the final hidden state $h_f^{(e)}$ and final cell state $c_f^{(e)}$ vector encodings. These final hidden and cell state vectors are expected to capture the entire summary of the source language sequence.

The decoder in the next step takes the hidden $h_f^{(e)}$ and cell state $c_f^{(e)}$ encodings as the context. This is done by setting the initial hidden $h_0^{(d)}$ and cell state of the decoder $c_{(0)}^{(d)}$ with the final hidden and cell states of the encoder.

The initial input to the decoder is the start of sequence token <START>. So, using the encoder context and the start of sequence token as input, the decoder LSTM predicts its first word. The predicted word is sent as input to the next step for the prediction of the second word, and the process continues until end of sequence word token is predicted.

During training, we have the actual inputs from the target language for the decoder, and hence we use the same for training instead of the predicted word from the earlier stage. The label for the target language at every instance is predicted through a SoftMax layer over all the words in the vocabulary of the target language.

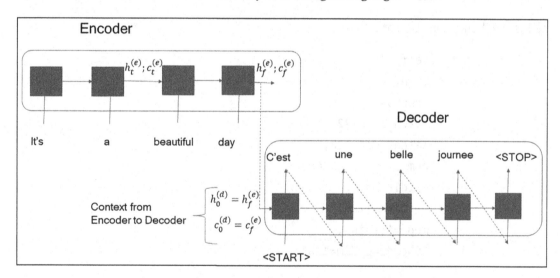

***Figure 4-16.** Seq2Seq neural machine translation model*

Please find the implementation of the Seq2Seq model in *Listing 4-6*.

Listing 4-6. Neural Machine Translation Using Seq2Seq Model

```python
import  tensorflow as tf
print(tf.__version__)
from tensorflow.keras.layers import Input, LSTM, Dense, Embedding
from tensorflow.keras import Model
from tensorflow.keras.models import load_model
import numpy as np
import codecs
import pandas as pd
import pickle
from elapsedtimer import ElapsedTimer
from  pathlib import Path
import os

# Machine Translation Model - English to French

class MachineTranslation:

    def __init__(self,
                path,
                epochs=20,
                batch_size=32,
                latent_dim=128,
                num_samples=40000,
                outdir=None,
                verbose=True,
                mode='train',
                embedding_dim=128):

        self.path = path
        self.epochs = epochs
        self.batch_size = batch_size
        self.latent_dim = latent_dim
        self.embedding_dim = embedding_dim
        self.num_samples = num_samples
```

```python
        self.outdir = outdir
        if self.outdir is not None:
            if not Path(self.outdir).exists():
                os.makedirs(Path(self.outdir))
        self.verbose = verbose
        self.mode = mode

    def read_input(self, path, num_samples=10e13):
        input_texts = []
        target_texts = []
        input_words = set()
        target_words = set()

        with codecs.open(path, 'r', encoding='utf-8') as f:
            lines = f.read().split('\n')

        for line in lines[: min(num_samples, len(lines) - 1)]:
            input_text, target_text = line.split('\t')  # \t as the start
            of sequence
            target_text = '\t ' + target_text + ' \n'  # \n as the end  of
            sequence
            input_texts.append(input_text)
            target_texts.append(target_text)
            for word in input_text.split(" "):
                if word not in input_words:
                    input_words.add(word)
            for word in target_text.split(" "):
                if word not in target_words:
                    target_words.add(word)

        return input_texts, target_texts, input_words, target_words

    def generate_vocab(self, path, num_samples, verbose=True):

        input_texts, target_texts, input_words, target_words =
        self.read_input(path, num_samples)
        input_words = sorted(list(input_words))
        target_words = sorted(list(target_words))
```

```python
        self.num_encoder_words = len(input_words)
        self.num_decoder_words = len(target_words)
        self.max_encoder_seq_length = max([len(txt.split(" ")) for txt in
        input_texts])
        self.max_decoder_seq_length = max([len(txt.split(" ")) for txt in
        target_texts])

        if verbose == True:
            print('Number of samples:', len(input_texts))
            print('Number of unique input tokens:', self.num_encoder_words)
            print('Number of unique output tokens:',
            self.num_decoder_words)
            print('Max sequence length for inputs:',
            self.max_encoder_seq_length)
            print('Max sequence length for outputs:',
            self.max_decoder_seq_length)

        self.input_word_index = dict(
            [(word, i) for i, word in enumerate(input_words)])
        self.target_word_index = dict(
            [(word, i) for i, word in enumerate(target_words)])
        self.reverse_input_word_dict = dict(
            (i, word) for word, i in self.input_word_index.items())
        self.reverse_target_word_dict = dict(
            (i, word) for word, i in self.target_word_index.items())

    def process_input(self, input_texts, target_texts=None, verbose=True):

        encoder_input_data = np.zeros(
            (len(input_texts), self.max_encoder_seq_length),
            dtype='float32')

        decoder_input_data = np.zeros(
            (len(input_texts), self.max_decoder_seq_length),
            dtype='float32')

        decoder_target_data = np.zeros(
            (len(input_texts), self.max_decoder_seq_length, 1),
            dtype='float32')
```

```
if self.mode == 'train':
    for i, (input_text, target_text) in enumerate(zip(input_texts,
    target_texts)):
        for t, word in enumerate(input_text.split(" ")):
            try:
                encoder_input_data[i, t] =
                self.input_word_index[word]
            except:
                encoder_input_data[i, t] = self.num_encoder_words

        for t, word in enumerate(target_text.split(" ")):
            # decoder_target_data is ahead of decoder_input_data by
              one timestep
            try:
                decoder_input_data[i, t] =
                self.target_word_index[word]
            except:
                decoder_input_data[i, t] = self.num_decoder_words
            if t > 0:
                # decoder_target_data will be ahead by one timestep
                # and will not include the start character.
                try:
                    decoder_target_data[i, t - 1] =
                    self.target_word_index[word]
                except:
                    decoder_target_data[i, t - 1] =
                    self.num_decoder_words
print(self.num_encoder_words)
print(self.num_decoder_words)
print(self.embedding_dim)
self.english_emb = np.zeros((self.num_encoder_words + 1,
self.embedding_dim))
self.french_emb = np.zeros((self.num_decoder_words + 1,
self.embedding_dim))
return encoder_input_data, decoder_input_data,
decoder_target_data, np.array(input_texts), np.array(
```

```
                target_texts)
        else:
            for i, input_text in enumerate(input_texts):
                for t, word in enumerate(input_text.split(" ")):
                    try:
                        encoder_input_data[i, t] = self.input_word_
                        index[word]
                    except:
                        encoder_input_data[i, t] = self.num_encoder_words

            return encoder_input_data, None, None, np.array
            (input_texts), None

    def train_test_split(self, num_recs, train_frac=0.8):
        rec_indices = np.arange(num_recs)
        np.random.shuffle(rec_indices)
        train_count = int(num_recs * 0.8)
        train_indices = rec_indices[:train_count]
        test_indices = rec_indices[train_count:]
        return train_indices, test_indices

    def model(self):
        # Encoder Model
        encoder_inp = Input(shape=(None,), name='encoder_inp')
        encoder_inp1 = Embedding(self.num_encoder_words + 1, self.
        embedding_dim, weights=[self.english_emb])(
            encoder_inp)
        encoder = LSTM(self.latent_dim, return_state=True, name='encoder')
        encoder_out, state_h, state_c = encoder(encoder_inp1)
        encoder_states = [state_h, state_c]

        # Decoder Model
        decoder_inp = Input(shape=(None,), name='decoder_inp')
        decoder_inp1 = Embedding(self.num_decoder_words + 1, self.
        embedding_dim, weights=[self.french_emb])(decoder_inp)
        decoder_lstm = LSTM(self.latent_dim, return_sequences=True, return_
        state=True, name='decoder_lstm')
```

```
    decoder_out, _, _ = decoder_lstm(decoder_inp1,
    initial_state=encoder_states)
    decoder_dense = Dense(self.num_decoder_words, activation='softmax',
    name='decoder_dense')
    decoder_out = decoder_dense(decoder_out)
    print(np.shape(decoder_out))
    # Combined Encoder Decoder Model
    model = Model([encoder_inp, decoder_inp], decoder_out)
    # Encoder Model
    encoder_model = Model(encoder_inp, encoder_states)
    # Decoder Model
    decoder_inp_h = Input(shape=(self.latent_dim,))
    decoder_inp_c = Input(shape=(self.latent_dim,))
    decoder_inp_state = [decoder_inp_h, decoder_inp_c]
    decoder_out, decoder_out_h, decoder_out_c = decoder_lstm(
    decoder_inp1, initial_state=decoder_inp_state)
    decoder_out = decoder_dense(decoder_out)
    decoder_out_state = [decoder_out_h, decoder_out_c]
    decoder_model = Model(inputs=[decoder_inp] + decoder_inp_state,
    outputs=[decoder_out] + decoder_out_state)

    return model, encoder_model, decoder_model

def decode_sequence(self, input_seq, encoder_model, decoder_model):
    # Encode the input as state vectors.
    states_value = encoder_model.predict(input_seq)

    # Generate empty target sequence of length 1.
    target_seq = np.zeros((1, 1))
    # Populate the first character of target sequence with the start
      character.
    target_seq[0, 0] = self.target_word_index['\t']

    # Sampling loop for a batch of sequences
    stop_condition = False
    decoded_sentence = ''
```

```python
    while not stop_condition:
        output_word, h, c = decoder_model.predict(
            [target_seq] + states_value)

        # Sample a token
        sampled_word_index = np.argmax(output_word[0, -1, :])
        try:
            sampled_char = self.reverse_target_word_dict[
            |sampled_word_index]
        except:
            sampled_char = '<unknown>'
        decoded_sentence = decoded_sentence + ' ' + sampled_char

        # Exit condition: either hit max length
        # or find stop character.
        if (sampled_char == '\n' or
                len(decoded_sentence) > self.max_decoder_seq_length):
            stop_condition = True

        # Update the target sequence (of length 1).
        target_seq = np.zeros((1, 1))
        target_seq[0, 0] = sampled_word_index

        # Update states
        states_value = [h, c]

    return decoded_sentence

# Run training

def train(self, encoder_input_data, decoder_input_data,
decoder_target_data):
    print("Training...")

    print(np.shape(encoder_input_data))
    print(np.shape(decoder_input_data))
    print(np.shape(decoder_target_data))

    model, encoder_model, decoder_model = self.model()
```

```
    model.compile(optimizer='rmsprop', loss='sparse_categorical_
    crossentropy')
    model.fit([encoder_input_data, decoder_input_data],
    decoder_target_data,
            batch_size=self.batch_size,
            epochs=self.epochs,
            validation_split=0.2)
    # Save model
    model.save(self.outdir + 'eng_2_french_dumm.h5')
    return model, encoder_model, decoder_model

def inference(self, model, data, encoder_model, decoder_model,
in_text):
    in_list, out_list = [], []
    for seq_index in range(data.shape[0]):
        input_seq = data[seq_index: seq_index + 1]
        decoded_sentence = self.decode_sequence(input_seq,
        encoder_model, decoder_model)
        print('-')
        print('Input sentence:', in_text[seq_index])
        print('Decoded sentence:', decoded_sentence)
        in_list.append(in_text[seq_index])
        out_list.append(decoded_sentence)
    return in_list, out_list

def save_models(self, outdir):
    self.model.save(outdir + 'enc_dec_model.h5')
    self.encoder_model.save(outdir + 'enc_model.h5')
    self.decoder_model.save(outdir + 'dec_model.h5')

    variables_store = {'num_encoder_words': self.num_encoder_words,
                       'num_decoder_words': self.num_decoder_words,
                       'max_encoder_seq_length': self.max_encoder_
                       seq_length,
                       'max_decoder_seq_length': self.max_decoder_
                       seq_length,
                       'input_word_index': self.input_word_index,
```

```
                    'target_word_index': self.target_word_index,
                    'reverse_input_word_dict': self.reverse_input_
                    word_dict,
                    'reverse_target_word_dict': self.reverse_target_
                    word_dict
                    }
        with open(outdir + 'variable_store.pkl', 'wb') as f:
            pickle.dump(variables_store, f)
            f.close()

    def load_models(self, outdir):
        self.model = load_model(outdir + 'enc_dec_model.h5')
        self.encoder_model = load_model(outdir + 'enc_model.h5')
        self.decoder_model = load_model(outdir + 'dec_model.h5')

        with open(outdir + 'variable_store.pkl', 'rb') as f:
            variables_store = pickle.load(f)
            f.close()

        self.num_encoder_words = variables_store['num_encoder_words']
        self.num_decoder_words = variables_store['num_decoder_words']
        self.max_encoder_seq_length = variables_store['max_encoder_seq_
        length']
        self.max_decoder_seq_length = variables_store['max_decoder_seq_
        length']
        self.input_word_index = variables_store['input_word_index']
        self.target_word_index = variables_store['target_word_index']
        self.reverse_input_word_dict = variables_store['reverse_input_
        word_dict']
        self.reverse_target_word_dict = variables_store['reverse_target_
        word_dict']

    def main(self):

        if self.mode == 'train':
            self.generate_vocab(self.path, self.num_samples,
            self.verbose)  # Generate the vocabulary
```

```
input_texts, target_texts, _, _ = self.read_input(self.path,
self.num_samples)
encoder_input_data, decoder_input_data, decoder_target_data,
input_texts, target_texts = \
    self.process_input(input_texts, target_texts, True)
num_recs = encoder_input_data.shape[0]
train_indices, test_indices = self.train_test_split(
num_recs, 0.8)
encoder_input_train_X, encoder_input_test_X =
encoder_input_data[train_indices,], encoder_input_data[
    test_indices,]
decoder_input_train_X, decoder_input_test_X =
decoder_input_data[train_indices,], decoder_input_data[
    test_indices,]
decoder_target_train_y, decoder_target_test_y =
decoder_target_data[train_indices,], decoder_target_data[
    test_indices,]
input_text_train, input_text_test = input_texts[train_indices],
input_texts[test_indices]
self.model, self.encoder_model, self.decoder_model = self.
train(encoder_input_train_X, decoder_input_train_X,
decoder_target_train_y)
in_list, out_list = self.inference(self.model,
encoder_input_test_X, self.encoder_model, self.decoder_model,
input_text_test)
out_df = pd.DataFrame()
out_df['English text'] = in_list
out_df['French text'] = out_list
out_df.to_csv(self.outdir + 'hold_out_results_val.csv',
index=False)
self.save_models(self.outdir)

else:
    self.load_models(self.outdir)
    input_texts, _, _, _ = self.read_input(self.path,
    self.num_samples)
```

```
            encoder_input_data, _, _, input_texts, _ = \
                self.process_input(input_texts, '', True)
            in_list, out_list = self.inference(self.model,
            encoder_input_test_X, self.encoder_model,
                                        self.decoder_model,
                                        input_text_test)
            out_df = pd.DataFrame()
            out_df['English text'] = in_list
            out_df['French text'] = out_list
            out_df.to_csv(self.outdir + 'results_test.csv', index=False)

translator = MachineTranslation(path='/home/santanu/ML_DS_Catalog-/Machine
Translation/fra-eng/fra.txt', outdir='/home/santanu/machine_translation/
output',epochs=10)
translator.main()

---output--

2.9.1
Number of samples: 40000
Number of unique input tokens: 8658
Number of unique output tokens: 16297
Max sequence length for inputs: 7
Max sequence length for outputs: 16
-
Input sentence: It's really bright.
Decoded sentence:  C'est vraiment bon.
```

Limitation of the Seq2Seq Model for Machine Translation

One of the limitations of Seq2Seq model for machine translation stems from the fact that the entire source language sequence is encoded into final hidden state $h_f^{(e)}$ and final cell state $c_f^{(e)}$ of the encoder LSTM. While this works well for short input sequences of length up to 15 to 20, the dimension of the hidden and cell state might not be enough to store the entire context of longer sequences. In general context from words that appear earlier in the source language sequence might not be captured well by the encoder as the sequence gets long due to the vanishing-gradient problems during training.

This is where attention comes in handy to provide additional context to the decoder from the source sequence processing. In attention, the intermediate hidden states corresponding to each time step is not thrown away but are used as context for predicting the target words at every time step. Attention figures out which source language words are related to a target word by making a direct connection and hence avoids loss of context from source to target. We discuss about attention and its various forms in the next section.

Attention

Although recurrent neural networks such as LSTMs and GRUs have been the default choice in modeling sequential data such as in language modeling and language translation, they do come with their set of limitations. Some of the limitations are outlined in the following:

- Recurrent neural network units cannot take advantage of any sort of parallelism because of the inherent sequential nature of processing. Hence processing very long sequences using recurrent neural network units can become prohibitory slow. In the age of GPUs for parallel processing, this comes as a major setup.

- Recurrent neural networks in application such as machine translation does not provide a natural mechanism of alignment between source words and phrases to target words and phrases. In such cases, the entire source sequence context passed is used for decoding the target sequence. This is limiting in a sense that the entire source sequence might be hard to encode in a fixed size hidden state of the encoder.

Both problems above are addressed by attention which provides a way to capture global dependencies between sequence elements within the same sequence (e.g., next-word prediction) or different sequences (e.g., machine translation). Thus, attention provides a mechanism of capturing dependencies between sequence elements without regard to their distances.

There are primarily two forms of attention—**self-attention** and **cross-attention.**

In the context of sequence of words in a sentence, self-attention would gather for each word context from other words in a sentence based on their similarities. Generally, these similarities are captured through dot products of the word with the other words in the sentence.

Cross-attention tries to capture similarity between elements of two different sequences. In the context of Seq2Seq model in machine translation, attention can be used to get the relevant context from source language word embeddings by attending to the current target language word embedding (mostly has the summarized information of all words prior to it through self-attention) for predicting the next word.

Now that we have intuitively defined attention using references to sequences of words, let us try to put some mathematical structure of attention.

There are three key components in the attention—keys (**K**), value (**V**), and query (**Q**). Attention can be defined as a mapping from the keys, value, and query tuple to an output which is the weighted sum of the values.

Scaled Dot Product Attention

The mapping for scaled dot product attention can be written as follows:

$$Attention(K,Q,V) = softmax\left(\frac{QK^{T}}{\sqrt{k}}\right)V$$

Here Q stands for the vector embeddings of queries, and hence Q is a matrix. Similarly, K stands for the vector embeddings of keys, while V stands for the vector of values. The dimension k of the key embeddings and the query embeddings must be the same for the scaled dot product attention to work.

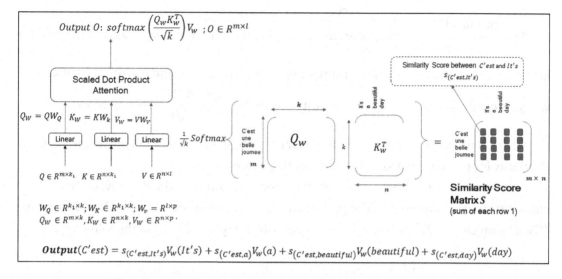

Figure 4-17. *Illustration of attention mechanism*

The very common scaled dot product attention mechanism is illustrated in Figure 4-17.

The keys (K), queries (Q), and values (V) prior to the attention are generally conditioned using projection matrices W_K, W_Q and W_V through linear layers. Their representation post the conditioning can be represented as follows:

$$Q_W = QW_Q$$

$$K_W = KW_K$$

$$V_W = VW_V$$

Again, coming to the machine translation example of translating the English sentence **"It's a beautiful day"** to the French sentence **"C'est une belle journee,"** the queries are each of the words of the French sentence, while the keys are each of the words of the English sentence **"It's a beautiful day."**

Through the cross-attention mechanism, we would like to see how we can represent query sequence words as a weighted average of the key sequence word representation. For NLP applications such as machine translation, the key sequence word representations also act as the value representation. We call these representation embeddings too.

The $\dfrac{softmax\left(Q_W K_W^T\right)}{\sqrt{k}}$ captures the similarity of each query sequence words with the

key sequence words. Hence if the number of query words is m and the number of key

words is n, the dimension of the similarity matrix $S = \dfrac{softmax\left(Q_W K_W^T\right)}{\sqrt{k}}$ is $m \times n$.

So, each row of the similarity matrix can be thought of a probability distribution of similarity for each query word over the key sequence words.

If we go back to the English to French translation example, the first row of this similarity matrix S would give the probability distribution of similarity of the query word "C'est" with each of English words **"It's a beautiful day"** as shown in the following.

$$S[0,:] = \left[\, s_{(C'est,It's)} \quad s_{("C'est,a)} \quad s_{("C'est,beatiful)} \quad s_{("C'est,day)} \,\right]$$

The multiplication of the similarity matrix S with the value representation of the keys V_k expresses each query word as a linear sum of the value representations or embeddings of the key words. Hence the output O of attention can be expressed as follows:

$$O = SV_W = \dfrac{softmax\left(Q_W K_W^T\right)}{\sqrt{k}} V_W$$

The first row of the output would correspond to **"C_est"** and can be expressed as follows:

$$O[0,:] = \boldsymbol{Output}(C'est)$$

$$= s_{(C'est,It's)} V_w\left(It's\right) + s_{(C'est,a)} V_w\left(a\right) + s_{(C'est,beautiful)} V_w\left(beautiful\right) + s_{(C'est,day)} V_w\left(day\right)$$

Multihead Attention

Instead of just a single attention, which can often be limiting, multihead attention performs multiple parallel attentions with different projection matrices. If there are r parallel attentions each having projection weights $W_Q^{(i)}$, $W_K^{(i)}$, and W_V^i, then the representation from the attention head i can be expressed as follows:

$$head(i) = Attention\left(QW_Q^{(i)}, KW_K^{(i)}, VW_V^{(i)}\right)$$

The output representation from each of the head is concatenated together and then projected through one final linear layer W_O to give it the output O its final dimension.

$$O = concat\left[head(0), head(1)..head(r-1)\right]W_O$$

We reduce the output dimension of each of the attention head by a factor of r to get the desired output dimension from the multihead attention. So, if we want the final dimension to be l, we would choose $W_V^{(i)} \in R^{n \times t}$ such that $t = \dfrac{l}{r}$. This ensures that the output dimension of the query words post the output concatenation sums to l. Multihead attention in general helps to represent the words in multiple different subspaces and thus fosters a richer representation.

Transformers

In this section, we will discuss about transformers which was first introduced in the paper "Attention Is All You Need" (`https://arxiv.org/abs/1706.03762?context=cs`). As is obvious from the paper name, the transformers are based on the attention mechanism that we have discussed in some details in the earlier section. Most of the sequential problems in deep learning uses an encoder-decoder architecture, and the transformer is no different. If we take the neural language translation problem Seq2Seq architecture, the input sequence of words $x = (x_1, x_2, ... x_m)$ are encoded to a sequence of latent representations $z = (z_1, z_2, ... z_m)$ using a RNN architecture such as LSTM. Using the z as the context, a decoder RNN then outputs the target sequence $(y_1, y_2, ... y_n)$ in an autoregressive fashion. Hence for the prediction of the word y_k, the decoder uses the input context z as well as the context from the first (k-1) predicted words $y_1, y_2...., y_{k-1}$.

The transformer follows the similar principle, but instead of using a RNN to capture the sequential context, the transformer instead captures context through attention.

The transformer architecture is illustrated in Figure 4-18.

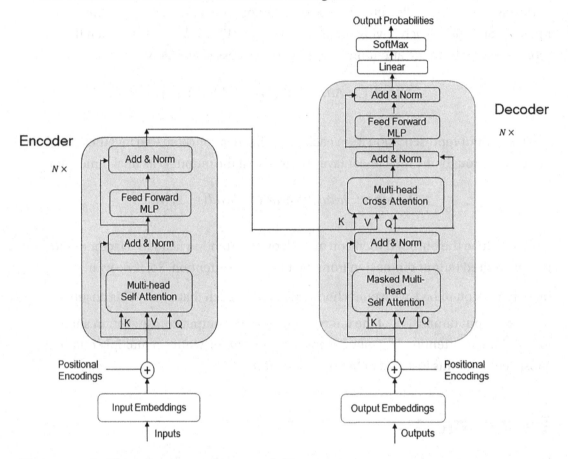

Figure 4-18. *Transformer architectural diagram*

As we can see from Figure 4-18, the transformer has an encoder and decoder module as its main components which we discuss in the following:

Encoder

The encoder has N identical units each of which has a multi-layer self-attention sublayer and a feed-forward MLP sublayer. A residual connection like that of ResNet is applied to each of the sublayers followed by layer normalization. The keys, value, and query are the same for the self-attention layer, and the idea is to represent the embedding of each word in the input sequence as weighted combination of the embedding of the rest of the words, where the weights are determined through self-attention.

The input to the first of the N encoder units can be pretrained word embeddings or embeddings through a learnable embeddings layer plus the positional encodings. Through recurrent neural network and convolution, we encode some form of order of the sequence. Since attention cannot encode the same, the transformers use the positional encodings to provide some notion of the relative or absolute position of the words in the sequence. We will discuss more about positional encodings in subsequent sections.

Decoder

The decoder has the same number of identical units as that of the encoder. Each of the units in decoder has three sublayers—the masked self-attention unit, the cross-attention unit, and feed-forward MLP unit.

For a given word at a location t in the output sequence, the masked self-attention tries to capture context from all words till the location t during training. The masked version of the attention prevents the scheme from getting information from words post the location t. The cross-attention helps extract relevant input sequence context at each given output position t for the prediction of the $t + 1$ word of the output sequence. For the cross-attention, the query consists of the target sequence word embeddings from the self-attention, while the keys and values are the input sequence word embeddings from the encoder. Each of the three sublayers has a residual connection followed by layer normalization like that of the encoder.

The input to the first of the N decoder is either pretrained word embeddings of the output sequence or embeddings through a learnable embeddings layer followed by positional encoding.

Embeddings from each of the N encoder units cross attend to the corresponding decoder unit.

Positional Encoding

As discussed earlier, the function of the positional encodings is to provide some notion of the relative or absolute position of the words in the sequence as we do not have the luxury of recurrent neural network or convolution in an attention-based mechanism.

The positional encodings used in the transformers described in "Attention Is All You Need" uses the following functions to encode position.

$$PE(p,2i) = \sin\left(\frac{p}{10000^{\left(\frac{2i}{dmodel}\right)}}\right)$$

$$PE(p,2i+1) = \cos\left(\frac{p}{10000^{\left(\frac{2i}{dmodel}\right)}}\right)$$

In the preceding equations, p stands for the position of the word in the sequence, and i denotes the dimension. So, each dimension of the positional encoding is represented as a sinusoidal.

Final Output

The final linear and SoftMax takes in the output of the decoder at the last stage and predicts the probability of a given target word at location t based on the $t-1$ output sequence word embeddings as well as the input sequence word embeddings.

Now that we have gone through all the relevant concepts of the transformer architecture, let us look at the TensorFlow code for each of the functionalities.

We begin with positional encoding in Listing 4-7.1. As discussed earlier, the positional encoding function should be able to take the position of the element in the sequence and the specific dimension of the representation and give out the positional encoding accordingly. Even dimensions follow a sin profile, while odd dimensions follow a cos profile in the positional encoding scheme.

Listing 4-7.1. Positional Encoding

```python
def get_angles(pos, i, dim):
    angle_rates = 1 / np.power(10000, (2 * (i//2)) / np.float32(dim))
    return pos * angle_rates
def positional_encoding(position, dim):
    angle_rads = get_angles(np.arange(position)[:, np.newaxis],
                            np.arange(dim)[np.newaxis, :],
                            dim)
    # apply sin to even indices in the array; 2i
    angle_rads[:, 0::2] = np.sin(angle_rads[:, 0::2])
    # apply cos to odd indices in the array; 2i+1
    angle_rads[:, 1::2] = np.cos(angle_rads[:, 1::2])
    pos_encoding = angle_rads[np.newaxis, ...]
    return tf.cast(pos_encoding, dtype=tf.float32)
# Sample Forward pass
n, d = 2048, 512

pos_encoding = positional_encoding(n, d)
print(f"Shape of the positional encodings : {np.shape(pos_encoding)}")
n, d = 2048, 512

pos_encoding = positional_encoding(n, d)
print(f"Shape of the positional encodings : {np.shape(pos_encoding)}")
--output--
Shape of the positional encodings (1. 2048, 512)
```

When dealing with encoder and decoder in transformer, we assume a fixed size sequence length. Hence, it is imperative that we perform masking to mask out entries that are beyond the sequence length. This ensures the model does not consider padding as input. Similarly for the decoder, it is important to mask out target words at timestamp $(t + 1)$ and beyond so that the current time step prediction at t is not influenced by the words that appear after it.

The logic for padding-specific masking and the look ahead masking is outlined in Listing 4-7.2.

Listing 4-7.2. Masking Routines for Handling Padding and to Remove Visibility of Future Words in Target

```
def create_padding_mask(seq):

    seq = tf.cast(tf.math.equal(seq, 0), tf.float32)

    # add extra dimensions to add the padding

    # to the attention logits.

    return seq[:, tf.newaxis, tf.newaxis, :]  # (batch_size, 1, 1, seq_len)

def create_look_ahead_mask(size):

    mask = 1 - tf.linalg.band_part(tf.ones((size, size)), -1, 0)

    return mask  # (seq_len, seq_len)

# Sample forward pass
x = tf.constant([[1, 2, 0, 0, 1], [1, 6, 7, 0, 0], [0, 1, 1, 4, 5]])
print(f"Padding mask: {create_padding_mask(x)}")

print(create_look_ahead_mask(3))

--output--
```
 # Padding Mask
```
tf.Tensor(
[[[[0. 0. 1. 1. 0.]]]

 [[[0. 0. 0. 1. 1.]]]
```

 # Mask to prevent look ahead

```
[[[1. 0. 0. 0. 0.]]]], shape=(3, 1, 1, 5), dtype=float32)
tf.Tensor(
[[0. 1. 1.]
 [0. 0. 1.]
 [0. 0. 0.]], shape=(3, 3), dtype=float32)
```

Next, we implement the scaled dot product scheme like what has been illustrated in "Attention Is All You Need" paper. The same has been implemented in Listing 4-7.3. Beyond the query, keys, and value inputs associated with attention, we take another input mask. Mask helps not to attend to unwanted tokens/words in the sequence.

Listing 4-7.3. Scaled Dot Product for Attention

```
def attention_scaled_dot(Q, K, V, mask):
    qk = tf.matmul(Q, K, transpose_b=True)
    _dim_ = tf.cast(tf.shape(K)[-1], tf.float32)
    scaled_qk = qk /tf.math.sqrt(_dim_)

    if mask is not None:
        scaled_qk += (mask * -1e9)

    attention_wts = tf.nn.softmax(scaled_qk, axis=-1)

    out = tf.matmul(attention_wts, V)

    return out, attention_wts
```

The multihead attention splits up the input dimension into multiple heads as defined by num_heads. It takes as input the embedding dimension given by "dim" and "num_heads." The detailed listing for multi_head_attention is illustrated in Listing 4-7.4.

Listing 4-7.4. Multihead Attention

```
class multi_head_attention(Layer):

    def __init__(self,*, dim, num_heads):
        super(multi_head_attention, self).__init__()
        self.num_heads = num_heads
        self.dim = dim
```

```python
        assert self.dim % self.num_heads == 0

        self.head_dim = self.dim // self.num_heads

        self.Wq = layers.Dense(self.dim)
        self.Wk = layers.Dense(self.dim)
        self.Wv = layers.Dense(self.dim)

        self.dense = layers.Dense(self.dim)

    def split_heads(self, x, batch_size):
        """Split the last dimension into (num_heads, depth).
        Transpose the result such that the shape is (batch_size, num_heads,
        seq_len, depth)
        """
        x = tf.reshape(x, (batch_size, -1, self.num_heads, self.head_dim))

        return tf.transpose(x, perm=[0, 2, 1, 3])

    def call(self, V, K, Q, mask):

        batch_size = tf.shape(Q)[0]

        Q = self.Wq(Q)  # (batch_size, seq_len, dim)
        K = self.Wk(K)  # (batch_size, seq_len, dim)
        V = self.Wv(V)  # (batch_size, seq_len, dim)

        Q = self.split_heads(Q, batch_size)  # (batch_size, num_heads,
        seq_len_q, head_dim)
        K = self.split_heads(K, batch_size)  # (batch_size, num_heads,
        seq_len_k, head_dim)
        V = self.split_heads(V, batch_size)  # (batch_size, num_heads,
        seq_len_v, head_dim)

        scaled_attention, attention_weights = attention_scaled_dot(
            Q, K, V, mask)

        scaled_attention = tf.transpose(scaled_attention, perm=[0, 2, 1, 3])
        # (batch_size, seq_len_q, num_heads, head_dim
```

```
        concat_attention = tf.reshape(scaled_attention, (batch_size, -1,
                                       self.dim))
                                       # (batch_size, seq_len_q, dim)

        output = self.dense(concat_attention)  # (batch_size, seq_
        len_q, dim)

        return output, attention_weights

# Sample forward pass
mha_layer = multi_head_attention(dim=512, num_heads=8)
x = tf.random.uniform((1, 30, 512))  # (batch_size, sequence_len,dim)
out, attn = mha_layer(V=x, K=x, Q=x, mask=None)
print(out.shape, attn.shape)

--output-

(TensorShape([1, 30, 512]), TensorShape([1, 8, 30, 30]))
```

The feed-forward pointwise MLP for both the encoder and decoder can be defined as in Listing 4-7.5.

Listing 4-7.5. Pointwise Feed-Forward Network

```
def pointwise_mlp(dim, hidden_dim,activation='relu'):

    return tf.keras.Sequential([

        layers.Dense(hidden_dim, activation=activation),  # (batch_size,
        seq_len, hidden)

        layers.Dense(dim)  # (batch_size, seq_len, dim)

    ])
```

Next, we implement the encoder layer as illustrated in Listing 4-7.6. As discussed earlier, the encoder layer consists of two major blocks—self-attention and pointwise feed-forward network. In general stack of encoder layer forms the transformer encoder.

Listing 4-7.6. The Encoder Layer Illustration

```
class encoder_layer(Layer):
    def __init__(self,*, dim, num_heads, hidden_dim, dropout=0.1):
        super(encoder_layer, self).__init__()

        self.mha = multi_head_attention(dim=dim, num_heads=num_heads)
        self.mlp = pointwise_mlp(dim,hidden_dim=hidden_dim)

        self.layernorm_1 = layers.LayerNormalization(epsilon=1e-6)
        self.layernorm_2 = layers.LayerNormalization(epsilon=1e-6)

        self.dropout_1 = layers.Dropout(dropout)
        self.dropout_2 = layers.Dropout(dropout)

    def call(self, x, training, mask):
        # Self Attention
        attn_output, _ = self.mha(x, x, x, mask)  # (batch_size,
        input_seq_len, dim)
        attn_output = self.dropout_1(attn_output, training=training)
        out_1 = self.layernorm_1(x + attn_output)  # (batch_size,
        input_seq_len, dim)

        mlp_output = self.mlp(out_1)  # (batch_size, input_seq_len, dim)
        mlp_output = self.dropout_2(mlp_output, training=training)
        out_2 = self.layernorm_2(out_1 + mlp_output)  # (batch_size,
        input_seq_len, dim)

        return out_2

# Sample Forward pass
enc_layer = encoder_layer(dim=512, num_heads=8, hidden_dim=2048)
out_e = enc_layer(tf.random.uniform((32, 30, 512)), False, None)
print(f"Encoder Shape:{out_e.shape}")  # (batch_size, input_seq_len, dim)

--output-

Encoder Shape:(32, 30, 512)
```

The decoder layer while very similar to the encoder layer has an extra block pertaining to cross-attention to extract information from the encoder inputs. The detailed implementation of the decoder layer is illustrated in the following (see Listing 4-7.7).

Listing 4-7.7. The Decoder Layer Illustration

```python
class decoder_layer(Layer):

    def __init__(self,*, dim, num_heads, hidden_dim, dropout=0.1):

        super(decoder_layer, self).__init__()

        self.mha_1 = multi_head_attention(dim=dim, num_heads=num_heads)
        # For self attention

        self.mha_2 = multi_head_attention(dim=dim, num_heads=num_heads)
        # For Cross attention

        self.mlp = pointwise_mlp(dim,hidden_dim=hidden_dim)

        self.layernorm_1 = layers.LayerNormalization(epsilon=1e-6)

        self.layernorm_2 = layers.LayerNormalization(epsilon=1e-6)

        self.layernorm_3 = layers.LayerNormalization(epsilon=1e-6)

        self.dropout_1 = layers.Dropout(dropout)

        self.dropout_2 = layers.Dropout(dropout)

        self.dropout_3 = layers.Dropout(dropout)

    def call(self, x, encoder_out, training, look_ahead_mask,
    padding_mask):

        # Self attention

        attn_1, attn_wts_block_1 = self.mha_1(x, x, x, mask=look_ahead_
        mask)  # (batch_size, target_seq_len, dim)

        attn_1 = self.dropout_1(attn_1, training=training)

        out_1 = self.layernorm_1(attn_1 + x)
```

```
        # Cross attention

        attn_2, attn_wts_block_2 = self.mha_2(encoder_out,encoder_
        out,out_1, mask=padding_mask)  # (batch_size, target_seq_len, dim)

        attn_2 = self.dropout_2(attn_2, training=training)

        out_2 = self.layernorm_2(attn_2 + out_1)  # (batch_size, target_
        seq_len, dim)

        # Feed forward MLP

        mlp_output = self.mlp(out_2)  # (batch_size, target_seq_len, dim)

        mlp_output = self.dropout_3(mlp_output, training=training)

        out_3 = self.layernorm_3(mlp_output + out_2)  # (batch_size,
        target_seq_len, dim)

        return out_3, attn_wts_block_1, attn_wts_block_2

# Sample forward run

dec_layer = decoder_layer(dim=512, num_heads=8, hidden_dim=2048)

out_d, _, _ = dec_layer(tf.random.uniform((32, 30, 512)), out_e, False,
None, None)

print("Decoder output shape",out_d.shape)  # (batch_size, target_seq_
len, dim)


--ouptut-
Decoder output shape (32, 30, 512)
```

The encoder block defined next (see below) in Listing 4-7.8 is nothing but a stack of encoder layers.

Listing 4-7.8. Encoder as a Stack of Encoder Layers

```python
class encoder(Layer):
    def __init__(self,*, num_layers,dim, num_heads, hidden_dim,
    input_vocab_size, dropout=0.1,max_tokens=2048):

        super(encoder, self).__init__()

        self.dim = dim

        self.num_layers = num_layers

        self.max_tokens = max_tokens

        self.num_heads = num_heads

        self.hidden_dim = hidden_dim

        self.embedding = layers.Embedding(input_vocab_size, dim)

        self.pos_encoding = positional_encoding(self.max_tokens, self.dim)

        self.encoder_layers = [

            encoder_layer(dim=self.dim, num_heads=self.num_heads,
            hidden_dim=hidden_dim,dropout=dropout)

            for _ in range(self.num_layers)]

        self.dropout = layers.Dropout(dropout)

    def call(self, x, training, mask):

        input_seq_len = tf.shape(x)[1]

        x = self.embedding(x)  # (batch_size, input_seq_len, dim)

        x *= tf.math.sqrt(tf.cast(self.dim, tf.float32))

        x += self.pos_encoding[:, :input_seq_len, :]

        x = self.dropout(x, training=training)

        for i in range(self.num_layers):
```

```
        x = self.encoder_layers[i](x, training, mask)

    return x  # (batch_size, input_seq_len, dim)

# Sample forward run
enc = encoder(num_layers=6, dim=512, num_heads=8, hidden_dim=2048,

                    input_vocab_size=1000)

X = tf.random.uniform((32, 30), dtype=tf.int64, minval=0, maxval=200)

encoder_out = enc(X, training=False, mask=None)

print(f"Encoder output shape: {encoder_out.shape}")  # (batch_size,
input_seq_len, dim)


--output -
Encoder output shape: (32, 30, 512)
```

Similar to the encoder block next, we illustrate the decoder block in Listing 4-7.9 (see below). Again, the decoder block is nothing but a stack of decoder layers.

Listing 4-7.9. Decoder as a Stack of Decoder Layers

```
class decoder(Layer):
    def __init__(self,*, num_layers, dim, num_heads, hidden_dim,
    target_vocab_size, dropout=0.1,max_tokens=2048):
        super(decoder, self).__init__()

        self.dim = dim
        self.num_layers = num_layers
        self.max_tokens = max_tokens
        self.num_layers = num_layers

        self.embedding = layers.Embedding(target_vocab_size, self.dim)
        self.pos_encoding = positional_encoding(self.max_tokens, self.dim)
```

```
        self.decoder_layers = [
            decoder_layer(dim=dim, num_heads=num_heads, hidden_dim=hidden_
            dim,dropout=dropout)
            for _ in range(num_layers)]

        self.dropout = layers.Dropout(dropout)

    def call(self, x, encoder_out, training, look_ahead_mask,
    padding_mask):

        output_seq_len = tf.shape(x)[1]

        attention_wts_dict = {}

        x = self.embedding(x)  # (batch_size, target_seq_len, dim)
        x *= tf.math.sqrt(tf.cast(self.dim, tf.float32))
        x += self.pos_encoding[:, :output_seq_len, :]

        x = self.dropout(x, training=training)

        for i in range(self.num_layers):
            x, block_1, block_2 = self.decoder_layers[i](x, encoder_out,
            training, look_ahead_mask, padding_mask)

            attention_wts_dict[f'decoder_layer{i}_block1'] = block_1
            attention_wts_dict[f'decoder_layer{i}_block2'] = block_2

        # x.shape == (batch_size, target_seq_len, dim)
        return x, attention_wts_dict

# Sample Forward pass
dec = decoder(num_layers=6, dim=512, num_heads=8, hidden_dim=2048,
                        target_vocab_size=1200)
X_target = tf.random.uniform((32, 45), dtype=tf.int64, minval=0,
maxval=200)
```

```
out_decoder, attn_wts_dict = dec(X_target, encoder_out=encoder_out,
                                 training=False, look_ahead_mask=None,
                                 padding_mask=None)

print(f"Decoder output shape: {out_decoder.shape}")
print(f"Cross Attention shape :{attn_wts_dict['decoder_layer0_block2'].
shape}")

--output -

Decoder output shape: (32, 45, 512)
Cross Attention shape :(32, 8, 45, 30)
```

Finally, we put the implemented positional encodings, encoder, decoder, and masking routines to form the transformer architecture. See Listing 4-7.10 for the detailed implementation.

Listing 4-7.10. Putting It All Together to Create Transformer

```
from tensorflow.keras import Model

class transformer(Model):
    def __init__(self,*, num_layers, dim, num_heads, hidden_dim, input_
    vocab_size,
                target_vocab_size, dropout=0.1,max_tokens_input=20,max_
                tokens_output=20):
        super(transformer,self).__init__()
        self.encoder = encoder(num_layers=num_layers, dim=dim,
                               num_heads=num_heads, hidden_dim=hidden_dim,
                               input_vocab_size=input_vocab_size,
                               dropout=dropout,max_tokens=max_tokens_input)

        self.decoder = decoder(num_layers=num_layers, dim=dim,
                               num_heads=num_heads, hidden_dim=hidden_dim,
                               target_vocab_size=target_vocab_size,
                               dropout=dropout,max_tokens=max_
                               tokens_output)

        self.final_layer = layers.Dense(target_vocab_size)
```

```python
    def call(self, inputs, training):

        input_, target_ = inputs

        padding_mask, look_ahead_mask = self.create_masks(input_,target_)

        encoder_output = self.encoder(input_, training, padding_mask)
        # (batch_size, inp_seq_len, dim)

        decoder_output, attn_wts_dict = self.decoder(
            target, encoder_output, training, look_ahead_mask,
            padding_mask)

        final_output = self.final_layer(decoder_output)  # (batch_size,
        target_seq_len, target_vocab_size)

        return final_output, attn_wts_dict

    def create_masks(self, input_, target_):
        padding_mask = create_padding_mask(input_)

        look_ahead_mask = create_look_ahead_mask(tf.shape(target_)[1])
        decoder_target_padding_mask = create_padding_mask(target_)
        look_ahead_mask = tf.maximum(decoder_target_padding_mask,
        look_ahead_mask)

        return padding_mask, look_ahead_mask

# Sample Forward pass
model = transformer(
    num_layers=3, dim=512, num_heads=8, hidden_dim=2048,
    input_vocab_size=1000, target_vocab_size=1200,max_tokens_input=30,max_
    tokens_output=45)

input = tf.random.uniform((32, 30), dtype=tf.int64, minval=0, maxval=200)
target = tf.random.uniform((32, 45), dtype=tf.int64, minval=0, maxval=200)

model_output, _ = model([input,target], training=False)
```

```
print(f"Transformer Output :{model_output.shape}")# (batch_size,
target_seq_len, target_vocab_size)
print(model.summary())

--output--

Transformer Output :(32, 45, 1200)
Model: "transformer_16"
```

Layer (type)	Output Shape	Param #
encoder_20 (encoder)	multiple	9969152
decoder_19 (decoder)	multiple	13226496
dense_1256 (Dense)	multiple	615600

```
Total params: 23,811,248
Trainable params: 23,811,248
Non-trainable params: 0
```

Hope the code implementation of the major components of the transformer helps the readers better understand the various implementation subtleties associated with them. Readers are further advised to apply the transformer solution to different domain problems to acquire an in-depth understanding. In general, pretrained version of the transformer architectures is readily available in several deep learning frameworks and offerings.

Summary

After this chapter, the reader is expected to have gained significant insights into the working principles of recurrent neural networks and their variants. Also, the reader should be able to implement RNN networks using TensorFlow with relative ease. Vanishing- and exploding-gradient problems with RNNs pose a key challenge in training them effectively, and thus many powerful versions of RNNs have evolved that take care of the problem. LSTM, being a powerful RNN architecture, is widely used in the community and has almost replaced the basic RNN. The reader is expected to know the uses and advantages of these advanced techniques, such as LSTM, GRU, and so forth, so that they can be accordingly implemented based on the problem. Pretrained Word2Vec and GloVe word-vector embeddings are used by several RNN, LSTM, and other networks so that the input word to the model at each sequence step can be represented by its pretrained Word2Vec or GloVe vector instead of learning these word-vector embeddings within the recurrent neural network itself.

In the next chapter, we will take up restricted Boltzmann machines (RBMs), which are energy-based neural networks, and various autoencoders as part of unsupervised deep learning. Also, we will discuss deep-belief networks, which can be formed by stacking several restricted Boltzmann machines and training such networks in a greedy manner, and collaborative filtering through RBMs. I look forward to your participation in the next chapter.

## CHAPTER 5

Unsupervised Learning with Restricted Boltzmann Machines and Autoencoders

Unsupervised learning is a branch of machine learning that tries to find hidden structures within unlabeled data and derive insights from it. Clustering, data dimensionality-reduction techniques, noise reduction, segmentation, anomaly detection, fraud detection, and other rich methods rely on unsupervised learning to drive analytics. Today, with so much data around us, it is impossible to label all data for supervised learning. This makes unsupervised learning even more critical. Restricted Boltzmann machines and autoencoders are unsupervised methods that are based on artificial neural networks. They have a wide range of uses in data compression and dimensionality reduction, noise reduction from data, anomaly detection, generative modeling, collaborative filtering, and initialization of deep neural networks, among other things. We will go through these topics in detail and then touch upon a couple of unsupervised preprocessing techniques for images, namely, PCA (principal component analysis) whitening and ZCA whitening. Also, since restricted Boltzmann machines use sampling techniques during training, I have briefly touched upon Bayesian inference and Markov chain Monte Carlo sampling for reader's benefit.

© Santanu Pattanayak 2023
S. Pattanayak, *Pro Deep Learning with TensorFlow 2.0*, https://doi.org/10.1007/978-1-4842-8931-0_5

Boltzmann Distribution

Restricted Boltzmann machines are energy models based on the Boltzmann distribution law of classical physics, where the state of particles of any system is represented by their generalized coordinates and velocities. These generalized coordinates and velocities form the phase space of the particles, and the particles can be in any location in the phase space with specific energy and probability. Let us consider a classical system with N gas molecule particles, and let the generalized position and velocity of any particle be represented by $r \in R^{3 \times 1}$ and $v \in R^{3 \times 1}$, respectively. The location of the particle in the phase space can be represented by (r, v). Every such possible value of (r, v) that the particle can take is called a *configuration* of the particle. Further, all the N particles are identical in the sense that they are equally likely to take up any state. Given such a system at thermodynamic temperature T, the probability of any such configuration is as follows:

$$P(r,v) \propto e^{-\frac{E(r,v)}{KT}}$$

$E(r, v)$ is the energy of any particle at configuration (r, v), and K is the Boltzmann constant. Hence, we see that the probability of any configuration in the phase space is proportional to the exponential of the negative of the energy divided by the product of the Boltzmann constant and the thermodynamic temperature. To convert the relationship into an equality, the probability needs to be normalized by the sum of the probabilities of all the possible configurations. If there are M possible phase-space configurations for the particles, then the probability of any generalized configuration (r, v) can be expressed as follows:

$$P(r,v) = \frac{e^{-\frac{E(r,v)}{KT}}}{Z}$$

where Z is the partition function given by the following:

$$Z = \sum_{i=1}^{M} e^{-\frac{E((r,v)_i)}{KT}}$$

There can be several values of r and v separately. However, M denotes the total number of possible unique combinations of r and v. In the preceding equation, we have in general represented the i-th unique combination of r and v as $(r, v)_i$. If r can take up n distinct coordinate values whereas v can take up m distinct velocity values, then the total number of possible configurations $M = n \times m$. In such cases, the partition function can also be expressed as follows:

$$Z = \sum_{j=1}^{m} \sum_{i=1}^{n} e^{-\frac{E\left(r_i, v_j\right)}{KT}}$$

The thing to note here is that the probability of any configuration is higher when its associated energy is low. For the gas molecules, it's intuitive as well given that high-energy states are always associated with unstable equilibrium and hence are less likely to retain the high-energy configuration for long. The particles in the high-energy configuration will always be in a pursuit to occupy much more stable low-energy states.

If we consider two configurations $s_1 = (r_1, v_1)$ and $s_2 = (r_2, v_2)$, and if the number of gas molecules in these two states are N_1 and N_2, respectively, then the probability ratio of the two states is a function of the energy difference between the two states:

$$\frac{N_1}{N_2} = \frac{P\left(r_1, v_1\right)}{P\left(r_2, v_2\right)} = e^{-\frac{E(r_1, v_1) - (r_2, v_2)}{KT}}$$

We will digress a little now and briefly discuss Bayesian inference and Markov chain Monte Carlo (MCMC) methods since restricted Boltzmann machines use sampling through MCMC techniques, especially Gibbs sampling, and some knowledge of these would go a long way toward helping the readers appreciate the working principles of restricted Boltzmann machines.

Bayesian Inference: Likelihood, Priors, and Posterior Probability Distribution

As discussed in Chapter 1, whenever we get data, we build a model by defining a likelihood function over the data conditioned on the model parameters and then try to maximize that likelihood function. Likelihood is nothing but the probability of the seen or observed data given the model parameters:

$$Likelihood = P(Data \,/\, Model)$$

To get the model defined by its parameters, we maximize the likelihood of the seen data:

$$Model = \underbrace{Arg\ \mathrm{Max}}_{Model}\, P(Data \,/\, Model)$$

Since we are only trying to fit a model based on the observed data, there is a high chance of overfitting and not generalizing to new data if we go for simple likelihood maximization.

If the data size is huge, the seen data is likely to represent the population well, and so maximizing the likelihood may suffice. On the other hand, if the seen data is small, there is a high chance that it might not represent the overall population well, and thus the model based on likelihood would not generalize well to new data. In that case, having a certain prior belief over the model and constraining the likelihood by that prior belief would lead to better results. Let's say our prior belief is in the form of knowing the probability distribution over the model parameters, i.e., P(*Model*) is known. We can in that case update our likelihood by the prior information to get a distribution over the model given the data. As per Bayes' theorem of conditional probability,

$$P(Model \,/\, Data) = \frac{P(Data \,/\, Model)\,P(Model)}{P(Data)}$$

P(Model/Data) is called the *posterior distribution* and is generally more informative since it combines one's prior knowledge about the data or model. Since this probability of data is independent of the model, the posterior is directly proportional to the product of the likelihood and the prior as shown in the following:

$$P(Model / Data) \propto P(Data / Model)P(Model)$$

One can build a model by maximizing the posterior probability distribution instead of the likelihood. This method of obtaining the model is called maximize a posterior, or MAP. Both likelihood and MAP are point estimates for the models and thus don't cover the whole uncertainty space. Taking the model that maximizes the posterior means taking the mode of the probability distribution of the model. Point estimates given by the maximum likelihood function don't correspond to any mode, since likelihood is not a probability distribution function. If the probability distribution turns out to be multimodal, these point estimates are going to perform even more poorly.

A better approach is to take the average of the model over the whole uncertainty space, i.e., to take the mean of the model based on the posterior distribution, as follows:

$$Model = E[Model / Data] = \int_{Model} Model \ P(Model / Data)d(Model)$$

To motivate the ideas of likelihood and posterior and how they can be used to derive model parameters, let's get back to the coin problem again.

Suppose we toss a coin six times, out of which heads appear five times. If one is supposed to estimate the probability of heads, what would the estimate be?

Here, the model for us is to estimate the probability of heads θ in a throw of a coin. Each toss of a coin can be treated as an independent Bernoulli trial with the probability of heads being θ. The likelihood of the data, given the model, is given by the following:

$$P(Data / \theta) = P(x_1 x_2 x_3 x_4 x_5 x_6 / \theta)$$

where $x_i \ \forall \ i \in \{1, 2, 3, 4, 5, 6\}$ denotes the event of either heads (H) or tails (T).

Since the throws of coins are independent, the likelihood can be factorized as follows:

$$P(Data / \theta) = P(x_1 x_2 x_3 x_4 x_5 x_6 / \theta) = \prod_{i=1}^{6} P(x_i / \theta) \qquad (5.1.1)$$

Each throw of dice follows the Bernoulli distribution; hence the probability of heads is θ, and the probability of tails is $(1 - \theta)$, and in general its probability mass function is given by the following:

$$P(x = j/\theta) = \theta^j (1-\theta)^{(1-j)} \ \forall j \in \{0,1\} \tag{5.1.2}$$

where $j = 1$ denotes heads and $j = 0$ denotes tails.

Since there are five heads and one tail combining (1) and (2), the likelihood L as a function of θ can be expressed as follows:

$$L(\theta) = \theta^5 (1-\theta) \tag{5.1.3}$$

Maximum likelihood methods treat the $\hat{\theta}$ that minimizes $L(\theta)$ as the model parameter. Hence,

$$\hat{\theta} = \underbrace{Arg \ \text{Max} \, L(\theta)}_{\theta}$$

If we take the derivative of the computed likelihood in (5.1.3) and set it to zero, we will arrive at the likelihood estimate for θ:

$$\frac{dL(\theta)}{d\theta} = 5\theta^4 - 6\theta^5 = 0 => \theta = \frac{5}{6}$$

In general, if someone asks us our estimate of θ without our doing a similar maximization of likelihood, we at once answer the probability to be $\frac{5}{6}$ by the basic definition of probability that we learned in high school, i.e.,

$$P(event = "a") = \frac{Number \ of \ occurences \ of \ the \ event "a"}{Total \ number \ of \ occurences \ of \ all \ events}$$

In a way, our head is thinking about likelihood and relying on the data seen thus far.

Now, let's suppose had we not seen the data and someone asked us to determine the probability of heads; what would have been a logical estimate?

Well, it depends on any prior belief that we may have about the coin. If we assume a fair coin, which in general is the most obvious assumption to make given that we have no information about the coin, $\theta = \frac{1}{2}$ would have been a good estimate. However, when we are assuming instead of doing a point estimate of prior for θ, it's better to have a probability distribution over θ with the probability maximum at $\theta = \frac{1}{2}$. The prior probability distribution is a distribution over the model parameter θ.

A beta distribution with parameters $\alpha = 2$, $\beta = 2$ would be a good prior distribution in this case since it has a maximum probability at $\theta = \frac{1}{2}$ and is symmetrical around it.

$$P(\theta) = Beta(\alpha = 2, \beta = 2) = \frac{\theta^{\alpha-1}(1-\theta)^{\beta-1}}{B(\alpha,\beta)} = \frac{\theta(1-\theta)}{B(\alpha,\beta)}$$

For fixed values of α and β, $B(\alpha, \beta)$ is constant and is the normalizing or partition function to this probability distribution. It can be computed as follows:

$$B(\alpha,\beta) = \frac{\tau(\alpha)\tau(\beta)}{\tau(\alpha+\beta)} = \frac{\tau(2)\tau(2)}{\tau(4)} = \frac{1!1!}{3!} = \frac{1}{6}$$

Even if one doesn't remember the formula, it can be found out by just integrating $\theta(1 - \theta)$ and taking the reciprocal of the same as the normalizing constant since the integral of the probability distribution should be 1.

$$P(\theta) = \frac{\theta\,(1-\theta)}{6} \tag{5.1.4}$$

If we combine the likelihood and the prior, we get the posterior probability distribution as follows:

$$P(\theta/D) \propto \theta^5(1-\theta)\frac{\theta\,(1-\theta)}{6} = \frac{\theta^6(1-\theta)^2}{6}$$

The proportional sign comes since we have ignored the probability of data. In fact, we can take the 6 out as well and express the posterior as follows:

$$P(\theta/D) \propto \theta^6(1-\theta)^2$$

Now, $0 \leq \theta \leq 1$ since θ is a probability. Integrating $\theta^6(1 - \theta)^2$ in the range of 0 to 1 and taking the reciprocal would give us the normalizing factor of the posterior, which comes out to be 252. Hence, the posterior can be expressed as follows:

$$P(\theta / D) = \frac{\theta^6 (1-\theta)^2}{252} \tag{5.1.5}$$

Now that we have the posterior, there are two ways we can estimate θ. We can maximize the posterior and get a MAP estimate of θ as follows:

$$\theta_{MAP} = \underbrace{Arg \ \mathrm{Max}}_{\theta} P(\theta / D)$$

$$\frac{dP(\theta / D)}{d\theta} = 0 => \theta = \frac{3}{4}$$

We see the MAP estimate of $\frac{3}{4}$ is more conservative than the likelihood estimates of $\frac{5}{6}$ since it takes the prior into consideration and doesn't blindly believe the data.

Now, let's look at the second approach, the pure Bayesian approach, and take the mean of the posterior distribution to average over all the uncertainties for θ:

$$E[\theta / D] = \int_{\theta=0}^{1} \theta P(\theta / D) d\theta$$

$$= \int_{\theta=0}^{1} \frac{\theta^7 (1-\theta)^2}{252} d\theta$$

$$= 0.7$$

Plotted in Figure 5-1a through Figure 5-1c are the likelihood function and the prior and posterior probability distributions for the coin problem. One thing to note is the fact that the likelihood function is not a probability density function or a probability mass function, whereas the prior and posteriors are probability mass or density functions.

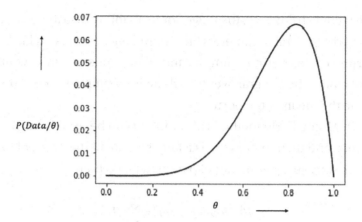

Figure 5-1a. *Likelihood function plot*

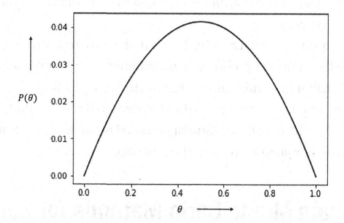

Figure 5-1b. *Prior probability distribution*

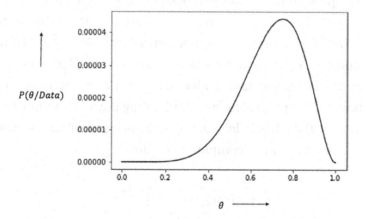

Figure 5-1c. *Posterior probability distribution*

For complicated distributions, the posterior probability distribution can turn out to be very complex with several parameters and is unlikely to represent known probability distribution forms such as normal, gamma, and so on. Thus, it may become seemingly impossible to compute the integral over the whole uncertainty space of the model in order to compute the mean of the posterior.

Markov chain Monte Carlo methods of sampling can be used in such cases to sample model parameters, and then their mean is a fair estimate of the mean of the posterior distribution. If we sample n sets of model parameters M_i, then

$$E[Model / Data] \approx \sum_{i=1}^{n} M_i$$

Generally, the mean of the distribution is taken since it minimizes the squared error. Basically, the expectation

$E[(y - c)^2]$ is minimized when $c = E[y]$. Given that we are trying to represent the probability of the distribution by a single representative such that the squared error over the probability distribution is minimized, mean is the best candidate.

However, one can take the median of the distribution if the distribution is skewed or there is more noise in the data in the form of potential outliers. This estimated median can be based on the samples drawn from the posterior.

Markov Chain Monte Carlo Methods for Sampling

Markov chain Monte Carlo methods, or MCMC, are some of the most popular techniques for sampling from complicated posterior probability distributions or in general from any probability distribution for multivariate data. Before we get to MCMC, let's talk about Monte Carlo sampling methods in general. Monte Carlo sampling methods try to compute the area under a curve based on sampled points.

For example, the area of the transcendental number $Pi(\pi)$ can be computed by sampling points within a square of radius 1 and noting down the number of sampled points within one-fourth of the circle of diameter 2 enclosed within the square. As shown in Figure 5-2, the area of Pi can be computed as follows:

$$\frac{4^* Area(OAC)}{Area(OABC)} = 4^* \frac{\left(\frac{1}{4}\right)\pi r^2}{r^2} = \pi$$

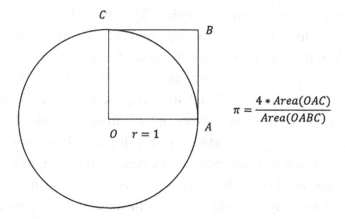

Figure 5-2. *Area of Pi*

In Listing 5-1, the Monte Carlo method for computing the value of *Pi* is illustrated. As we can see, the value comes out to nearly the value of *Pi*. The accuracy can be improved by sampling more points.

Listing 5-1. Computation of Pi Through Monte Carlo Sampling

```
import numpy as np
number_sample = 100000
inner_area,outer_area = 0,0

for i in range(number_sample):
    x = np.random.uniform(0,1)
    y = np.random.uniform(0,1)
    if (x**2 + y**2) < 1 :
        inner_area += 1
    outer_area += 1

print("The computed value of Pi:",4*(inner_area/float(outer_area)))

--Output--
('The computed value of Pi:', 3.142)
```

The simple Monte Carlo method is highly inefficient if the dimension space is large since the larger the dimensionality is, the more prominent the effects of correlation are. Markov chain Monte Carlo methods are efficient in such scenarios since they spend more time collecting samples from high-probability regions than from lower-probability regions. The normal Monte Carlo method explores the probability space uniformly and hence spends as much time exploring low-probability zones as it does high-probability zones. As we know, the contribution of a low-probability zone is insignificant when computing the expectation of functions through sampling, and hence when an algorithm spends a lot of time in such a zone, it leads to significantly higher processing time. The main heuristic behind the Markov chain Monte Carlo method is to explore the probability space not uniformly but rather to concentrate more on the high-probability zones. In high-dimensional space, because of correlation, most of the space is sparse, with high density found only at specific areas. So, the idea is to spend more time and collect more samples from those high-probability zones and spend as little time as possible exploring low-probability zones.

Markov chain can be thought of as a stochastic/random process to generate a sequence of random samples evolving over time. The next value of the random variable is only determined by the prior value of the variable. Markov chain, once it enters a high-probability zone, tries to collect as many points with a high-probability density as possible. It does so by generating the next sample, conditioned on the current sample value, so that points near the current sample are chosen with high probability and points far away are chosen with low probability. This ensures that the Markov chain collects as many points as possible from a current high-probability zone. However, occasionally a long jump from the current sample is required to explore other potential high-probability zones far from the current zone where the Markov chain is working.

The Markov chain concept can be illustrated with the movement of gas molecules in an enclosed container at a steady state. A few parts of the container have a higher density of gas molecules than the other areas, and since the gas molecules are at a steady state, the probabilities of each state (determined by the position of a gas molecule) would remain constant even though there might be gas molecules moving from one position to another.

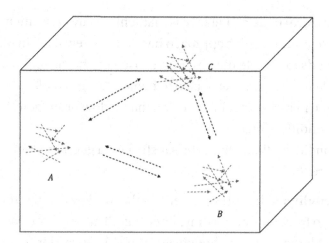

Figure 5-3. *Movement of gases in an enclosed container at steady state with only three states: A, B, and C*

For the sake of simplicity, let us assume there are only three states (position of the gas molecules, in this case) for the gas molecules, as shown in Figure 5-3. Let us denote those states by A, B, and C and their corresponding probabilities by P_A, P_B, and P_C.

Since the gas molecules are in steady state, if there are gas molecules transitioning to other states, equilibrium needs to be maintained to keep the probability distribution stationary. The simplest assumption to consider is that probability mass going from state A to state B should come back to A from B; i.e., pairwise, the states are in equilibrium.

Let's say $P(B/A)$ determines the transition probability from A to B. So, probability mass going from A to B is given by the following:

$$P(A)(B/A) \tag{5.2.1}$$

Likewise, probability mass coming to A from B is given by the following:

$$P(B)P(A/B) \tag{5.2.2}$$

So, in steady state from (5.2.1) and (5.2.2), we have the following:

$$P(A)(B/A) = P(B)P(A/B) \tag{5.2.3}$$

to maintain the stationarity of the probability distribution. This is called a *detailed balance condition*, and it is a sufficient but not necessary condition for the stationarity of a probability distribution. The gas molecules can be in equilibrium in more complex ways, but since this form of detail balance is mathematically convenient when the possible state space is infinite, this approach has been widely used in Markov chain Monte Carlo methods to sample the next point based on the current point. In short, movement of the Markov chain is expected to behave like gas molecules at steady state spending more time in high-probability region than in low probability keeping the detailed balance condition intact.

A few other conditions that need to be satisfied for a good implementation of Markov chain are listed here:

> **Irreducibility:** A desirable property of the Markov chain is that we can go from one state to any other state. This is important since in Markov chain, although we want to keep exploring nearby states of a given state with high probability, at times we might have to take a jump and explore some far neighborhood with the expectation that the new zone might be another high-probability zone.

> **Aperiodicity:** The Markov chain shouldn't repeat too often, as otherwise it won't be possible to traverse the whole space. Imagine a space with 20 states. If, after exploring five states, the chain repeats, it would not be possible to traverse all 20 states, thus leading to suboptimal sampling.

Metropolis Algorithm

The Metropolis algorithm is a Markov chain Monte Carlo method that uses the current accepted state to determine the next state. A sample at time $(t + 1)$ is conditionally dependent upon the sample at time t. The proposed state at time $(t + 1)$ is drawn from a normal distribution with a mean equal to the current sample at time t with a specified variance. Once drawn, the ratio of the probability is checked between the sample at time $(t + 1)$ and time t. If $\dfrac{P\left(x^{(t+1)}\right)}{P\left(x^{(t)}\right)}$ is greater than or equal to 1, then the sample $x^{(t+1)}$ is chosen with a probability of 1; if it is less than 1, then the sample is chosen randomly.

Mentioned next are the detailed implementation steps.

- Start with any random sample point $X^{(1)}$.

- Choose the next point $X^{(2)}$ that is conditionally dependent on $X^{(1)}$. You can choose $X^{(2)}$ from a normal distribution with a mean of $X^{(1)}$ and some finite variance; let's say S^2. So,

  $X^{(2)} \sim Normal\ (X^{(1)}, S^2)$. A key deciding factor for good sampling is to choose the variance S^2 very judicially. The variance shouldn't be too large, since in that case the next sample $X^{(2)}$ has less of a chance of staying near the current sample $X^{(1)}$, in which case a high-probability region might not be explored as much since the next sample is selected far away from the current sample most of the time. At the same time, the variance shouldn't be too small. In such cases, the next samples would almost always stay near the current point, and hence the probability of exploring a different high-probability zone far from the current zone would reduce.

- Some special heuristics are used in determining whether to accept $X^{(2)}$ once it has been generated from the preceding step.

  - If the ratio $P(X^{(2)})/P(X^{(1)}) \geq 1$, then accept $X^{(2)}$ and keep it as a valid sample point. The accepted sample becomes the $X^{(1)}$ for generating the next sample.

  - The ratio $P(X^{(2)})/P(X^{(1)}) < 1$, $X^{(2)}$ is accepted if the it is greater than a randomly generated number from the uniform distribution between 0 and 1, i.e., $U[0, 1]$.

As we can see, if we move to a higher-probability sample, then we accept the new sample, and if we move to a lower-probability sample, we sometimes accept and sometimes reject the new sample. The probability of rejection increases if the ratio $P(X^{(2)})|P(X^{(1)})$ is small. Let's say the ratio of $P(X^{(2)})|P(X^{(1)}) = 0.1$. When we generate a random number r_u between 0 and 1 from a uniform distribution, then the probability of $r_u > 0.1$ is 0.9, which in turn implies that the probability of the new samples getting rejected is 0.9. In general,

$$P(r_u > r) = 1 - r$$

where r is the ratio of the probability of the new sample and the old sample.

Let's try to intuit why such heuristics work for Markov chain Monte Carlo methods. As per detailed balance,

$$P\left(X^{(1)}\right)P\left(X^{(2)}/X^{(1)}\right)=P\left(X^{(2)}\right)P\left(X^{(1)}/X^{(2)}\right)$$

We have conveniently taken the transition probabilities to follow normal distribution without validating whether these transition probabilities adhere to the detailed balanced conditions for stationarity of the probability distribution that we wish to sample from. Let us consider that the ideal transition probabilities between the two states X_1 and X_2 to maintain the stationarity of the distribution are given by $P(X_1|X_2)$ and $P(X_2|X_1)$. Hence, as per detailed balance, the following condition must be satisfied:

$$P\left(X_1/X_2\right)P\left(X_1\right)=P\left(X_2/X_1\right)P\left(X_2\right)$$

However, discovering such an ideal transition probability function that ensures stationarity by imposing a detailed balance condition is hard. We start off with a suitable transition probability function; let's say $T(x/y)$, where y denotes the current state and x denotes the next state to be sampled based on y. For the two states X_1 and X_2, the assumed transition probabilities are thus given by $T(X_1/X_2)$ for a move from state X_2 to X_1 and by $T(X_2/X_1)$ for a move from state X_1 to X_2. Since the assumed transition probabilities are different than the ideal transition probabilities required to maintain stationarity through detailed balance, we get the opportunity to accept or reject samples based on how good the next move is. To cover up this opportunity, an acceptance probability for the transition of states is considered such that for a transition of a state from X_1 to X_2

$$P\left(X_2/X_1\right)=T\left(X_2/X_1\right)A\left(X_2/X_1\right)$$

where $A(X_2/X_1)$ is the acceptance probability of the move from X_1 to X_2.
As per detailed balance,

$$P\left(X_1/X_2\right)P\left(X_1\right)=P\left(X_2/X_1\right)P\left(X_2\right)$$

Replacing the ideal transition probability as the product of the assumed transition probability and the acceptance probability, we get the following:

$$T\left(X_2/X_1\right)A\left(X_2/X_1\right)P\left(X_1\right)=T\left(X_1/X_2\right)A\left(X_1/X_2\right)P\left(X_2\right)$$

Rearranging this, we get the acceptance probability ratio as follows:

$$\frac{A(X_2 / X_1)}{A(X_1 / X_2)} = \frac{T(X_1 / X_2)P(X_2)}{T(X_2 / X_1)P(X_1)}$$

One simple proposal that satisfies this is given by the Metropolis algorithm as follows:

$$A(X_2 / X_1) = \min\left(1, \frac{T(X_1 / X_2)P(X_2)}{T(X_2 / X_1)P(X_1)}\right)$$

In the Metropolis algorithm, the assumed transitional probability is generally assumed to be a normal distribution that is symmetric, and hence $T(X_1/X_2) = T(X_2/X_1)$. This simplifies the acceptance probability of the move from X_1 to X_2 as follows:

$$A(X_2 / X_1) = \min\left(1, \frac{P(X_2)}{P(X_1)}\right)$$

If the acceptance probability is 1, then we accept the move with probability 1, while if the acceptance probability is less than 1, let's say r, then we accept the new sample with probability r and reject the sample with probability $(1 - r)$. This rejection of samples with the probability $(1 - r)$ is achieved by comparing the ratio with the randomly generated sample r_u from a uniform distribution between 0 and 1 and rejecting the sample in cases where $r_u > r$. This is because for a uniform distribution probability $P(r_u > r) = 1 - r$, which ensures the desired rejection probability is maintained.

In Listing 5-2, we illustrate the sampling from a bivariate Gaussian distribution through the Metropolis algorithm.

Listing 5-2. Bivariate Gaussian Distribution Through Metropolis Algorithm

```
import numpy as np
import matplotlib.pyplot as plt
%matplotlib inline
import time
start_time = time.time()
```

```
"""
Generate samples from Bivariate Guassian Distribution with mean (0,0) and
covariance of 0.7 using
Markov Chain Monte Carlo(MCMC) method called Metropolis Hastings
algorithm.

We will use a Independent Gaussian transition probability distribution
with covariaace 0.2
So the next point X_next is going to be sampled from a Gaussian
Distribution with current point X_curr as the mean
and the Transition Covariance of 0.2

X_next ~ N(X_curr,covariance=[[0.2 , 0],[0,0.2]])

"""

def metropolis_hastings(target_dist,cov_trans,num_samples=100000):

    _mean_,_cov_ = target_dist[0],target_dist[1]

    x_list,y_list = [],[]
    accepted_samples_count = 0

    # Start with Initial Point (0,0)
    x_init, y_init = 0,0

    x_curr,y_curr = x_init, y_init

    for i in range(num_samples):

    # Set up the Transition(Conditional) Probability distribution taking
      the current point
    # as the mean and a small variance (we pass it through cov_trans) so
      that points near
    # the existing point have a high chance of getting sampled.

        mean_trans = np.array([x_curr,y_curr])
        # Sample next point using the Transition Probability distribution
        x_next, y_next = np.random.multivariate_normal(mean_trans,
        cov_trans).T
```

```
    X_next = np.array([x_next,y_next])
    X_next = np.matrix(X_next)
    X_curr = np.matrix(mean_trans)

    # Compute the probability density of the existing point and the
      newly sampled
    # point. We can ignore the normalizer as it would cancel
    # out when we take density ratio of next and curr point

    mahalnobis_dist_next = (X_next - _mean_)*np.linalg.inv(_cov_)*
    (X_next - _mean_).T
    prob_density_next = np.exp(-0.5*mahalnobis_dist_next)
    mahalnobis_dist_curr = (X_curr - _mean_)*np.linalg.inv(_cov_)*
    (X_curr - _mean_).T
    prob_density_curr = np.exp(-0.5*mahalnobis_dist_curr)

    # This is the heart of the algorithm.  Compute the acceptance
      ratio(r) as the
    # Probability density of the new point to existing point.
      Select the new point
    # 1. acceptance ratio(r) >= 1
    # 2. If acceptance ratio(r) < 1  chose the new point randomly
      proportional to its acceptance ratio(r)

    acceptance_ratio = prob_density_next[0,0] /
    float(prob_density_curr[0,0])

    if (acceptance_ratio >= 1) | ((acceptance_ratio < 1) and
    (acceptance_ratio >= np.random.uniform(0,1)) ) :
        x_list.append(x_next)
        y_list.append(y_next)
        x_curr = x_next
        y_curr = y_next
        accepted_samples_count += 1
end_time = time.time()
```

```
    print(f"Time taken to sample {accepted_samples_count} points :
    {str(end_time - start_time)} seconds")
    print(f"Acceptance ratio : {accepted_samples_count/float(
    num_samples)}")
    print(f"Mean of the Sampled Points: ({np.mean(x_list)} ,
    {np.mean(y_list)})")
    print(f"Covariance matrix of the Sampled Points\n")
    print(np.cov(x_list,y_list))

    plt.xlabel('X')
    plt.ylabel('Y')
    plt.title(f"Scatter plot for the Sampled Points")
    plt.scatter(x_list,y_list,color='black')
# Let's trigger some MCMC
num_samples=100000
_cov_ = np.array([[1,0.7],[0.7,1]])
_mean_ = np.matrix(np.array([0,0]))
metropolis_hastings(target_dist=(_mean_,_cov_),cov_trans=np.
array([[1,0.2],[0.2,1]]))

-Output-
Time taken to sample 71538 points ==> 30.3350000381 seconds
Acceptance ratio ===>  0.71538
Mean of the Sampled Points
-0.0090486292629 -0.008610932357
Covariance matrix of the Sampled Points
[[ 0.96043199  0.66961286]
 [ 0.66961286  0.94298698]]
```

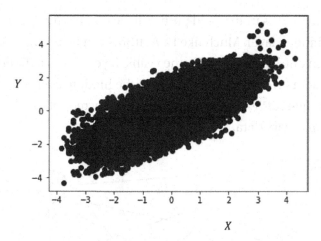

Figure 5-4. *Plot of sampled points from multivariate Gaussian distribution using Metropolis algorithm*

We see from the output that the mean and the covariance of the sampled points closely represent the mean and covariance of the bivariate Gaussian distribution from which we are sampling. Also, the scatter plot in Figure 5-4 closely resembles the bivariate Gaussian distribution.

Now that we are aware of the Markov chain Monte Carlo methods of sampling from probability distribution, we will learn about another MCMC method called Gibbs sampling while examining restricted Boltzmann machines.

Restricted Boltzmann Machines

Restricted Boltzmann machines (RBMs) belong to the unsupervised class of machine-learning algorithms that utilize the Boltzmann equation of probability distribution. Illustrated in Figure 5-5 is a two-layer restricted Boltzmann machine architecture that has a hidden layer and a visible layer. There are weight connections between all the hidden and visible layers' units. However, there are no hidden-to-hidden or visible-to-visible unit connections. The term *restricted* in RBM refers to this constraint on the network. The hidden units of an RBM are conditionally independent from one another given the set of visible units. Similarly, the visible units of an RBM are conditionally independent from one another given the set of hidden units. Restricted Boltzmann machines are most often used as a building block for deep networks rather than as an individual network itself. In terms of probabilistic graphic models, restricted Boltzmann machines can be defined as undirected probabilistic graphic models containing a visible layer and a single hidden layer. Much like PCA, RBMs can be thought of as a way of representing data in one space (given by the visible layer v) into a different space (given by the hidden or latent layer h). When the size of the hidden layer is less than the size of the visible layer, then RBMs perform a dimensionality reduction of data. RBMs are generally trained on binary data.

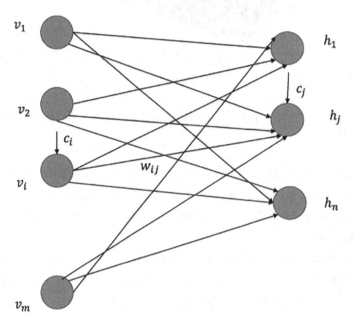

Figure 5-5. *Restricted Boltzmann machine visible and hidden layer architecture*

Let the visible units of the RBM be represented by vector $v = [v_1\ v_2 ... v_m]^T \in R^{m \times 1}$ and the hidden units be represented by $h = [h_1 h_2 ... h_n]^T \in R^{n \times 1} h = [h_1\ h_2 ... h_n]^T \in R^{n \times 1}$. Also, let the weight connecting the *i-th* visible unit to the *j-th* hidden unit be represented by w_{ij}, $\forall i \in \{1, 2, ..m\}$, $\forall j \in \{1, 2, ..n\}$. We can represent the matrix containing the weights w_{ij} as $W \in R^{m \times n}$.

The energy of the joint probability distribution having hidden state h and visible state v is given by the following:

$$P(v, h) = \frac{e^{-E(h, v)}}{Z} \tag{5.3.1}$$

where $E(v, h)$ is the energy of the joint configuration (v, h) and Z is the normalizing factor, commonly known as the partition function. This probability is based on the Boltzmann distribution and assumes the Boltzmann constant and thermal temperature as 1.

$$Z = \sum_v \sum_h e^{-E(v,h)} \tag{5.3.2}$$

The energy $E(v, h)$ of the joint configuration (v, h) is given by the following:

$$E(v,h) = -b^T v - c^T h - v^T W h \tag{5.3.3}$$

$$E(v,h) = -\sum_{i=1}^{m} b_i v_i - \sum_{j=1}^{n} c_j h_j - \sum_{j=1}^{n} \sum_{i=1}^{m} v_i w_{ij} h_j \tag{5.3.4}$$

The vectors $b = [b_1\ b_2 ... b_m]^T \in R^{m \times 1}$ and $c = [c_1\ c_2 .... c_n]^T \in R^{n \times 1}$ are biases at the visible and hidden units, respectively, as we will see later.

In any graphical probabilistic model, the idea is to compute the joint probability distribution over various sets of events. Combining (5.3.1) and (5.3.3), we get the following:

$$P(v,h) = \frac{e^{b^T v + c^T h + v^T W h}}{Z} \tag{5.3.5}$$

The partition function Z is hard to compute, which makes the computation of $P(v, h)$ hard to compute. For a small set of events, it is possible to compute the partition function. If there are many variables in v and h, the number of possible joint events will be exceedingly large, considering all such combinations become hard.

However, the conditional probability distribution $P(h/v)$ is easy to compute and sample from. The following deduction will justify this:

$$P(h/v) = \frac{P(v,h)}{P(v)}$$

$$= \frac{P(v,h)}{\sum_h P(v,h)}$$

$$= \frac{\dfrac{e^{b^T v + c^T h + v^T Wh}}{Z}}{\sum_h \dfrac{e^{b^T v + c^T h + v^T Wh}}{Z}}$$

$$= \frac{e^{b^T v + c^T h + v^T Wh}}{\sum_h e^{b^T v + c^T h + v^T Wh}}$$

$$= \frac{e^{b^T v} e^{c^T h} e^{v^T Wh}}{e^{b^T v} \sum_h e^{c^T h} e^{v^T Wh}}$$

$$= \frac{e^{c^T h} e^{v^T Wh}}{\sum_h e^{c^T h} e^{v^T Wh}}$$

We can expand the numerator and denominator in terms of the components of the different vectors involved, as follows:

$$e^{c^T h} e^{v^T Wh} = e^{c^T h + v^T Wh}$$

$$= e^{\sum_{j=1}^{n} \left(c_j h_j + v^T W[:,j] h_j \right)}$$

Since the exponential of a sum is equal to the product of the exponentials, the preceding equation can be written in product form as follows:

$$e^{c^T h} e^{v^T Wh} = \prod_{j=1}^{n} e^{c_j h_j + v^T W[:,j] h_j} \tag{5.3.6}$$

Now, let's look at the denominator, which looks similar to the numerator only with a sum over all possible hidden states h. Using the expression for $e^{c^T h} e^{v^T W h}$ in (5.3.6), the denominator can be expressed as follows:

$$\sum_h e^{c^T h} e^{v^T W h} = \sum_h \prod_{j=1}^{n} e^{c_j h_j + v^T W[:,j] h_j} \qquad (5.3.7)$$

The sum over vector means the sum over all combinations of its components. Each hidden unit h_i can have a binary state of 0 or 1, and hence $h_j \in \{0, 1\}\ \forall j \in (1, 2, 3, ..n)$. So, the summation over vector h in (5.3.7) can be expanded into multiple summations corresponding to each of its components:

$$\sum_h e^{c^T h} e^{v^T W h} = \sum_{h_1=0}^{1} \sum_{h_2=0}^{1} .. \sum_{h_n=0}^{1} \prod_{j=1}^{n} e^{c_j h_j + v^T W[:,j] h_j}$$

$$= \sum_{h_1=0}^{1} \sum_{h_2=0}^{1} .. \sum_{h_n=0}^{1} \left(e^{c_1 h_1 + v^T W[:,1] h_1} \right) \left(e^{c_2 h_2 + v^T W[:,2] h_2} \right) ... \left(e^{c_n h_n + v^T W[:,n] h_n} \right) \qquad (5.3.8)$$

Now, let's look at a very simple manipulation involving products and sums with two discrete variables a and b:

$$\sum_{j=1}^{2} \sum_{i=1}^{3} a_i b_j = a_1 b_1 + a_2 b_1 + a_3 b_1 + a_1 b_2 + a_2 b_2 + a_3 b_2$$

$$= b_1 (a_1 + a_2 + a_3) + b_2 (a_1 + a_2 + a_3)$$

$$= (a_1 + a_2 + a_3)(b_1 + b_2)$$

$$= \left(\sum_{i=1}^{3} a_i \right) \left(\sum_{j=1}^{2} b_j \right)$$

So, we see that when we take elements of variables with independent indices, the sum of the products of the variables can be expressed as the product of the sum of the variables. Similar to this example, the elements of h (i.e., h_i) in general are involved in the product $\left(e^{c_1 h_1 + v^T W[:,1] h_1} \right) \left(e^{c_2 h_2 + v^T W[:,2] h_2} \right) ... \left(e^{c_n h_n + v^T W[:,n] h_n} \right)$ independently, and hence the expression in (5.3.8) can be simplified as follows:

$$\sum_h e^{c^T h} e^{v^T W h} = \sum_{h_1=0}^{1} \left(e^{c_1 h_1 + v^T W[:,1] h_1} \right) \sum_{h_1=0}^{1} \left(e^{c_2 h_2 + v^T W[:,2] h_2} \right) .. \sum_{h_n=0}^{1} \left(e^{c_n h_n + v^T W[:,n] h_n} \right)$$

$$\sum_h e^{c^T h} e^{v^T W h} = \prod_{j=1}^{n} \sum_{h_j=0}^{1} \left(e^{c_j h_j + v^T W[:,j]h_j} \right) \tag{5.3.9}$$

Combining the expressions for numerator and denominator from (5.3.6) and (5.3.9), we have the following:

$$P(h/v) = \frac{\displaystyle\prod_{j=1}^{n} e^{c_j h_j + v^T W[:,j]h_j}}{\displaystyle\prod_{j=1}^{n} \sum_{h_j=0}^{1} e^{c_j h_j + v^T W[:,j]h_j}}$$

Simplifying this in terms of components of h on both sides, we get the following:

$$P(h_1 h_2 .. h_n / v) = \prod_{j=1}^{n} \left(\frac{e^{c_j h_j + v^T W[:,j]h_j}}{\displaystyle\sum_{h_j=0}^{1} e^{c_j h_j + v^T W[:,j]h_j}} \right)$$

$$= \left(\frac{e^{c_1 h_1 + v^T W[:,1]h_1}}{\displaystyle\sum_{h_1=0}^{1} e^{c_1 h_1 + v^T W[:,1]h_1}} \right) \left(\frac{e^{c_2 h_2 + v^T W[:,1]h_2}}{\displaystyle\sum_{h_2=0}^{1} e^{c_2 h_2 + v^T W[:,2]h_2}} \right) \left(\frac{e^{c_n h_n + v^T W[:,n]h_n}}{\displaystyle\sum_{h_n=0}^{1} e^{c_n h_n + v^T W[:,n]h_{1n}}} \right) \tag{5.3.10}$$

The joint probability distribution of elements of h conditioned on v has factored into the product of expressions independent of each other conditioned on v. This leads to the fact that the components of h (i.e., $h_i \,\forall\, i \in \{1, 2, ...n\}$) are conditionally independent of each other given v. This gives us the following:

$$P(h_1 h_2 .. h_n / v) = P(h_1 / v) P(h_2 / v) .. P(h_n / v) \tag{5.3.11}$$

$$P(h_j / v) = \frac{e^{c_j h_j + v^T W[:,j]h_j}}{\displaystyle\sum_{h_j=0}^{1} e^{c_j h_j + v^T W[:,j]h_j}} \tag{5.3.12}$$

Replacing $h_j = 1$ and $h_j = 0$ in (5.3.12), we get the following:

$$P\left(h_j = 1/v\right) = \frac{e^{c_j + v^T W[:,j]}}{1 + e^{c_j + v^T W[:,j]}} \qquad (5.3.13)$$

$$P\left(h_j = 0/v\right) = \frac{1}{1 + e^{c_j + v^T W[:,j]}} \qquad (5.3.14)$$

The expressions for (5.3.13) and (5.3.14) illustrate the fact that the hidden units $h_i \, \forall \, i \in \{1, 2, \ldots n\}$ are independent sigmoid units:

$$P\left(h_j = 1/v\right) = \sigma\left(c_j + v^T W[:, j]\right) \qquad (5.3.15)$$

Expanding the components of v and $W[:, j]$, we can rewrite (5.3.15) as follows:

$$P\left(h_j = 1/v\right) = \sigma\left(c_j + \sum_{i=1}^{m} v_i w_{ij}\right) \qquad (5.3.16)$$

where $\sigma(.)$ represents the sigmoid function such that

$$\sigma(x) = \frac{1}{\left(1 + e^{-x}\right)}$$

Proceeding in a similar fashion, it can be proven that

$$P\left(v_1 v_2 .. v_m / h\right) = P\left(v_1 / h\right) P\left(v_2 / h\right) .. P\left(v_m / h\right)$$

which means the hidden units are conditionally independent of each other given the visible states. Since RBM is a symmetrical undirected network, like the visible units, the probability of the visible units given the hidden states can be similarly expressed as follows:

$$P\left(v_i = 1/h\right) = \sigma\left(b_i + \sum_{j=1}^{n} h_j w_{ij}\right) \qquad (5.3.17)$$

From (5.3.16) and (5.3.17), we can clearly see that the visible and hidden units are actually binary sigmoid units with the vectors b and c providing the biases at the visible and hidden units, respectively. This symmetrical and independent conditional dependence of the hidden and visible units can be useful while training the model.

Training a Restricted Boltzmann Machine

We need to train the Boltzmann machine in order to derive the model parameters b, c, W, where b and c are the bias vectors at the visible and hidden units, respectively, and W is the weight connection matrix between the visible and hidden layers. For ease of reference, the model parameters can be collectively referred to as follows:

$$\theta = [b;c;W]$$

The model can be trained by maximizing the log likelihood function of the input data points with respect to the model parameters. The input is nothing but the data corresponding to the visible units for each data point. The likelihood function is given by the following:

$$L(\theta) = P\left(v^{(1)}v^{(2)}..v^{(m)} / \theta\right)$$

Assuming the data points are independent given the model,

$$L(\theta) = P\left(v^{(1)} / \theta\right)P(v^{(2)} / \theta)..P\left(v^{(m)} / \theta\right) = \prod_{t=1}^{m} P\left(v^{(t)} / \theta\right) \tag{5.3.18}$$

Taking the log on both sides to get the log likelihood expression of the function in (5.3.18), we have the following:

$$C = \log L(\theta) = \sum_{t=1}^{m} \log P\left(v^{(t)} / \theta\right) \tag{5.3.19}$$

Expanding the probabilities in (5.3.19) by its joint probability form, we get the following:

$$
\begin{aligned}
C &= \sum_{t=1}^{m} \log P\!\left(v^{(t)}/\theta\right) \\
&= \sum_{t=1}^{m} \log \sum_{h} P\!\left(v^{(t)},h/\theta\right) \\
&= \sum_{t=1}^{m} \log \sum_{h} \frac{e^{-E\left(v^{(t)},h\right)}}{Z} \\
&= \sum_{t=1}^{m} \log \frac{\sum_{h} e^{-E\left(v^{(t)},h\right)}}{Z} \\
&= \sum_{t=1}^{m} \log \sum_{h} e^{-E\left(v^{(t)},h\right)} - \sum_{t=1}^{m} \log Z \\
&= \sum_{t=1}^{m} \log \sum_{h} e^{-E\left(v^{(t)},h\right)} - m\log Z
\end{aligned}
\tag{5.3.20}
$$

The partition function Z is not constrained by visible-layer inputs $v^{(t)}$ unlike the first term in (5.3.20). Z is the sum of the negative exponentials of the energies over all possible combinations of v and h and so can be expressed as follows:

$$
Z = \sum_{v}\sum_{h} e^{-E(v,h)}
$$

Replacing Z with this expression in (5.3.20), we get the following:

$$
C = \sum_{t=1}^{m} \log \sum_{h} e^{-E\left(v^{(t)},h\right)} - m\log \sum_{v}\sum_{h} e^{-E(v,h)}
\tag{5.3.21}
$$

Now, let's take the gradient of the cost function with respect to the combined parameter θ. We can think of C as comprising two components, ρ^{+} and ρ^{-}, as shown in the following:

$$
\rho^{+} = \sum_{t=1}^{m}\sum_{h} e^{-E\left(v^{(t)},h\right)}
$$

$$
\rho^{-} = m\log \sum_{v}\sum_{h} e^{-E(v,h)}
$$

Taking the gradient of ρ^+ with respect to θ, we have the following:

$$\nabla_\theta\left(\rho^+\right)=\sum_{t=1}^{m}\frac{\sum_h e^{-E\left(v^{(t)},h\right)}\nabla_\theta\left(-E\left(v^{(t)},h\right)\right)}{\sum_h e^{-E\left(v^{(t)},h\right)}} \tag{5.3.22}$$

Now, let's simplify $\dfrac{\sum_h e^{-E\left(v^{(t)},h\right)}\nabla_\theta\left(-E\left(v^{(t)},h\right)\right)}{\sum_h e^{-E\left(v^{(t)},h\right)}}$ by dividing both the numerator and the denominator by Z:

$$\nabla_\theta\left(\rho^+\right)=\sum_{t=1}^{m}\frac{\sum_h\dfrac{e^{-E\left(v^{(t)},h\right)}}{Z}\nabla_\theta\left(-E\left(v^{(t)},h\right)\right)}{\dfrac{\sum_h e^{-E\left(v^{(t)},h\right)}}{Z}} \tag{5.3.23}$$

$\dfrac{e^{-E\left(v^{(t)},h\right)}}{Z}=P\left(v^{(t)},h/\theta\right)$ and $\dfrac{\sum_h e^{-E\left(v^{(t)},h\right)}}{Z}=P\left(v^{(t)}/\theta\right)$. Using these expressions for the probabilities in (5.3.23), we get the following:

$$\begin{aligned}
\nabla_\theta\left(\rho^+\right)&=\sum_{t=1}^{m}\frac{\sum_h P\left(v^{(t)},h/\theta\right)\nabla_\theta\left(-E\left(v^{(t)},h\right)\right)}{P\left(v^{(t)}/\theta\right)}\\
&=\sum_{t=1}^{m}\sum_h\frac{P\left(v^{(t)},h/\theta\right)}{P\left(v^{(t)}/\theta\right)}\nabla_\theta\left(-E\left(v^{(t)},h\right)\right)\\
&=\sum_{t=1}^{m}\sum_h P\left(h/v^{(t)},\theta\right)\nabla_\theta\left(-E\left(v^{(t)},h\right)\right)
\end{aligned} \tag{5.3.24}$$

One can remove the θ from the probability notations, such as $P(v^{(t)}, h/\theta)$, $P(v^{(t)}, h/\theta)$, and so forth, for ease of notation if one wishes to, but it is better to keep them since it makes the deductions more complete, which allows for better interpretability of the overall training process.

Let us look at the expectation of functions, which gives us the expression seen in (5.3.24) in a more meaningful form ideal for training purposes. The expectation of $f(x)$, given x, follows probability mass function $P(x)$ and is given by the following:

$$E\big[f(x)\big]=\sum_{x}P(x)f(x)$$

If $x = [x_1\ x_2\ldots x_n]^T \in R^{n \times 1}$ is multivariate, then the preceding expression holds true and

$$E\big[f(x)\big]=\sum_{x}P(x)f(x)=\sum_{x_1}\sum_{x_2}..\sum_{x_n}P(x_1,x_2,..,x_n)f(x_1,x_2,..,x_n)$$

Similarly, if $f(x)$ is a vector of functions such that $f(x) = [f_1(x)f_2(x)]^T$, one can use the same expression as for expectation. Here, one would get a vector of expectations, as follows:

$$E\big[f(x)\big]=\sum_{x}P(x)f(x)=\begin{bmatrix}\sum_{x_1}\sum_{x_2}..\sum_{x_n}P(x_1,x_2,..,x_n)f_1(x_1,x_2,..,x_n)\\ \sum_{x_1}\sum_{x_2}..\sum_{x_n}P(x_1,x_2,..,x_n)f_2(x_1,x_2,..,x_n)\end{bmatrix} \qquad (5.3.25)$$

To explicitly mention the probability distribution in the expectation notation, one can rewrite the expectation of functions or expectation of vector of functions whose variables x follow probability distribution $P(x)$ as follows:

$$E_{P(x)}\big[f(x)\big]=\sum_{x}P(x)f(x)$$

Since we are working with gradients, which are vectors of different partial derivatives, and each of the partial derivatives is a function of h for given values of θ and v, the equation in (5.3.24) can be expressed in terms of expectation of the gradient $\nabla_\theta(-E(v^{(t)}, h))$ with respect to the probability distribution $P(h/v^{(t)}, \theta)$ as follows:

$$\nabla_\theta\big(\rho^+\big)=\sum_{t=1}^{m}E_{P(h/v^{(t)},\theta)}\Big[\nabla_\theta\big(-E\big(v^{(t)},h\big)\big)\Big] \qquad (5.3.26)$$

Note that the expectation $E_{P(h/v^{(t)},\theta)}\Big[\nabla_\theta\big(-E\big(v^{(t)},h\big)\big)\Big]$ is a vector of expectations, as has been illustrated in (5.3.25).

Now, let's get to the gradient of $\rho^- = m \log \sum_v \sum_h e^{-E(v,h)}$ with respect to the θ:

$$
\begin{aligned}
\nabla_\theta(\rho^-) &= m \frac{\sum_v \sum_h e^{-E(v,h)} \nabla_\theta(-E(v,h))}{\sum_v \sum_h e^{-E(v,h)}} \\
&= m \frac{\sum_v \sum_h e^{-E(v,h)} \nabla_\theta(-E(v,h))}{Z} \\
&= m \sum_v \sum_h \frac{e^{-E(v,h)}}{Z} \nabla_\theta(-E(v,h)) \\
&= m \sum_v \sum_h P(v,h/\theta) \nabla_\theta(-E(v,h)) \\
&= m E_{P(h,v/\theta)} \left[\nabla_\theta(-E(v,h)) \right]
\end{aligned}
\tag{5.3.27}
$$

The expectation in (5.3.27) is over the joint distribution of h and v, whereas the expectation in (5.3.26) is over the h given a seen v. Combining (5.3.26) and (5.3.27), we get the following:

$$
\nabla_\theta(C) = \sum_{t=1}^{m} E_{P\left(h/v^{(t)},\theta\right)} \left[\nabla_\theta\left(-E\left(v^{(t)},h\right)\right) \right] - m E_{P(h,v/\theta)} \left[\nabla_\theta(-E(v,h)) \right]
\tag{5.3.28}
$$

If we look at the gradient with respect to all the parameters in (5.3.28), it has two terms. The first term is dependent on the seen data $v^{(t)}$, while the second term depends on samples from the model. The first term increases the likelihood of the given observed data, while the second term reduces the likelihood of data points from the model.

Now, let's do some simplification of the gradient for each of the parameter sets in θ, i.e., b, c, and W.

$$
\nabla_b(-E(v,h)) = \nabla_b\left(b^T v + c^T h + v^T W h\right) = v
\tag{5.3.29}
$$

$$
\nabla_c(-E(v,h)) = \nabla_c\left(b^T v + c^T h + v^T W h\right) = h
\tag{5.3.30}
$$

$$
\nabla_W(-E(v,h)) = \nabla_W\left(b^T v + c^T h + v^T W h\right) = v h^T
\tag{5.3.31}
$$

Using (5.3.28) through (5.3.31), the expression for the gradient with respect to each of the parameter sets is given by the following:

$$\nabla_b(C) = \sum_{t=1}^{m} E_{P\left(h/v^{(t)},\theta\right)}\left[v^{(t)}\right] - mE_{P(h,v/\theta)}[v] \tag{5.3.32}$$

Since the probability distribution of the first term is conditioned on $v^{(t)}$, the expectation of $v^{(t)}$ with respect to $P(h/v^{(t)}, \theta)$ is $v^{(t)}$.

$$\nabla_b(C) = \sum_{t=1}^{m} v^{(t)} - mE_{P(h,v/\theta)}[v] \tag{5.3.33}$$

$$\nabla_c(C) = \sum_{t=1}^{m} E_{P\left(h/v^{(t)},\theta\right)}[h] - mE_{P(h,v/\theta)}[h] \tag{5.3.34}$$

The expectation of h over the probability distribution $P(h/v^{(t)}, \theta)$ can be easily computed since each of the units of h (i.e., h_j) given $v^{(t)}$ is independent. Each of them is a sigmoid unit with two possible outcomes, and their expectation is nothing but the output of the sigmoid units, i.e.,

$$E_{P\left(h/v^{(t)},\theta\right)}[h] = \hat{h}^{(t)} = \sigma\left(c + W^T v^{(t)}\right)$$

If we replace the expectation with $\hat{h}$, then the expression in (5.3.34) can be written as follows:

$$\nabla_c(C) = \sum_{t=1}^{m} \hat{h}^{(t)} - mE_{P(h,v/\theta)}[h] \tag{5.3.35}$$

Similarly,

$$\nabla_W(C) = \sum_{t=1}^{m} E_{P\left(h/v^{(t)},\theta\right)}\left[v^{(t)}h^T\right] - mE_{P(h,v/q)}[h]$$

$$= \sum_{t=1}^{m} v^{(t)}\hat{h}^{(t)T} - mE_{P(h,v/\theta)}[h] \tag{5.3.36}$$

So, the expressions in (5.3.33), (5.3.35), and (5.3.36) represent the gradients with respect to the three parameter sets. For easy reference,

$$
\begin{cases}
\nabla_b(C) = \sum_{t=1}^{m} v^{(t)} - mE_{P(h,v/\theta)}[v] \\[2ex]
\nabla_c(C) = \sum_{t=1}^{m} \hat{h}^{(t)} - mE_{P(h,v/\theta)}[h] \\[2ex]
\nabla_W(C) = \sum_{t=1}^{m} v^{(t)} \hat{h}^{(t)^T} - mE_{P(h,v/\theta)}[h]
\end{cases}
\qquad (5.3.37)
$$

Based on these gradients, gradient-descent techniques can be invoked to iteratively get the parameter values that maximize the likelihood function. However, there is a little complexity involved to compute the expectations with respect to the joint probability distribution $P(h, v/\theta)$ at each iteration of gradient descent. The joint distribution is hard to compute because of the seemingly large number of combinations for h and v in cases where they are moderate to large dimensionality vectors. Markov chain Monte Carlo sampling (MCMC) techniques, especially Gibbs sampling, can be used to sample from the joint distribution and compute the expectations in (5.3.37) for the different parameter sets. However, MCMC techniques take a long time to converge to a stationary distribution, after which they provide good samples. Hence, to invoke MCMC sampling at each iteration of gradient descent would make the learning very slow and impractical.

Gibbs Sampling

Gibbs sampling is a Markov chain Monte Carlo method that can be used to sample observations from a multivariate probability distribution. Suppose we want to sample from a multivariate joint probability distribution $P(x)$ where $x = [x_1 x_2 .. x_n]^T$.

Gibbs sampling generates the next value of a variable x_i conditioned on all the current values of the other variables. Let the t-th sample drawn be represented by $x^{(t)} = [x_1^{(t)} x_2^{(t)} .. x_n^{(t)}]^T$. To generate the $(t + 1)$ sample seen next, follow this logic:

- Draw the variable $x_j^{(t+1)}$ by sampling it from a probability distribution conditioned on the rest of the variables. In other words, draw $x_j^{(t+1)}$ from

$$
P\left(x_j^{(t+1)} / x_1^{(t+1)} x_2^{(t+1)} .. x_{j-1}^{(t+1)} x_{j+1}^{(t)} .. x_n^{(t)}\right)
$$

So basically, for sampling x_j conditioned on the rest of the variables, for the $j - 1$ variables before x_j, their values for the $(t + 1)$ instance are considered since they have already been sampled, while for the rest of the variables, their values at instance t are considered since they are yet to be sampled. This step is repeated for all the variables.

If each x_j is discrete and can take, let's say, two values 0 and 1, then we need to compute the probability $p_1 = P(x_j^{(t+1)} = 1/x_1^{(t+1)}x_2^{(t+1)} .. x_{j-1}^{(t+1)}x_{j+1}^{(t)} .. x_n^{(t)})$. We can then draw a sample u from a uniform probability distribution between 0 and 1 (i.e., $U[0, 1]$), and if $p_1 \geq u$ set $x_j^{(t+1)} = 1$, else set $x_j^{(t+1)} = 0$. This kind of random heuristics ensures that the higher the probability p_1 is, the greater the chances are of $x_j^{(t+1)}$ getting selected as 1. However, it still leaves room for 0 getting selected with very low probability, even if p_1 is relatively large, thus ensuring that the Markov chain doesn't get stuck in a local region and can explore other potential high-density regions as well. There is the same kind of heuristics that we saw for the Metropolis algorithm as well.

- If one wishes to generate m samples from the joint probability distribution $P(x)$, the preceding step has to be repeated m times.

The conditional distributions for each variable based on the joint probability distribution are to be determined before the sampling can proceed. If one is working on Bayesian networks or restricted Boltzmann machines, there are certain constraints within the variables that help determine these conditional distributions in an efficient manner.

As an example, if one needs to do Gibbs sampling from a bivariate normal distribution with mean [0 0] and covariance matrix $\begin{bmatrix} 1 & \rho \\ \rho & 1 \end{bmatrix}$, then the conditional probability distributions can be computed as follows:

$$P(x_2 / x_1) = P(x_1, x_2) / P(x_1)$$

$$P(x_1 / x_2) = P(x_1, x_2) / P(x_2)$$

If one derives the marginal distributions $P(x_1)$ and $P(x_2)$ as $\int_{x_2} P(x_1, x_2) dx_2$ and $\int_{x_1} P(x_1, x_2) dx_1$, then

$$x_2 / x_1 \sim Normal(\rho x_1, 1 - p^2)$$

$$x_1 / x_2 \sim Normal(\rho x_2, 1 - p^2)$$

Block Gibbs Sampling

There are several variants of Gibbs sampling. Block Gibbs sampling is one of them. In block Gibbs sampling, more than one variable is grouped together, and then the group of variables is sampled together conditioned on the rest of the variables, as opposed to sampling for individual variables separately. For example, in a restricted Boltzmann machine, the hidden unit state variables $h = [h_1 \ h_2.... \ h_n]^T$ can be sampled together conditioned on the visible unit states $=[v_1 \ v_2...v_m]^T$ and vice versa. Hence, for sampling from the joint probability distribution over $P(v, h)$, through block Gibbs sampling, one can sample all the hidden states given the visible unit states through the conditional distribution $P(h/v)$ and sample all the visible unit states given the hidden unit states through the conditional distribution $P(v/h)$. The samples at the $(t + 1)$ iteration of Gibbs sampling can be generated as follows:

$$h^{(t+1)} \sim P\left(h|v^{(t)}\right)$$

$$v^{(t+1)} \sim P(v|h^{(t+1)})$$

Therefore, $(v^{(t+1)}, h^{(t+1)})$ is the combined sample at iteration $(t + 1)$.

Based on samples collected by the above sampling method, the expectation of a function $f(h, v)$ can be computed as follows:

$$E\left[f(h,v)\right] \approx \frac{1}{M}\sum_{t=1}^{M}f\left(h^{(t)},v^{(t)}\right)$$

where M denotes the number of samples generated from the joint probability distribution $P(v, h)$.

Burn-in Period and Generating Samples in Gibbs Sampling

To consider the samples as independent as possible for expectation computation generally the samples are picked up at an interval of k samples. The larger the value of k, the better at removing the autocorrelation among the generated samples. Also, the samples generated at the beginning of the Gibbs sampling are ignored. These ignored samples are said to have been generated in the burn-in period.

The burn-in period uses the Markov chain to settle down to an equilibrium distribution before we can start drawing samples from it. This is required because we generate the Markov chain from an arbitrary sample that might be a low-probability zone with respect to the actual distribution, and so we can throw away those unwanted samples. The low-probability samples don't contribute much to the actual expectation, and thus having plenty of them in the sample would obscure the expectation. Once the Markov chain has run long enough, it will sample from the high-probability zones frequently, at which point we can start collecting the samples.

Using Gibbs Sampling in Restricted Boltzmann Machines

Block Gibbs sampling can be used to compute the expectations with respect to the joint probability distribution $P(v, h/\theta)$ as mentioned in the equations in (5.3.37) for computing the gradient with respect to the model parameters b, c, and W. Here are the equations from (5.3.37) for easy reference:

$$
\begin{cases}
\nabla_b (C) = \sum_{t=1}^{m} v^{(t)} - m E_{P(h,v/\theta)}[v] \\
\nabla_c (C) = \sum_{t=1}^{m} \hat{h}^{(t)} - m E_{P(h,v/\theta)}[h] \\
\nabla_W (C) = \sum_{t=1}^{m} v^{(t)} \hat{h}^{(t)^T} - m E_{P(h,v/\theta)}[h]
\end{cases}
$$

The expectations $E_{P(h,v/\theta)}[v]$, $E_{P(h,v/\theta)}[h]$, and $E_{P(h,v/\theta)}[vh^T]$, all require sampling from the joint probability distribution $P(v, h/\theta)$. Through block Gibbs sampling, samples (v, h) can be drawn as follows based on their conditional probabilities, where t denotes the iteration number of Gibbs sampling:

$$
h^{(t+1)} \sim P\left(h/v^{(t)}, \theta\right)
$$

$$
v^{(t+1)} \sim P\left(v/h^{(t+1)}, \theta\right)
$$

What makes sampling even easier is the fact that the hidden units are independent given the visible unit states, and vice versa:

$$P(h/v) = P(h_1/v)P(h_2/v)..P(h_n/v) = \prod_{j=1}^{n} P(h_j/v)$$

This allows each of the individual hidden unit h_j to be sampled independently of the other in parallel given the visible unit states. The parameter θ has been removed from the preceding notation since θ would remain constant for a step of gradient descent when we perform Gibbs sampling.

Now, each of the hidden unit output state h_j can be either 0 or 1, and its probability of assuming state 1 is given by (5.3.16) as follows:

$$P(h_j = 1/v) = \sigma\left(c_j + \sum_{i=1}^{m} v_i w_{ij}\right)$$

This probability can be computed based on the current value of $v = v^{(t)}$ and model parameters c, $W \in \theta$. The computed probability $P(h_j = 1/v^{(t)})$ is compared to a random sample u generated from a uniform distribution $U[0, 1]$. If $P(h_j = 1/v^{(t)}) > u$, then the sampled $h_j = 1$, else $h_j = 0$. Each of the hidden unit h_j sampled in this manner forms the combined hidden unit state vector $h^{(t+1)}$.

Similarly, the visible units are independent given the hidden unit states:

$$P(v/h) = P(v_1/h)P(v_2/h)..P(v_n/h) = \prod_{i=1}^{m} P(v_i/h)$$

Each of the visible units can be sampled independently given $h^{(t+1)}$ to get the combined $v^{(t+1)}$ in the same way as for the hidden units. The required sample thus generated in the $(t + 1)$ iteration is given by $(v^{(t+1)}, h^{(t+1)})$.

All the expectations $E_{P(h,v/\theta)}[v]$, $E_{P(h,v/\theta)}[h]$, and $E_{P(h,v/\theta)}[vh^T]$ can be computed by taking the averages of the samples generated through Gibbs sampling. Through Gibbs sampling, if we take N samples after considering burn-in periods and autocorrelation as discussed earlier, the required expectation can be computed as follows:

$$E_{P(h,v/\theta)}[v] \approx \frac{1}{N}\sum_{i=1}^{N} v^{(i)}$$

$$E_{P(h,v/\theta)}[h] \approx \frac{1}{N}\sum_{i=1}^{N}h^{(i)}$$

$$E_{P(h,v/\theta)}\left[vh^{T}\right] \approx \frac{1}{N}\sum_{i=1}^{N}v^{(i)}h^{(i)T}$$

However, performing Gibbs sampling for the joint distribution to generate N samples in each iteration of gradient descent becomes a tedious task and is often impractical. There is an alternate way of approximating these expectations, called *contrastive divergence*, which we will discuss in the next section.

Contrastive Divergence

Performing Gibbs sampling on the joint probability distribution $P(h, v|\theta)$ at each step of gradient descent becomes challenging since Markov chain Monte Carlo methods such as Gibbs sampling take a long time to converge, which is required in order to produce unbiased samples. These unbiased samples drawn from the joint probability distribution are used to compute the expectations terms $E_{P(h,v/\theta)}[v]$, $E_{P(h,v/\theta)}[h]$, and $E_{P(h,v/\theta)}[vh^{T}]$, which are nothing but components of the term $E_{P(h,v/\theta)}[\nabla_{\theta}(-E(v,h))]$ in the combined expression for gradients as deduced in (5.3.28).

$$\nabla_{\theta}(C) = \sum_{t=1}^{m}E_{P(h/v^{(t)},\theta)}\left[\nabla_{\theta}\left(-E\left(v^{(t)},h\right)\right)\right] - mE_{P(h,v/\theta)}\left[\nabla_{\theta}\left(-E(v,h)\right)\right]$$

The second term in the preceding equation can be rewritten as a summation over the m data points, and hence

$$\nabla_{\theta}(C) = \sum_{t=1}^{m}E_{P(h/v^{(t)},\theta)}\left[\nabla_{\theta}\left(-E\left(v^{(t)},h\right)\right)\right] - \sum_{t=1}^{m}E_{P(h,v/\theta)}\left[\nabla_{\theta}\left(-E(v,h)\right)\right]$$

Contrastive divergence approximates the overall expectation $E_{P(h,v/\theta)}[\nabla_{\theta}(-E(v,h))]$ by a point estimate at a candidate sample $(\bar{v},\bar{h})$ obtained by performing Gibbs sampling for only a couple of iterations.

$$E_{P(h,v/\theta)}\left[\nabla_{\theta}\left(-E(v,h)\right)\right] \approx \nabla_{\theta}\left(-E\left(\bar{v},\bar{h}\right)\right)$$

This approximation is done for every data point $v^{(t)}$, and hence the expression for the overall gradient can be rewritten as follows:

$$\nabla_{\theta}(C) \approx \sum_{t=1}^{m} E_{P(h/v^{(t)},\theta)}\left[\nabla_{\theta}\left(-E\left(v^{(t)},h\right)\right)\right] - \sum_{t=1}^{m}\left[\nabla_{\theta}\left(-E\left(\bar{v}^{(t)},\bar{h}^{(t)}\right)\right)\right]$$

Figure 5-6 illustrates how Gibbs sampling is performed for each input data point $v^{(t)}$ to get the expectation approximation over the joint probability distribution by a point estimate. The Gibbs sampling starts with $v^{(t)}$, and based on the conditional probability distribution $P(h/v^{(t)})$, the new hidden state h' is obtained. As discussed earlier, each of the hidden unit h_j can be sampled independently and combined to form the hidden state vector h'. Then v' is sampled based on the conditional probability distribution $P(v/h')$. This iterative process is generally run for couple of iterations, and the final v and h sampled are taken as the candidate sample $(\bar{v},\bar{h})$.

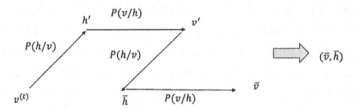

Figure 5-6. *Gibbs sampling for two iterations to get one sample for contrastive divergence*

The contrastive divergence method makes the gradient descent faster since the Gibbs sampling in each step of gradient descent is limited to only a few iterations, mostly one or two per data point.

A Restricted Boltzmann Implementation in TensorFlow

In this section, we will go through the implementation of restricted Boltzmann machines using the MNIST dataset. Here, we try to model the structure of the MNIST images by defining a restricted Boltzmann machine network that consists of the image pixels as the visible units and 500 hidden layers in order to decipher the internal structure of each image. Since the MNIST images are 28 × 28 in dimension, when flattened as a vector, we have 784 visible units. We try to capture the hidden structures properly by training the Boltzmann machines. Images that represent the same digit should have similar hidden states, if not the same, when said hidden states are sampled given the visible representations of the input images. When the visible units are sampled, given their hidden structure, the visible unit values when structured in an image form should correspond to the label of the image. The detailed code is illustrated in Listing 5-3a.

Listing 5-3a. Restricted Boltzmann Machine Implementation with MNIST Dataset

```
##Import the Required libraries
import numpy as np
import pandas as pd
import tensorflow as tf
from tensorflow.keras import layers,Model
print(f"Tensorflow version: {tf.__version__}")
import matplotlib.pyplot as plt
%matplotlib inline

"""
    Restricted Boltzmann machines Class
"""

class rbm:

    def __init__(self,n_visible,n_hidden,lr=0.01,num_epochs=100,batch_
    size=256,weight_init='normal',k_steps=2):
        self.n_visible = n_visible
        self.n_hidden = n_hidden
        self.lr = lr
        self.num_epochs = num_epochs
        self.batch_size = batch_size
        self.weight_init = weight_init
        self.k = k_steps

    def model(self):
        if self.weight_init == 'glorot':
            self.W = tf.Variable(
                tf.random.normal([self.n_visible, self.n_hidden], mean=0.0,
                stddev=0.1, dtype=tf.float32) * tf.cast(tf.sqrt(
```

```
                2 / (self.n_hidden + self.n_visible)), tf.float32),
                tf.float32, name="weights")

        elif self.weight_init == 'normal':
            self.W = tf.Variable(
                tf.random.normal([self.n_visible, self.n_hidden], mean=0.0,
                stddev=0.1,  dtype=tf.float32),
                tf.float32, name="weights")

        self.b_v = tf.Variable(tf.random.uniform([1, self.n_visible],
        0, 0.1,  dtype=tf.float32), tf.float32, name="visible_biases")

        self.b_h = tf.Variable(tf.random.uniform([1, self.n_hidden],
        0, 0.1, dtype=tf.float32), tf.float32, name="hidden_biases")

        self.model_params = {'weights': self.W, 'visible_biases': self.b_v,
                        'hidden_biases': self.b_h}

# Converts the probability into discrete binary states i.e. 0 and 1
@staticmethod
def sample(probs):
    return tf.floor(probs + tf.random.uniform(tf.shape(probs), 0, 1))
# Sample hidden activations from  visible inputs
def visible_to_hidden(self,x):
    h = self.sample(tf.sigmoid(tf.matmul(x, self.W) + self.b_h))
    return h
# Sample visible activations from hidden inputs
def hidden_to_visible(self,h):
    x = self.sample(tf.sigmoid(tf.matmul(h, tf.transpose(self.W)) +
    self.b_v))
    return x

# Gibbs sampling step to sample hidden state given visible and then
  visible given hidden.
def gibbs_step(self,x):
    h = self.visible_to_hidden(x)
    x = self.hidden_to_visible(h)
    return x
```

```python
# Multiple Gibbs sampling steps to extract a constrastive visible
  sample starting from
# a given visible sample
def gibbs_sample(self,x,k):
    for i in range(k):
        x = self.gibbs_step(x)
    # Returns the gibbs sample after k iterations
    return x

# Ensure the data is converted to binary form
def data_load(self):
    (train_X, train_Y), (test_X, test_Y) = tf.keras.datasets.mnist.
    load_data()
    train_X, test_X , = train_X.reshape(-1,28*28), test_X.reshape
    (-1,28*28)
    train_X, test_X = train_X/255.0, test_X/255.0
    train_X[train_X < 0.5], test_X[test_X < 0.5] = 0.0, 0.0
    train_X[train_X >= 0.5], test_X[test_X >= 0.5] = 1.0, 1.0
    return np.float32(train_X), train_Y, np.float32(test_X), test_Y

# Train step to update model weights for a batch
def train_step(self,x,k):

    # Constrastive Sample
    x_s = self.gibbs_sample(x,k)
    h_s = self.sample(self.visible_to_hidden(x_s))
    h = self.sample(self.visible_to_hidden(x))
    # Update weights
    batch_size = tf.cast(tf.shape(x)[0], tf.float32)
    W_add  = tf.multiply(self.lr/batch_size, tf.subtract(tf.matmul(
    tf.transpose(x), h), tf.matmul(tf.transpose(x_s), h_s)))
    bv_add = tf.multiply(self.lr/batch_size, tf.reduce_sum
    (tf.subtract(x, x_s), 0, True))
    bh_add = tf.multiply(self.lr/batch_size, tf.reduce_sum(tf.subtract
    (h, h_s), 0, True))
    self.W.assign_add(W_add)
    self.b_h.assign_add(bh_add)
```

```python
        self.b_v.assign_add(bv_add)

    def validation(self,x,y):

        x_reconst = self.gibbs_step(tf.constant(x))
        x_reconst = x_reconst.numpy()

        plt.figure(1)

        for k in range(20):
            plt.subplot(4, 5, k+1)
            image = x[k,:].reshape(28,28)
            image = np.reshape(image,(28,28))
            plt.imshow(image,cmap='gray')

        plt.figure(2)

        for k in range(20):
            plt.subplot(4, 5, k+1)
            image = x_reconst[k,:].reshape(28,28)
            plt.imshow(image,cmap='gray')

    def train_model(self):
        # Initialize the Model parameters
        self.model()
        train_X, train_Y, test_X, test_Y = self.data_load()
        num_train_recs = train_X.shape[0]

        batches = num_train_recs//self.batch_size
        order = np.arange(num_train_recs)
        for i in range(self.num_epochs):
            np.random.shuffle(order)
            train_X, train_Y = train_X[order], train_Y[order]

            for batch in range(batches):
                batch_xs = train_X[batch*self.batch_size:(batch+1)*self.
                batch_size]
                #print(batch_xs.shape)
```

```
        batch_xs = tf.constant(batch_xs)
        self.train_step(batch_xs,self.k)

    if i % 5 == 0 :
        print(f"Completed Epoch {i}")
        self.model_params = {'weights': self.W, 'visible_biases':
        self.b_v,
                    'hidden_biases': self.b_h}

    print("RBM training Completed !")

    self.validation(x=test_X[:20,:], y=test_Y[:20])
rbm = rbm(n_visible=28*28,num_epochs=100,n_hidden=500)
rbm.train_model()

--output --
```

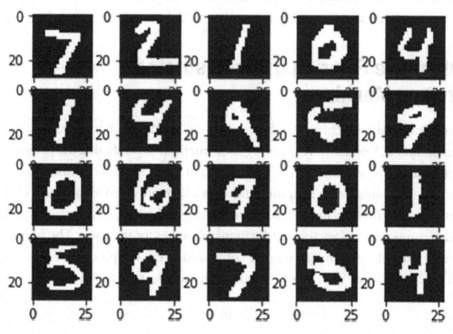

Figure 5-7. *Actual test images*

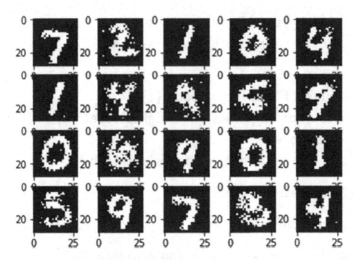

Figure 5-8. *Simulated images given the hidden states*

We can see from Figure 5-7 and Figure 5-8 that the RBM model did a good job of simulating the input images given their hidden representations. Hence, restricted Boltzmann machines can be used as generative models as well.

Collaborative Filtering Using Restricted Boltzmann Machines

Restricted Boltzmann machines can be used for collaborative filtering in making recommendations. Collaborative filtering is the method of predicting the preference of a user for an item by analyzing the preferences of many users for items. Given a set of items and users along with the ratings the users have provided for a variety of items, the most common method for collaborative filtering is the matrix factorization method, which determines a set of vectors for the items as well as for the users. The rating assigned by a user to an item can then be computed as the dot product of the user vector $u^{(j)}$ with the item vector $v^{(k)}$. Hence, the rating can be expressed as follows:

$$r^{(jk)} = u^{(j)T} v^{(k)}$$

where j and k represent the *j-th* user and the *kth* item, respectively. Once the vectors for each item and each user have been learned, the expected ratings a user would assign to a product they haven't rated yet can be found out by the same method. Matrix factorization can be thought of as decomposing a big matrix of ratings into user and item vectors.

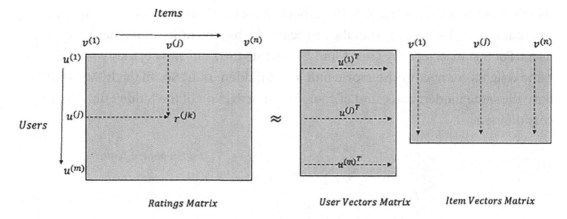

Figure 5-9. *Schematic diagram of a matrix factorization method for collaborative filtering*

Illustrated in Figure 5-9 is a schematic diagram of a matrix factorization method that decomposes a user item rating matrix into two matrices consisting of user vectors and item vectors. The dimensionality of the user and rating vectors must be equal for their dot product to hold true, which gives an estimate of what rating the user might assign to a particular item. There are several matrix factorization methods, such as singular value decomposition (SVD), nonnegative matrix factorization, alternating least squares, and so on. Any of the suitable methods can be used for matrix factorization depending on the use. Generally, SVD requires filling in the missing ratings (where the users have not rated the items) in the matrix, which might be a difficult task, and hence methods such as alternating least squares, which only take the provided ratings and not the missing values, work well for collaborative filtering.

Now, we will look at a different method of collaborative filtering that uses restricted Boltzmann machines. Restricted Boltzmann machines were used by the winning team in the Netflix Challenge of Collaborative Filtering, and so let's consider the items as movies for this discussion. The visible units for this RBM network would correspond to movie ratings, and instead of being binary, each movie would be a five-way SoftMax unit to account for the five possible ratings from 1 to 5. The number of hidden units can be chosen arbitrarily; we chose d here. There would be several missing values for different movies since all the movies would not be rated by all the users. The way to handle them is to train a separate RBM for each user based on only the movies that user has rated. The weights from the movies to the hidden units would be shared by all users. For instance, let's say *User A* and *User B* rate the same movie; they would use the same weights connecting the movie unit to the hidden units. So, all the RBMs would have the same hidden units, and of course their activation of the hidden units may be very different.

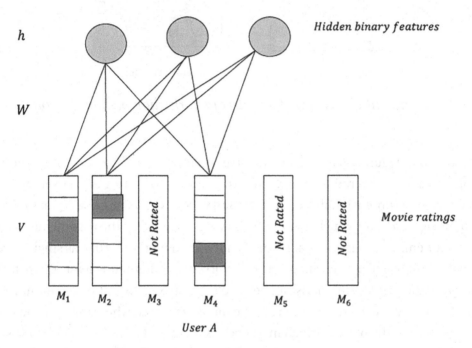

Figure 5-10. *Restricted Boltzmann view for User A*

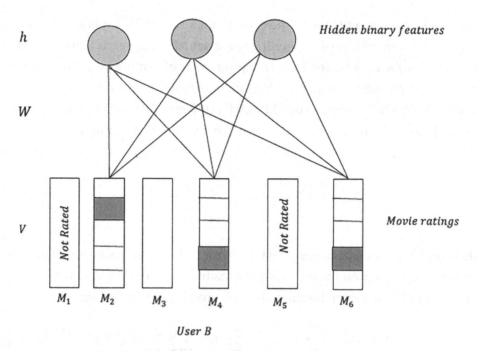

Figure 5-11. *Restricted Boltzmann view for User B*

As we can see from Figure 5-10 and Figure 5-11, the restricted Boltzmann views for User A and User B are different since they differ in the selection of movies they have rated. However, for the movies they both have rated, the weight connections are the same. This kind of architecture—where each user's RBM is trained separately while the RBMs share weights for same movies—helps overcome the problem of missing ratings and at the same time allows generalized weights for a movie to hidden layer connection for all the users. From each movie to the hidden unit and vice versa, there are actually five connections, one for each of the possible ratings for a movie. However, to keep the representation simple, only one combined connection has been shown in the diagrams. Each of the models can be trained separately through gradient descent using contrastive divergence, and the model weights can be averaged across the different RBMs so that all the RBMs share the same weights.

From 5.3.17, we have for the binary visible unit the following:

$$P(v_i = 1/h) = \sigma\left(b_i + \sum_{j=1}^{n} h_j w_{ij}\right)$$

Now that the visible units have K possible ratings, the visible units are K dimensional vectors with only one index corresponding to the actual rating set to 1, and the rest are all set to zero. So, the new expression of the probability of a rating over K possible ratings would be given by a SoftMax function. Also, do note that the m in this case is the number of movies a user has watched and would vary for different RBMs for different users. The constant n denotes the number of hidden units in the hidden layer for each RBM.

$$P\left(v_i^{(k)} = 1/h\right) = \frac{e^{\left(b_i^{(k)} + \sum_{j=1}^n h_j w_{ij}^{(k)}\right)}}{\sum_{l=1}^K e^{\left(b_i^{(l)} + \sum_{j=1}^n h_j w_{ij}^{(l)}\right)}} \tag{5.4.1}$$

where $w_{ij}^{(k)}$ is the weight connecting the kth rating index for visible unit i to the j-th hidden unit and $b_i^{(k)}$ represents the bias at the visible unit i for its k-th rating.

The energy of a joint configuration $E(v, h)$ is given by the following:

$$E(v,h) = -\sum_{k=1}^K \sum_{i=1}^m b_i^{(k)} v_i^{(k)} - \sum_{j=1}^n c_j h_j - \sum_{k=1}^K \sum_{j=1}^n \sum_{i=1}^m v_i^{(k)} w_{ij}^{(k)} h_j \tag{5.4.2}$$

So,

$$P(v,h) \propto e^{-E(v,h)} = e^{\sum_{k=1}^K \sum_{i=1}^m b_i^{(k)} v_i^{(k)} + \sum_{j=1}^n c_j h_j + \sum_{k=1}^K \sum_{j=1}^n \sum_{i=1}^m v_i^{(k)} w_{ij}^{(k)} h_j} \tag{5.4.3}$$

The probability of the hidden unit given the input v is as follows:

$$P(hj = 1/v) = \frac{e^{\left(c_j + \sum_{i=1}^m \sum_{k=1}^K v_i^{(k)} w_{ij}^{(k)}\right)}}{1 + e^{\left(c_j + \sum_{i=1}^m \sum_{k=1}^K v_i^{(k)} w_{ij}^{(k)}\right)}} \tag{5.4.4}$$

Now, the obvious question: How do we predict the rating for a movie a user has not seen? As it turns out, the computation for this is not that involved and can be computed in linear time. We need to compute the probability of the user rating r to an unknown movie q given the user's ratings on seen movies. Let the movie ratings the user has already provided be denoted by V. So, we need to compute the probability $P(v_q^{(k)}/V)$ as follows:

$$P\left(v_q^{(k)}/V\right) = \sum_h P\left(v_q^{(k)}, h/V\right) = \frac{\sum_h P\left(v_q^{(k)}, h, V\right)}{P(V)} \tag{5.4.5}$$

Since $P(V)$ is fixed for all movie ratings k, from (5.4.5) we have the following:

$$P\left(v_q^{(k)}/V\right) \propto \sum_h P\left(v_q^{(k)},h,V\right)$$

$$\propto \sum_h e^{-E\left(v_q^{(k)},h,V\right)}$$

(5.4.6)

This is a three-way energy configuration and can be computed easily by adding the contribution of $v_q^{(k)}$ in (5.4.2), as shown here:

$$E\left(v_q^{(k)},V,h\right) = -\sum_{k=1}^{K}\sum_{i=1}^{m}b_i^{(k)}v_i^{(k)} - \sum_{j=1}^{n}c_jh_j - \sum_{k=1}^{K}\sum_{j=1}^{n}\sum_{i=1}^{m}v_i^{(k)}w_{ij}^{(k)}h_j - \sum_{j=1}^{n}v_s^{(k)}w_{sj}^{(k)}h_j - v_s^{(k)}b^{(k)} \quad (5.4.7)$$

Substituting $v_q^{(k)} = 1$ in (5.4.7), one can find the value of $E(v_q^{(k)} = 1, V, h)$, which is proportional to the following:

$$P\left(v_q^{(k)} = 1, V, h\right)$$

For all K values of rating k, the preceding quantity $E(v_q^{(k)} = 1, V, h)$ needs to be computed and then normalized to form probabilities. One can then either take the value of k for which the probability is maximum or compute the expected value of k from the derived probabilities, as shown here:

$$\hat{k} = \underbrace{argmax}_{k} P\left(v_q^{(k)} = 1/V\right)$$

$$\hat{k} = \sum_{k=1}^{5} k \times P\left(v_q^{(k)} = 1/V\right)$$

The expectation way of deriving the rating turns out to give better predictions than a hard assignment of a rating based on the maximum probability.

Also, one simple way to derive the probability of the rating k for a specific unrated movie q by a user with rating matrix V is to first sample the hidden states h given the visible ratings input V; i.e., draw $h \sim P(h/V)$. The hidden units are common to all and hence carry information about patterns for all movies. From the sampled hidden units, we try to sample the value of $v_q^{(k)}$; i.e., draw $v_q^{(k)} \sim P(v_q^{(k)}/h)$. This back-to-back sampling, first from $V \to h$ and then from $h \to v_q^{(k)}$, is equivalent to sampling $v_q^{(k)} \sim P(v_q^{(k)}/V)$. I hope this helps in providing an easier interpretation.

Deep-Belief Networks (DBNs)

Deep-belief networks are based on the restricted Boltzmann machine, but, unlike an RBN, a DBN has multiple hidden layers. The weights in each hidden layer K are trained by keeping all the weights in the prior $(K-1)$ layers constant. The activities of the hidden units in the $(K-1)$ layer are taken as input for the Kth layer. At any time during training, two layers are involved in learning the weight connections between them. The learning algorithm is the same as that of restricted Boltzmann machines. Illustrated in Figure 5-12 is a high-level schematic diagram for a deep-belief network.

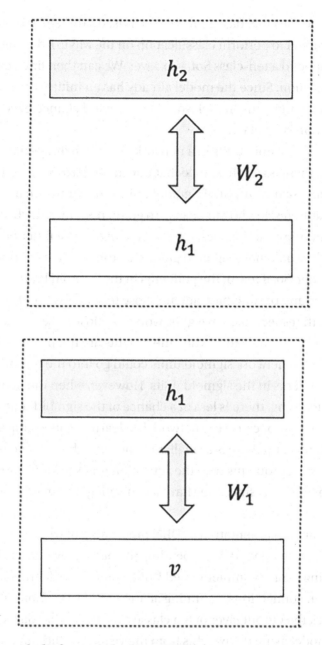

Figure 5-12. *Deep-belief network using RBMs*

Like RBM, in DBN each layer can be trained by gradient descent using contrastive divergence. The DBN learning algorithm is used to learn the initial weights of a deep network being used for supervised learning so that the network has a good set of initial weights to start with. Once the pretraining is done for the deep-belief network, we can add an output layer to the DBN based on the supervised problem at hand. Let's say we want to train a model to perform classification on the MNIST dataset. In that case, we would have to append a ten-class SoftMax layer. We can then fine-tune the model by using backpropagation. Since the model already has an initial set of weights from unsupervised DBN learning, the model would have a good chance of converging faster when backpropagation is invoked.

Whenever we have sigmoid units in a network, if the network weights are not initialized properly, there is a high chance that one might have a vanishing-gradient problem. This is because the output of sigmoid units is linear within a small range, after which the output saturates, leading to near-zero gradients. Since the backpropagation is essentially a chain rule of derivatives, the gradient of the cost function with respect to any weight would have sigmoid gradients from the layers prior to it from a backpropagation order. So, if few of the gradients in the sigmoid layers are operating in the saturated regions and producing gradients close to zero, the latter layers, gradients of the cost function with respect to the weights, would be close to zero, and there is a high chance that the learning would stop. When the weights are not properly initialized, there is a high chance that the network sigmoid units could go into the unsaturated region and lead to near-zero gradients in the sigmoid units. However, when the network weights are initialized by DBN learning, there is less of a chance of the sigmoid units operating in the saturated zone. This is because the network has learned something about the data while it was pretraining, and there is a smaller chance that the sigmoid units will operate in saturated zones. Such problems with the activation unit saturating are not present for ReLU activation functions since they have a constant gradient of 1 for input values greater than zero.

We now look at an implementation of DBN pretraining of weights followed by the training of a classification network by appending the output layer to the hidden layer of the RBM. In Listing 5-3a, we implemented RBM, wherein we learned the weights of the visible-to-hidden connections, assuming all the units are sigmoid. To that RBM, we are going to stack the output layer of ten classes for the MNIST dataset and train the classification model using the weights from the visible-to-hidden units learned as the initial weights for the classification network. Of course, we would have a new set of weights corresponding to the connection of the hidden layer to the output layer. See the detailed implementation in Listing 5-3b.

Listing 5-3b. Basic Implementation of DBN

```python
#Import the Required libraries
import numpy as np
import pandas as pd
import tensorflow as tf
from tensorflow.keras import layers,Model
print(f"Tensorflow version: {tf.__version__}")
import matplotlib.pyplot as plt
%matplotlib inline

"""

A MNIST classifier pretrained as a Deep Belief Network using RBM.
"""

class rbm_pretrained_classifier:

    def __init__(self,n_visible,n_hidden,lr=0.01,n_out=10,
    num_epochs=100,batch_size=256,weight_init='normal',k_steps=2):

        self.n_visible = n_visible
        self.n_hidden = n_hidden
        self.n_out = n_out
        self.lr = lr
        self.num_epochs = num_epochs
        self.batch_size = batch_size
        self.weight_init = weight_init
        self.k = k_steps

    def model(self):

        if self.weight_init == 'glorot':
            self.W = tf.Variable(
                tf.random.normal([self.n_visible, self.n_hidden], mean=0.0,
                stddev=0.1, dtype=tf.float32) * tf.cast(tf.sqrt(
                 2 / (self.n_hidden + self.n_visible)), tf.float32),
                tf.float32, name="weights")
```

```python
        self.Wf = tf.Variable(
            tf.random.normal([self.n_hidden,self.n_out], mean=0.0,
            stddev=0.1, dtype=tf.float32) * tf.cast(tf.sqrt(
             2 / (self.n_hidden + self.n_out)), tf.float32),
            tf.float32, name="weights_final")

    elif self.weight_init == 'normal':
        self.W = tf.Variable(
            tf.random.normal([self.n_visible, self.n_hidden], mean=0.0,
            stddev=0.1, dtype=tf.float32),
            tf.float32, name="weights")
        self.Wf = tf.Variable(
            tf.random.normal([self.n_hidden, self.n_out], mean=0.0,
            stddev=0.1, dtype=tf.float32),
            tf.float32, name="weights_final")

    self.b_v = tf.Variable(tf.random.uniform([1, self.n_visible],
    0, 0.1, dtype=tf.float32), tf.float32, name="visible_biases")

    self.b_h = tf.Variable(tf.random.uniform([1, self.n_hidden],
    0, 0.1, dtype=tf.float32), tf.float32, name="hidden_biases")

    self.b_f = tf.Variable(tf.random.uniform([1, self.n_out], 0, 0.1,
    dtype=tf.float32), tf.float32, name="final_biases")

    self.model_params = {'weights': self.W, 'visible_biases': self.b_v,
                        'hidden_biases': self.b_h, 'weights_final':self.
                        Wf, 'final_biases':self.b_f}

# Converts the probability into discrete binary states i.e. 0 and 1
@staticmethod
def sample(probs):
    return tf.floor(probs + tf.random.uniform(tf.shape(probs), 0, 1))

def visible_to_hidden(self,x):
    h = self.sample(tf.sigmoid(tf.matmul(x, self.W) + self.b_h))
    return h
```

```python
def hidden_to_visible(self,h):
    x = self.sample(tf.sigmoid(tf.matmul(h, tf.transpose(self.W)) +
    self.b_v))
    return x

# Gibbs sampling step
def gibbs_step(self,x):
    h = self.visible_to_hidden(x)
    x = self.hidden_to_visible(h)
    return x

# Run multiple gives Sampling step starting from an initital point
def gibbs_sample(self,x,k):
    for i in range(k):
        x = self.gibbs_step(x)
    # Returns the gibbs sample after k iterations
    return x

def data_load(self):
    (train_X, train_Y), (test_X, test_Y) = tf.keras.datasets.mnist.
    load_data()
    train_X, test_X , = train_X.reshape(-1,28*28), test_X.
    reshape(-1,28*28)
    train_X, test_X = train_X/255.0, test_X/255.0
    train_X[train_X < 0.5], test_X[test_X < 0.5] = 0.0, 0.0
    train_X[train_X >= 0.5], test_X[test_X >= 0.5] = 1.0, 1.0
    return np.float32(train_X), train_Y, np.float32(test_X), test_Y

def train_rbm_step(self,x,k):

    # Constrastive Sample
    x_s = self.gibbs_sample(x,k)
    h_s = self.sample(self.visible_to_hidden(x_s))
    h = self.sample(self.visible_to_hidden(x))
    # Update weights
    batch_size = tf.cast(tf.shape(x)[0], tf.float32)
    W_add  = tf.multiply(self.lr/batch_size, tf.subtract(tf.matmul(
    tf.transpose(x), h), tf.matmul(tf.transpose(x_s), h_s)))
```

```python
        bv_add = tf.multiply(self.lr/batch_size, tf.reduce_sum(
        tf.subtract(x, x_s), 0, True))
        bh_add = tf.multiply(self.lr/batch_size, tf.reduce_sum(
        tf.subtract(h, h_s), 0, True))
        self.W.assign_add(W_add)
        self.b_h.assign_add(bh_add)
        self.b_v.assign_add(bv_add)

    @tf.function
    def train_classifier_step(self,x):
        _h_ = tf.sigmoid(tf.matmul(x, self.W) + self.b_h)
        _out_ = tf.matmul(_h_, self.Wf) + self.b_f
        return _out_

    def rbm_validation(self,x,y):

        x_reconst = self.gibbs_step(tf.constant(x))
        x_reconst = x_reconst.numpy()

        plt.figure(1)

        for k in range(20):
            plt.subplot(4, 5, k+1)
            image = x[k,:].reshape(28,28)
            image = np.reshape(image,(28,28))
            plt.imshow(image,cmap='gray')

        plt.figure(2)

        for k in range(20):
            plt.subplot(4, 5, k+1)
            image = x_reconst[k,:].reshape(28,28)
            plt.imshow(image,cmap='gray')

    def classifier_test_accuracy(self,test_X,test_Y):
        test_accuracy = 0
        num_test_batches = test_X.shape[0]//self.batch_size
        for batch in range(num_test_batches):
```

```
                X_batch = test_X[batch*self.batch_size:(batch+1)*
                self.batch_size]
                y_batch = test_Y[batch*self.batch_size:(batch+1)*
                self.batch_size]
                X_batch, y_batch = tf.constant(X_batch),
                tf.constant(y_batch)
                y_pred_batch = self.train_classifier_step(X_batch)
                test_accuracy += np.sum(y_batch.numpy() == np.argmax
                (y_pred_batch.numpy(),axis=1))
        print(f"Test Classification Accuracy: {test_accuracy/float
        (test_X.shape[0])}")

    def train_model(self):
        # Initialize the Model parameters
        self.model()
        train_X, train_Y, test_X, test_Y = self.data_load()
        batches = int(train_X.shape[0]/self.batch_size)
        num_train_recs = train_X.shape[0]
        order = np.arange(num_train_recs)

        for i in range(self.num_epochs):
            np.random.shuffle(order)
            train_X, train_Y = train_X[order], train_Y[order]
            for batch in range(batches):
                X_batch = train_X[batch*self.batch_size:(batch+1)*
                self.batch_size]
                #print(batch_xs.shape)
                X_batch = tf.constant(X_batch)
                self.train_rbm_step(X_batch,self.k)

            if i % 5 == 0 :
                print(f"Completed Epoch {i}")
                self.model_params = {'weights': self.W, 'visible_biases':
                self.b_v, 'hidden_biases': self.b_h}
```

```python
print("RBM Pre-training Completed !")

self.rbm_validation(x=test_X[:20,:], y=test_Y[:20])

print(f"Classification training starts")

loss_fn = tf.keras.losses.SparseCategoricalCrossentropy(
from_logits=True,reduction=tf.keras.losses.Reduction.SUM)
optimizer = tf.keras.optimizers.Adam(self.lr)
for i in range(self.num_epochs):
    np.random.shuffle(order)
    train_X, train_Y = train_X[order], train_Y[order]
    loss_epoch = 0
    accuracy = 0
    for batch in range(batches):
        X_batch = train_X[batch*self.batch_size:(batch+1)*
        self.batch_size]
        y_batch = train_Y[batch*self.batch_size:(batch+1)*
        self.batch_size]
        X_batch, y_batch = tf.constant(X_batch),
        tf.constant(y_batch)
        with tf.GradientTape() as tape:
            y_pred_batch = self.train_classifier_step(X_batch)
            loss_ = loss_fn(y_batch,y_pred_batch)

        gradients = tape.gradient(loss_,(self.W,self.b_h,self.
        Wf,self.b_f))
        optimizer.apply_gradients(zip(gradients,(self.W,self.b_h,
        self.Wf,self.b_f)))
        loss_epoch += loss_.numpy()
        accuracy += np.sum(y_batch.numpy() == np.argmax(y_pred_
        batch.numpy(),axis=1))
    if i % 5 == 0:
        print(f"Epoch {i}, Training loss: {loss_epoch/float
        (train_X.shape[0])}, Training Accuracy: {accuracy/float
        (train_X.shape[0])}")
```

```
            self.model_params = {'weights': self.W, 'visible_biases':
            self.b_v,
                        'hidden_biases': self.b_h, 'weights_final':
                        self.Wf, 'final_biases':self.b_f}

        self.classifier_test_accuracy(test_X,test_Y)

rbm_pretrained_classifier = rbm_pretrained_classifier(n_visible=28*28,
num_epochs=6,n_hidden=500)
rbm_pretrained_classifier.train_model()

--Output--
RBM Pre-training Completed !
Classification training starts
Epoch 0, Training loss: 0.26531310059229535, Training Accuracy: 0.91905
Epoch 5, Training loss: 0.00810452818373839, Training Accuracy:
0.9969833333333333
Test Classification Accuracy: 0.9781
```

As we can see from the preceding output, with the pretrained weights from RBM used as initial weights for the classification network, we can get good accuracy of around 95% on the MNIST test dataset by just running it for few epochs. This is impressive given that the network does not have any convolutional layers.

Autoencoders

Autoencoders are unsupervised artificial neural networks that are used to generate a meaningful internal representation of the input data. The autoencoder network generally consists of three layers—the input layer, the hidden layer, and the output layer. The input layer and hidden layer combination acts as the encoder, while the hidden layer and output layer combination acts as the decoder. The encoder tries to represent the input as a meaningful representation at the hidden layer, while the decoder reconstructs the input back into its original dimension at the output layer. Typically, some cost function between the reconstructed input and the original input is minimized as part of the training process.

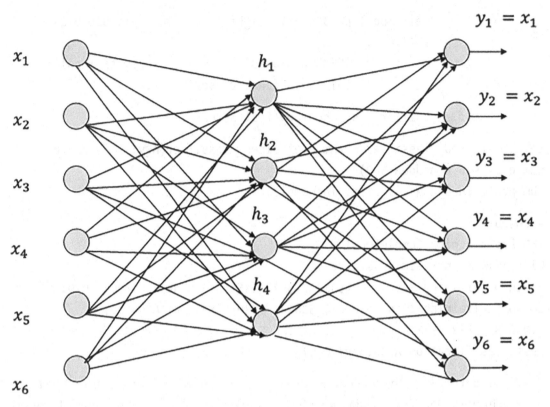

Figure 5-13. *Architecture of a basic autoencoder*

Figure 5-13 represents a basic autoencoder with one hidden layer and the input and the output layers. The input $x = [x_1 \, x_2...x_6]^T \in R^{6 \times 1}$, while the hidden layer $h = [h_1 \, h_2.. \, h_4]^T \in R^{4 \times 1}$. The output y is chosen to be equal to x so that the error between the reconstructed input $\hat{y}$ can be minimized to get a meaningful representation of the input in the hidden layer. For generality purposes, let's take the following:

$$x = \begin{bmatrix} x_1 \, x_2...x_n \end{bmatrix}^T \in R^{n \times 1}$$

$$h = \begin{bmatrix} h_1 \, h_2...h_d \end{bmatrix}^T \in R^{d \times 1}$$

$$y = x = \begin{bmatrix} y_1 \, y_2...y_6 \end{bmatrix}^T \in R^{n \times 1}$$

Let the weights from x to h be represented by the weight matrix $W \in R^{d \times n}$ and the biases at the hidden unit be represented by $b = [b_1 \, b_2...b_d]^T \in R^{d \times 1}$.

Similarly, let the weights from h to y be represented by the weight matrix $W' \in R^{n \times d}$ and the biases at the output units be represented by $b' = [b_1\ b_2...b_n]^T \in R^{n \times 1}$.

The output of the hidden unit can be expressed as follows:

$$h = f_1(Wx + b)$$

where f_1 is the element-wise activation function at the hidden layer. The activation function can be linear, ReLU, sigmoid, and so forth depending on its use.

Similarly, the output of the output layer can be expressed as follows:

$$\hat{y} = f_2(W'h + b')$$

If the input features are of a continuous nature, one can minimize a least square-based cost function as follows to derive the model weights and biases based on the training data:

$$C = \sum_{k=1}^{m} \|\hat{y}^{(k)} - y^{(k)}\|_2^2 = \sum_{k=1}^{m} \|\hat{y}^{(k)} - x^{(k)}\|_2^2$$

where $\|\hat{y}^{(k)} - x^{(k)}\|_2^2$ is the Euclidean or the l^2 norm distance between the reconstructed output vector and the original input vector and m is the number of data points on which the model is trained on.

If we represent all the parameters of the model by the vector, $\theta = [W; b; W'; b']$, then the cost function C can be minimized with respect to all the parameters of the model θ to derive the model

$$\hat{\theta} = \underbrace{\text{argmin}}_{\theta} C(\theta) = \underbrace{\text{argmin}}_{\theta} \sum_{k=1}^{m} \|\hat{y}^k - x^k\|_2^2$$

$$\hat{\theta} = \underset{\theta}{\text{argmin}}\ C(\theta) = \underset{\theta}{\text{argmin}} \sum_{k=1}^{m} \|\hat{y}^{(k)} - x^{(k)}\|_2^2$$

The learning rule of the model as per gradient descent is as follows:

$$\theta^{(t+1)} = \theta^{(t)} - \eta \nabla_\theta C(\theta^{(t)})$$

where η is the learning rate, t represents the iteration number, and $\nabla_\theta C(\theta^{(t)})$ is the gradient of the cost function with respect to θ at $\theta = \theta^{(t)}$.

Now, let us consider several cases as follows:

- When the dimension of the hidden layer is less than that of the input layer, i.e., $(d < n)$ where d is the hidden layer and n is the dimension of input layer, then the autoencoder works as a data-compression network that projects the data from a high-dimensional space to a lower-dimensional space given by the hidden layer. This is a lossy data-compression technique. It can also be used for noise reduction in the input signal.

- When $(d < n)$ and all the activation functions are linear, then the network learns to do a linear PCA (principal component analysis).

- When $(d \geq n)$ and the activation functions are linear, then the network may learn an identity function, which might not be of any use. However, if the cost function is regularized to produce a sparse hidden representation, then the network may still learn an interesting representation of the data.

- Complex nonlinear representations of input data can be learned by having more hidden layers in the network and by making the activation functions nonlinear. A schematic diagram of such a model is represented in Figure 5-14. When taking multiple hidden layers, it is a must that one takes nonlinear activation functions to learn nonlinear representations of data since several layers of linear activations are equivalent to a single linear activation layer.

Feature Learning Through Autoencoders for Supervised Learning

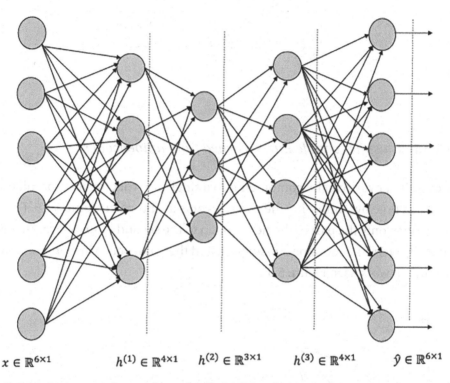

$$x \in \mathbb{R}^{6 \times 1} \qquad h^{(1)} \in \mathbb{R}^{4 \times 1} \quad h^{(2)} \in \mathbb{R}^{3 \times 1} \quad h^{(3)} \in \mathbb{R}^{4 \times 1} \quad \hat{y} \in \mathbb{R}^{6 \times 1}$$

Figure 5-14. *Autoencoder with multiple hidden layers*

When we deal with multiple hidden layers, as shown in Figure 5-14, and have a nonlinear activation function in the neural units, then the hidden layers learn nonlinear relations between the variables of the input data. If we are working on a classification-related problem of two classes where the input data is represented by $x \in R^{6 \times 1}$, we can learn interesting nonlinear features by training the autoencoder as in Figure 5-14 and then using the output of the second hidden layer vector $h^{(2)} \in R^{3 \times 1}$. This new nonlinear feature representation given by $h^{(2)}$ can be used as the input to a classification model, as shown in Figure 5-15. When the hidden layer whose output we are interested in has a dimensionality less than that of the input, it is equivalent to the nonlinear version of principal component analysis wherein we are just consuming the important nonlinear features and discarding the rest as noise.

$h^{(2)} \in \mathbb{R}^{3 \times 1}$

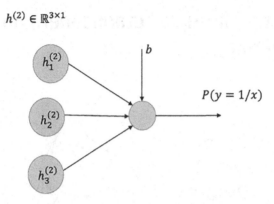

Figure 5-15. *Classifier with features learned from autoencoder*

The overall network can be combined into a single network for class-prediction purpose at test time by combining the two networks as shown in Figure 5-16. From the autoencoder, only the part of the network up to the second hidden layer that is producing output $h^{(2)}$ needs to be considered, and then it needs to be combined with the classification network as in Figure 5-15.

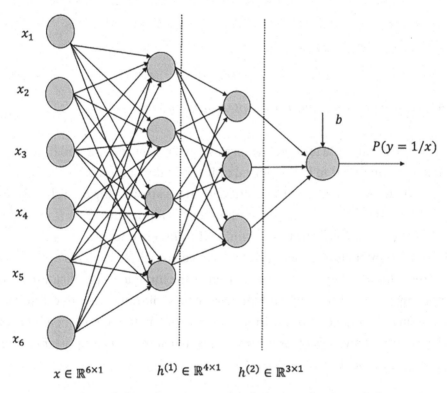

$x \in \mathbb{R}^{6 \times 1}$ $h^{(1)} \in \mathbb{R}^{4 \times 1}$ $h^{(2)} \in \mathbb{R}^{3 \times 1}$

Figure 5-16. *Combined classification network for prediction of classes*

One might ask the obvious question: Why would linear PCA not suffice for the task the autoencoder is performing in this example? Linear PCA or principal component analysis only takes care of capturing the linear relationship between input variables and tries to decompose the input variables into components that are not linearly dependent on each other. These components are called *principal components* and are uncorrelated with each other, unlike the original input variables. However, the input variables are not always going to be related in a linear fashion that leads to linear correlation. Input variables might be correlated in much more complex, nonlinear ways, and such nonlinear structures within the data can be captured only through nonlinear hidden units in the autoencoders.

Kullback-Leibler (KL) Divergence

The KL divergence measures the disparity or divergence between two random variables. If two random variables X_1 and X_2 have distributions $P(X_1 = x)$ and $Q(X_2 = x)$, then the KL divergence of Q with respect to P is given by the following:

$$KL(P\|Q) = E_{x \sim P(x)} \log \frac{P(x)}{Q(x)}$$

If P and Q are discrete probability distributions over a domain χ, we can express KL divergence in a sum form as the following:

$$KL(P\|Q) = \sum_{x \in \chi} P(x) \log \frac{P(x)}{Q(x)}$$

Alternately if P and Q are continuous probability distributions, we can write KL divergence in an integral form as follows:

$$KL(P\|Q) = \int_{\chi} P(x) \log \frac{P(x)}{Q(x)} dx$$

Now let us look at some of the properties of KL divergence:

- When the distribution of P and Q are same, i.e., $P(x) = Q(x)$; $\forall\, x \in \chi$, then KL divergence is zero:

$$KL(P\|Q) = \int_{\chi} P(x)\log\frac{P(x)}{Q(x)}dx = \int_{\chi} P(x)\log\frac{P(x)}{P(x)}dx = \int_{\chi} P(x)\log 1\,dx = 0$$

- At a given point x, if $P(x)$ is exceedingly larger than $Q(x)$, then it would increase the KL divergence. This is because the log ratio $\dfrac{P(x)}{Q(x)}$ is going to be large in this case weighted by a large value of the density $P(x)$. Hence KL divergence in some sense gives the disparity between two probability distributions over a common domain.

- At a given point x if $Q(x)$ is exceedingly larger than $P(x)$, then it would not have a very high impact on KL divergence since in this case the relatively smaller $P(x)$ would weigh the high negative log ratio $\dfrac{P(x)}{Q(x)}$ down.

- The KL divergence is not symmetric since for two different distributions,

$$\int_{\chi} P(x)\log\frac{P(x)}{Q(x)}dx \neq \int_{\chi} Q(x)\log\frac{Q(x)}{P(x)}dx$$

Hence $KL(P\|Q) \neq KL(Q\|P)$. The non-symmetry property of KL divergence prevents it from being a distance measure.

For Bernoulli random variables X and Y having means of ρ_1 and ρ_2, respectively, the KL divergence between the variables X and Y is given by the following:

$$KL(\rho_1\|\rho_2) = \rho_1\log\left(\frac{\rho_1}{\rho_2}\right) + (1-\rho_1)\log\left(\frac{1-\rho_1}{1-\rho_2}\right)$$

From the preceding expression, we can see that the KL divergence is 0 when

$\rho_1 = \rho_2$; i.e., when both distributions are identical. When $\rho_1 \neq \rho_2$, the KL divergence increases monotonically with the difference of the means. If ρ_1 is chosen to be 0.2, then the KL divergence versus the ρ_2 plot is as expressed in Figure 5-17.

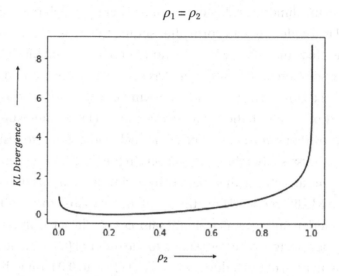

Figure 5-17. *KL divergence plot for mean $\rho_1 = 0.2$*

As we can see, the KL divergence is at its minimum at $\rho_2 = \rho_1 = 0.2$ and increases monotonically on either side of $\rho_2 = 0.2$. We will be using KL divergence to introduce sparsity in the hidden layer for sparse autoencoders in the next section.

Sparse Autoencoders

The purpose of autoencoders, as we discussed earlier, is to learn interesting hidden structures of the input data or, more specifically, to learn interesting relations among the different variables in the input. The most common way to derive these hidden structures is to make the hidden layer dimension smaller than the input data dimension so that the autoencoder is forced to learn a compressed representation of the input data. This compressed representation is forced to reconstruct the original data, and hence the compressed representation should have enough information to capture the input sufficiently well. This compressed representation will only be able to capture the input data efficiently if there is redundancy in the data in the form of correlation and other nonlinear associations between the input variables. If the input features are relatively independent, then such compression would not be able to represent the original data well. So, for the autoencoder to give an interesting low-dimensional representation of the input data, the data should have enough structure in it in the form of correlation and other nonlinear associations between input variables.

One thing that we touched upon earlier is that when the number of hidden layer units is larger than the dimensionality of the input, there is a high possibility that the autoencoder will learn identity transform after setting the weights corresponding to the extra hidden layers to zero. In fact, when the number of input and hidden layers is the same, the optimal solution for the weight matrix connecting the input to the hidden layer is the identity matrix. However, even when the number of hidden units is larger than the dimensionality of the input, the autoencoder can still learn interesting structures within the data, provided some constraints. One such constraint is to restrict the hidden layer output to be sparse so that those activations in the hidden layer units on average are close to zero. We can achieve this sparsity by adding a regularization term to the cost function based on KL divergence. Here, the ρ_1 will be very close to zero, and the average activation in the hidden unit over all training samples would act as the ρ_2 for that hidden unit. Generally, ρ_1 is selected to be very small, to the order of 0.04, and hence if the average activation in each of the hidden units is not close to 0.04, then the cost function would be penalized.

Let $h \in R^{d \times 1}$ be the hidden layer sigmoid activations of the input $x \in R^{n \times 1}$ where $d > n$. Further, let the weights connecting the inputs to the hidden layer be given by $W \in R^{d \times n}$ and the weights connecting the hidden layer to the output layer be given by $W' \in R^{n \times d}$. If the bias vectors at the hidden and output layers are given by b and b', respectively, then the following relationship holds true:

$$h^{(k)} = \sigma\left(Wx^{(k)} + b\right)$$

$$\hat{y}^{(k)} = f\left(W'h^{(k)} + b'\right)$$

where $h^{(k)}$ and $\hat{y}^{(k)}$ are the hidden layer output vector and the reconstructed input vector for the kth input training data point. The cost function to be minimized with respect to the parameters of the model (i.e., W, W', b, b') is given by the following:

$$C = \sum_{k=1}^{m} \|\hat{y}^{(k)} - x^{(k)}\|_2^2 + \lambda \sum_{j=1}^{d} KL(\rho \| \hat{\rho}_j)$$

where $\hat{\rho}_j$ is the average activation in the *j-th* unit of the hidden layer over all the training samples and can be represented as follows. Also, $h_j^{(k)}$ represents the hidden layer activation at unit *j* for the *kth* training sample.

$$\hat{\rho}_j = \frac{1}{m} \sum_{k=1}^{m} h_j^{(k)}$$

$$KL\left(\rho \| \hat{\rho}_j\right) = \rho \log\left(\frac{\rho}{\hat{\rho}_j}\right) + (1-\rho)\log\left(\frac{1-\rho}{1-\hat{\rho}_j}\right)$$

Generally, ρ is selected as 0.04 to 0.05 so that the model learns to produce average hidden layer unit activations very close to 0.04, and in the process the model learns sparse representation of the input data in the hidden layer.

Sparse autoencoders are useful in computer vision to learn low-level features that represent the different kinds of edges at different locations and orientations within the natural images. The hidden layer output gives the weight of each of these low-level features, which can be combined to reconstruct the image. If 10×10 images are processed as 100-dimensional input, and if there are 200 hidden units, then the weights connecting input to hidden units—i.e., W or hidden units to the output reconstruction layer W'— would comprise 200 images of size 100 (10×10). These images can be displayed to see the nature of the features they represent. Sparse encoding works well when supplemented with PCA whitening, which we will discuss briefly later in this chapter.

Sparse Autoencoder Implementation in TensorFlow

In this section, we will implement a sparse autoencoder that has more hidden units than the input dimensionality. The dataset for this implementation is the MNIST dataset. Sparsity has been introduced in the implemented network through KL divergence. Also, the weights of the encoder and decoder are used for L2 regularization to ensure that in the pursuit of sparsity, these weights don't adjust themselves in undesired ways. The autoencoder and decoder weights represent overrepresented basis, and each of these basis tries to learn some low-level feature representations of the images, as discussed earlier. The encoder and the decoder are taken to be the same. These weights that make up the low-level feature images have been displayed to highlight what they represent. The detailed implementation has been outlined in Listing 5-4.

Listing 5-4. Sparse Autoencoder implementation using Tensorflow

```python
import tensorflow as tf
import numpy as np
import matplotlib.pyplot as plt
%matplotlib inline
import time
from tensorflow.keras import layers, Model

class encoder(Model):
    def __init__(self,n_hidden):
        super(encoder,self).__init__()
        self.fc1 = layers.Dense(n_hidden,activation='sigmoid')

    def call(self,x):
        out = self.fc1(x)
        return out

class decoder(Model):
    def __init__(self,n_input):
        super(decoder,self).__init__()
        self.fc1 = layers.Dense(n_input,activation='sigmoid')

    def call(self,x):
        out = self.fc1(x)
        return out

class encoder_decoder(Model):
    def __init__(self,encoder,decoder):
        super(encoder_decoder,self).__init__()
        self.encoder = encoder
        self.decoder = decoder

    def call(self,x):
        h = self.encoder(x)
        # The average activation rho(bernoulli probability) should be
          constrained to be small
        # and similar to a target rho using KL divergence.
        rho = tf.reduce_mean(h,0)
```

```python
        out = self.decoder(x)
        return h, out, rho

def logfunc(x, x2):
    return tf.multiply( x, tf.math.log(tf.math.divide(x,x2)))

# KL divergence loss to make the per latent dimension Bernoulli activation
probability rho small and
# similar to a target rho rho_hat induce sparsity

def KL_Div(rho, rho_hat):
    invrho = tf.subtract(tf.constant(1.), rho)
    invrhohat = tf.subtract(tf.constant(1.), rho_hat)
    logrho = logfunc(rho,rho_hat) + logfunc(invrho, invrhohat)
    return logrho

def data_load():
    (train_X, train_Y), (test_X, test_Y) = tf.keras.datasets.mnist.
load_data()
    train_X, test_X , = train_X.reshape(-1,28*28), test_X.reshape(-1,28*28)
    train_X, test_X = train_X/255.0, test_X/255.0
    return np.float32(train_X), train_Y, np.float32(test_X), test_Y

def train(n_hidden,n_input,rho,lr,num_epochs,batch_size):

    enc = encoder(n_hidden=n_hidden)
    dec = decoder(n_input=n_input)
    model = encoder_decoder(encoder=enc, decoder=dec)
    model_graph = tf.function(model)

    train_X,train_Y, test_X, test_Y = data_load()
    num_train_recs = train_X.shape[0]
    order = np.arange(num_train_recs)
    num_batches = num_train_recs//batch_size

    optimizer = tf.keras.optimizers.Adam(lr)
```

```python
    for i in range(num_epochs):
        epoch_loss = 0
        np.random.shuffle(order)
        train_X = train_X[order]
        for batch in range(num_batches):
            X_batch = train_X[batch*batch_size:(batch+1)*batch_size]
            X_batch = tf.constant(X_batch)
            bsize = X_batch.shape[0]

            with tf.GradientTape() as tape:
                _, X_pred_batch, rho_hat = model_graph(X_batch)
                reconst_loss = tf.reduce_mean(tf.pow(X_batch - X_pred_
                batch, 2))
                sparsity_loss =  0.01*tf.reduce_sum(KL_Div(rho,rho_hat))
                l2_reg_loss = 0.0001*tf.nn.l2_loss(model_graph.encoder.fc1.
                variables[0])
                #print(reconst_loss,sparsity_loss)
                loss = reconst_loss + sparsity_loss + l2_reg_loss

            gradients = tape.gradient(loss, model.trainable_variables)
            optimizer.apply_gradients(zip(gradients, model.trainable_
            variables))
            epoch_loss += loss.numpy()*bsize
        if i % 5 == 0:
            print(f"Training Loss at epoch {i}: {epoch_loss/num_train_recs}")

    # Check the quality of the 1st 10 test image reconstruction
    _,X_pred_test,_ = model_graph(tf.constant(test_X[:10,:]))
    f, a = plt.subplots(2, 10, figsize=(10, 2))
    for i in range(10):
        a[0][i].imshow(np.reshape(test_X[i,:], (28, 28)),cmap='gray')
        a[1][i].imshow(np.reshape(X_pred_test[i], (28, 28)),cmap='gray')
    return model, model_graph

model, model_graph = train(n_hidden=1024,n_input=28*28,rho=0.2,lr=0.01,
num_epochs=100,batch_size=256)

-- Output --
```

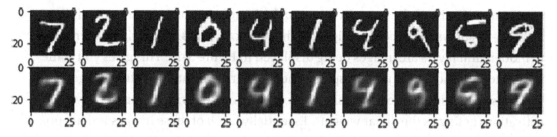

Figure 5-18. *Display of the original image followed by the reconstructed image*

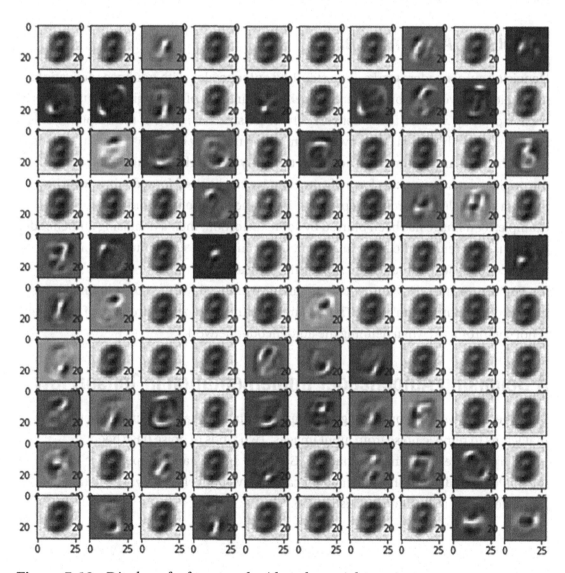

Figure 5-19. *Display of a few encoder/decoder weights as images*

Figure 5-18 shows the reconstruction of the images by sparse encoders, while Figure 5-19 shows the decoder weights in the form of images. The weights corresponding to hidden unit layers are images being displayed in Figure 5-19. This gives you some idea as to the kind of features the sparse encoders are learning. The final image that is reconstructed is the linear combination of these images, with the hidden layer activations acting as the linear weights. Essentially, each of these images is detecting low-level features in the form of hand strokes for the written digits. In terms of linear algebra, these images form a basis for representing the reconstructed images.

Denoising Autoencoder

Denoising autoencoders works like a standard autoencoder, with nonlinear activations in the hidden layers, the only difference being that instead of the original input x, a noisy version of x, say $\tilde{x}$, is fed to the network. The reconstructed image at the output layer is compared with the actual input x while computing the error in reconstruction. The idea is that the hidden structured learned from the noisy data is rich enough to reconstruct the original data. Hence, this form of autoencoder can be used for reducing noise in the data since it learns a robust representation of the data from the hidden layer. For example, if an image has been blurred by some distortion, then a denoising autoencoder can be used to remove the blur. An autoencoder can be converted into a denoising autoencoder by just introducing a stochastic noise addition unit.

For images, denoising autoencoders can have hidden layers as convolutional layers instead of standard neural units. This ensures that the topological structure of the image is not compromised when defining the autoencoder network.

A Denoising Autoencoder Implementation in TensorFlow

In this section, we will work through the implementation of a denoising autoencoder that learns to remove noise from input images. Two kinds of noise have been introduced to the input images—namely, Gaussian and salt and pepper noise—and the implemented denoising autoencoder can remove both efficiently. The detailed implementation is illustrated in Listing 5-5.

Listing 5-5. Denoising Autoencoder Using Convolution and
Deconvolution Layers

```python
# Import the required library
from tensorflow.keras import layers, Model
from skimage import transform
import numpy as np
import matplotlib.pyplot as plt

class encoder(Model):
    def __init__(self):
        super(encoder,self).__init__()
        self.conv1 = layers.Conv2D(filters=32,kernel_size=5,strides=(2,2),
        padding='SAME')
        self.batchnorm1 = layers.BatchNormalization()
        self.act1 = layers.LeakyReLU()
        self.conv2 = layers.Conv2D(filters=16,kernel_size=5,strides=(2,2),
        padding='SAME')
        self.batchnorm2 = layers.BatchNormalization()
        self.act2 = layers.LeakyReLU()
        self.conv3 = layers.Conv2D(filters=8,kernel_size=5,strides=(4,4),
        padding='SAME')
        self.batchnorm3 = layers.BatchNormalization()
        self.act3 = layers.LeakyReLU()

    def call(self,x):
        x = self.conv1(x)
        x = self.batchnorm1(x)
        x = self.act1(x)
        x = self.conv2(x)
        x = self.batchnorm2(x)
        x = self.act2(x)
        x = self.conv3(x)
        x = self.batchnorm3(x)
        x = self.act3(x)
        return x
```

```python
class decoder(Model):
    def __init__(self):
        super(decoder,self).__init__()
        self.conv1 = layers.Conv2DTranspose(filters=16,kernel_size=5,
        strides=(4,4),padding='SAME')
        self.batchnorm1 = layers.BatchNormalization()
        self.act1 = layers.LeakyReLU()
        self.conv2 = layers.Conv2DTranspose(filters=32,kernel_size=5,
        strides=(2,2),padding='SAME')
        self.batchnorm2 = layers.BatchNormalization()
        self.act2 = layers.LeakyReLU()
        self.conv3 = layers.Conv2DTranspose(filters=1,kernel_size=5,
        strides=(2,2),padding='SAME')

    def call(self,x):
        x = self.conv1(x)
        x = self.batchnorm1(x)
        x = self.act1(x)
        x = self.conv2(x)
        x = self.batchnorm2(x)
        x = self.act2(x)
        x = self.conv3(x)
        return x

class encoder_decoder(Model):
    def __init__(self,encoder,decoder):
        super(encoder_decoder,self).__init__()
        self.encoder = encoder
        self.decoder = decoder

    def call(self,x):
        x = self.encoder(x)
        x = self.decoder(x)
        return x
```

```python
def data_load():
    (train_X, train_Y), (test_X, test_Y) =
    tf.keras.datasets.mnist.load_data()
    train_X, test_X , = train_X.reshape(-1,28*28), test_X.reshape(-1,28*28)
    train_X, test_X = train_X/255.0, test_X/255.0
    return np.float32(train_X), train_Y, np.float32(test_X), test_Y

def resize_batch(imgs):
    # Resize to 32 for convenience
    imgs = imgs.reshape((-1, 28, 28, 1))
    resized_imgs = np.zeros((imgs.shape[0], 32, 32, 1))
    for i in range(imgs.shape[0]):
        resized_imgs[i, ..., 0] = transform.resize(imgs[i, ..., 0], (32, 32))
    return np.float32(resized_imgs)

# Introduce Gaussian Noise
def gaussian_noise(image):
    row,col= image.shape
    mean,var = 0, 0.1
    sigma = var**0.5
    gauss = np.random.normal(mean,sigma,(row,col))
    gauss = gauss.reshape(row,col)
    noisy_image = image + gauss
    return np.float32(noisy_image)

def create_noisy_batch(image_arr,noise_profile='gaussian'):
    image_arr_out = []

    for i in range(image_arr.shape[0]):
        img = image_arr[i,:,:,0]
        if noise_profile == 'gaussian':
            img = gaussian_noise(img)
        else:
            img = salt_pepper_noise(img)
        image_arr_out.append(img)
    image_arr_out = np.array(image_arr_out)
    image_arr_out = image_arr_out.reshape(-1,32,32,1)
    return image_arr_out
```

```
# Introduce Salt and Pepper Noise
def salt_pepper_noise(image):
    row,col = image.shape
    s_vs_p = 0.5
    amount = 0.05
    out = np.copy(image)
    # Salt mode
    num_salt = np.ceil(amount * image.size * s_vs_p)
    coords = [np.random.randint(0, i - 1, int(num_salt)) for i in
    image.shape]
    out[coords] = 1
    # Pepper mode
    num_pepper = np.ceil(amount* image.size * (1. - s_vs_p))
    coords = [np.random.randint(0, i - 1, int(num_pepper)) for i in
    image.shape]
    out[coords] = 0
    return out

def train(lr,num_epochs=100,batch_size=256,noise_profile='gaussian'):
    enc = encoder()
    dec = decoder()
    model = encoder_decoder(encoder=enc, decoder=dec)
    model_graph = tf.function(model)

    train_X,train_Y, test_X, test_Y = data_load()
    num_train_recs = train_X.shape[0]
    order = np.arange(num_train_recs)
    num_batches = num_train_recs//batch_size

    optimizer = tf.keras.optimizers.Adam(lr)
    loss_fn = tf.keras.losses.MeanSquaredError(reduction=tf.keras.losses.
    Reduction.SUM)

    for i in range(num_epochs):
        epoch_loss = 0
        np.random.shuffle(order)
        train_X = train_X[order]
```

```python
    for batch in range(num_batches):
        X_batch = train_X[batch*batch_size:(batch+1)*batch_size]
        #X_batch = X_batch.resize(-1,28,28,1)
        X_batch = resize_batch(X_batch) # Output as its the clean image
        X_batch_noisy = create_noisy_batch(X_batch,noise_profile=noise_
        profile) # Noisy input image

        X_batch_noisy , X_batch = tf.constant(X_batch_noisy),
        tf.constant(X_batch)

        with tf.GradientTape() as tape:
            X_pred_batch = model_graph(X_batch_noisy)
            loss = loss_fn(X_batch,X_pred_batch)

        gradients = tape.gradient(loss, model.trainable_variables)
        optimizer.apply_gradients(zip(gradients, model.trainable_
        variables))
        epoch_loss += loss.numpy()
    if i % 5 == 0:
        print(f"Training Loss at epoch {i}: {epoch_loss/num_
        train_recs}")

# Check the quality of the 1st 50 test image reconstruction
X_batch = test_X[:50,:]
X_batch_resize = resize_batch(X_batch)
X_batch_resize_noisy = create_noisy_batch(X_batch_resize,noise_
profile=noise_profile)
X_pred_test = model_graph(tf.constant(X_batch_resize_noisy))
# plot the reconstructed images and their ground truths (inputs)
plt.figure(1)
plt.title('Reconstructed Images')
for i in range(50):
    plt.subplot(5, 10, i+1)
    plt.imshow(X_pred_test[i, ..., 0], cmap='gray')
plt.figure(2)
plt.title('Input Images with Gaussian Noise')
for i in range(50):
```

```
        plt.subplot(5, 10, i+1)
        plt.imshow(X_batch_resize_noisy[i, ..., 0], cmap='gray')
    plt.show()
    return model, model_graph
```

```
model, model_graph = train(lr=0.01,num_epochs=20,batch_size=256,
noise_profile='gaussian')
```

--Output--

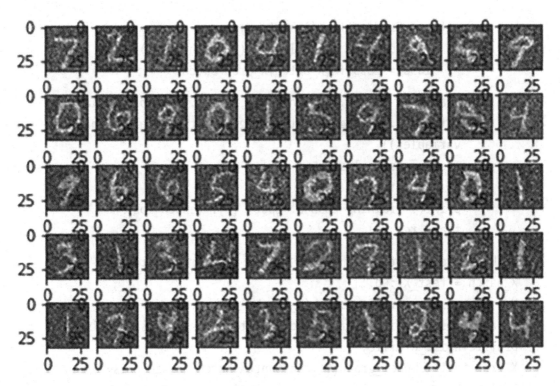

Figure 5-20. *Images with Gaussian noise*

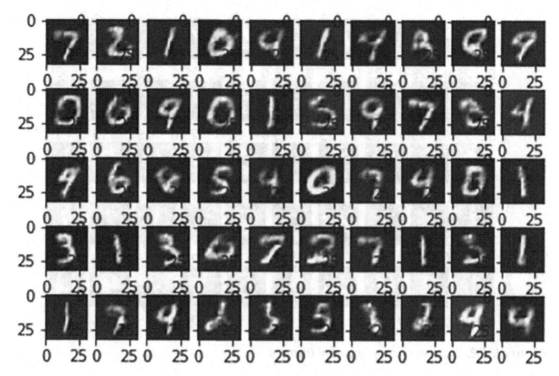

Figure 5-21. *Reconstructed images (without Gaussian noise) generated by the denoising autoencoder*

We can see from Figure 5-20 and Figure 5-21 that the Gaussian noise has been removed by the denoising autoencoders.

```
model, model_graph = train(lr=0.01,num_epochs=20,batch_size=256,
noise_profile='salt_and_pepper')
```

--Output--

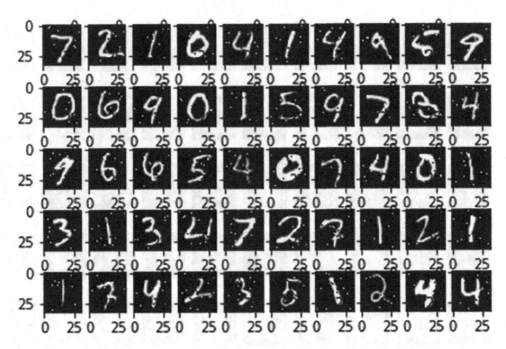

Figure 5-22. *Salt and pepper noisy images*

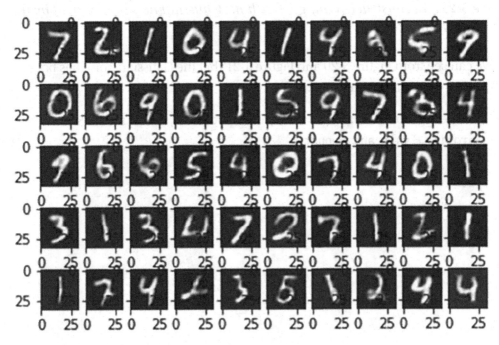

Figure 5-23. *Reconstructed images without the salt and pepper noise generated by the denoising autoencoder*

From Figure 5-22 and Figure 5-23, it is evident that the denoising autoencoder does a good job of removing the salt and pepper noise. Do note that the autoencoders are trained separately, once for handling Gaussian noise and once for handling salt and pepper noise.

Variational Autoencoders

Variational autoencoders are a special form of autoencoders that can be used as generative models and are based on variational inference. Generally, in an autoencoder, the encoder has a fixed mapping from an observable x to a latent variable z, while in a variational autoencoder, the encoder maps the observable x to a probability distribution $P(z|x)$. To appreciate variation autoencoder, let us spend some time on variational inference.

Variational Inference

To understand variational inference, we assume a generative process where the latent variable z generates the observable x through some probability distribution $p(x|z)$. We want to learn about the latent variable z by observing x. In other words, we are interested in the posterior distribution $p(z|x)$.

The posterior $p(z|x)$ can be expressed as in the following using Bayes rule:

$$p(z|x) = \frac{p(x,z)}{p(x)}$$

The evidence $p(x)$ can be expressed as follows:

$$p(x) = \int_z p(x|z)p(z)dz$$

Computing this integral is often intractable especially when z is very high dimensional. This makes computing $p(x)$ intractable.

One of the ways to get rid of the problem is to compute the integral using Markov chain Monte Carlo methods such as Metropolis-Hastings algorithm or Gibbs sampling. The integral computation in such cases would have low bias but high variance.

The other way to deal with the problem is to resort to variational inference where we try to approximate the posterior $p(z|x)$ by another distribution $q_\theta(z|x)$ by minimizing the KL(Kullback-Leibler) divergence between $q_\theta(z|x)$ and $p(z|x)$. Remember we have studied earlier in this chapter that KL (Kullback-Leibler) divergence captures the discrepancy or distance between two probability distributions. The variational inference premises is captured in Figure 5-24 (see below).

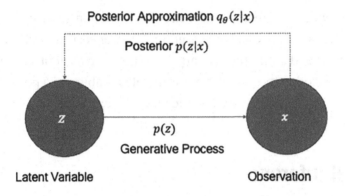

Understand Latent Variable z from observation x through the posterior $p(z|x)$ or through its approximation $q_\theta(z|x)$ using Variational Inference

Figure 5-24. *Variational inference setup*

The KL divergence between the approximate posterior distribution $q_\theta(z|x)$ and the true posterior distribution is given by the following:

$$KL(q\|p) = E_{z \sim q_\theta(z|x)} \log \frac{q_\theta(z|x)}{p(z|x)}$$

We would like to minimize the KL divergence with respect to the parameter set θ. Before we work on the minimization, let us try to simplify the KL divergence expression that would be convenient to have.

$$KL(q\|p) = E_{z \sim q_\theta(z|x)} \log \frac{q_\theta(z|x)}{p(z|x)}$$

$$= E_{z \sim q_\theta(z|x)} \log \frac{q_\theta(z|x)p(x)}{p(x,z)}$$

$$= E_{z \sim q_\theta(z|x)} \log \frac{q_\theta(z|x)p(x)}{p(x|z)p(z)}$$

$$= E_{z \sim q_\theta(z|x)} \log \frac{q_\theta(z|x)}{p(x|z)p(z)} + E_{z \sim q_\theta(z|x)} \log p(x)$$

The $\log p(x)$ does not depend on the latent variable z and hence is a constant. This makes the KL divergence expression as follows:

$$KL(q\|p) = E_{z \sim q_\theta(z|x)} \log \frac{q_\theta(z|x)}{p(z|x)}$$

$$= E_{z \sim q_\theta(z|x)} \log \frac{q_\theta(z|x)}{p(x|z)p(z)} + \log p(x) E_{z \sim q_\theta(z|x)}(1)$$

$$= E_{z \sim q_\theta(z|x)} \log \frac{q_\theta(z|x)}{p(x|z)p(z)} + \log p(x)$$

If we rearrange the terms,

$$\log p(x) = KL(q\|p) - E_{z \sim q_\theta(z|x)} \log \frac{q_\theta(z|x)}{p(x|z)p(z)}$$

Since $\log p(x)$ is not dependent on z and is constant, minimization of the $KL(q\|p)$ which we had set out to do initially would simplify to maximization of the following:

$$-E_{z \sim q_\theta(z|x)} \log \frac{q_\theta(z|x)}{p(x|z)p(z)}$$

Also, since the KL divergence is always greater than zero,

$$\log p(x) \geq -E_{z \sim q_\theta(z|x)} \log \frac{q_\theta(z|x)}{p(x|z)p(z)}$$

Because the above holds true, $-E_{z \sim q_\theta(z|x)} \log \dfrac{q_\theta(z|x)}{p(x|z)p(z)}$ is called the lower bound on the evidence $\log p(x)$. This lower bound on the evidence is also abbreviated as ELBO. Minimizing the KL divergence is equivalent to maximizing the ELBO objective.

So, in variation inference, we can maximize $-E_{z \sim q_\theta(z|x)} \log \dfrac{q_\theta(z|x)}{p(x|z)p(z)}$ or rather minimize the cost $E_{z \sim q_\theta(z|x)} \log \dfrac{q_\theta(z|x)}{p(x|z)p(z)}$ with an objective to learn the encoder $q_\theta(z|x)$. We can represent the ELBO cost to be minimized as L and hence

$$L = E_{z \sim q_\theta(z|x)} \log \frac{q_\theta(z|x)}{p(x|z)p(z)}$$

Variational Autoencoder Objective from ELBO

In a variational autoencoder, the posterior approximation $q_\theta(z|x)$ can be modeled as an encoder that takes x as input and outputs a conditional distribution $q_\theta(z|x)$ over the latent space z. We want to have a latent space of z from which we can sample x using a generative process $p(x|z)$. Hence, we would like to constraint the latent space over z to a known distribution that we can sample from. An independent Gaussian distribution over the latent space z would be a good choice. Hence, we can constraint $z \in R^d$ to have an independent Gaussian prior with zero mean and unit variance.

$$p(z) = N(0, I_{d \times d})$$

Coming back to the generative process, we can define $p_\phi(x|z)$ through a decoder block which can take in z sampled from Gaussian prior $P(z)$ and give out a distribution $p_\phi(x|z)$. Given an observation x if z happens to be sampled from $q_\theta(z|x)$, then x should be highly probable under $p_\phi(x|z)$.

The constraint on z to belong to a zero mean unit variance independent Gaussian prior also enforces the posterior approximation distribution $q_\theta(z|x)$ to give out z as if it has been sampled from the prior $p(z)$. The variational autoencoder setup we just described is illustrated in Figure 5-25.

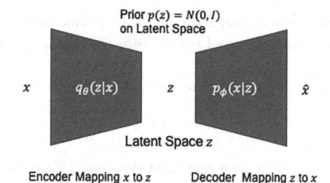

Figure 5-25. *Variational autoencoder block diagram*

Now let us see if we can rearrange the ELBO cost to give the desired capabilities to the variational autoencoders as we have discussed earlier in this section:

- $z \sim q_\theta(z|x)$ should look like samples from the prior distribution $p(z)$ so that we can use decoder $p_\phi(x|z)$ as a generative process that maps z to x.

- If z is sampled from $q_\theta(z|x)$, then $p_\phi(x|z)$ should be highly probable.

The ELBO cost that we last saw and would like to minimize can be simplified as follows:

$$L = E_{z \sim q_\theta(z|x)} \log \frac{q_\theta(z|x)}{p(x|z)p(z)}$$

$$= E_{z \sim q_\theta(z|x)} \log \frac{q_\theta(z|x)}{p(z)} - E_{z \sim q_\theta(z|x)} \log p(x|z)$$

The first term $E_{z \sim q_\theta(z|x)} \log \dfrac{q_\theta(z|x)}{p(z)}$ is nothing but the KL divergence of $q_\theta(z|x)$ with that of prior $p(z)$. So, to minimize the ELBO cost, we should minimize $E_{z \sim q_\theta(z|x)} \log \dfrac{q_\theta(z|x)}{p(z)}$ which would basically make the distribution $q_\theta(z|x)$ biased toward the prior $p(z)$.

The minimization of the second term $- E_{z \sim q_\theta(z|x)} \log p(x|z)$ would mean, given a latent variable z sampled from the approximate posterior $q_\theta(z|x)$ with x as input, the same (latent variable z) should be highly probable under $p_\phi(x|z)$. The distribution $p_\phi(x|z)$

495

is represented by a mean prediction $\hat{x}$ and constant covariance matrix $\sigma^2 I$. Hence, $-E_{z \sim q_\theta(z|x)} \log p(x|z)$ can be expressed as a reconstruction loss $L_R(x,\hat{x})$ in going from x to $\hat{x}$.

Hence the overall ELBO cost can be rewritten as follows:

$$L = KL\big(q_\theta(z|x) \| p(z)\big) + L_R(x,\hat{x})$$

In general, we use a scalar λ to weight the two losses in ELBO, and hence we can write the overall ELBO cost as follows:

$$L = KL\big(q_\theta(z|x) \| p(z)\big) + \lambda L_R(x,\hat{x})$$

If we take M training data points summing the loss for all M data points, we can express the loss as follows:

$$L = \sum_{i=0}^{M-1} KL\big(q_\theta(z_i|x_i) \| p(z_i)\big) + \lambda \sum_{i=0}^{M-1} L_R(x_i,\hat{x}_i)$$

The optimal estimates of encoder parameters θ and decoder parameters ϕ can be determined by minimizing L with respect to θ and ϕ.

$$\hat{\theta}, \hat{\phi} = \operatorname*{argmin}_{\theta,\phi} L = \operatorname*{argmin}_{\theta,\phi} \sum_{i=0}^{M-1} KL\big(q_\theta(z_i|x_i)\| p(z_i)\big) + \lambda \sum_{i=0}^{M-1} L_R(x_i,\hat{x}_i)$$

Implementation Details of the Variational Autoencoder

The reconstruction loss is generally taken as either the L2 or the L1 norm of the difference of the original input x and the reconstructed input $\hat{x}$, and hence in a generalized form, we can write the reconstruction loss as follows:

$$L_R(x_i,\hat{x}_i) = \| x - \hat{x} \|_p^p$$

where p=1 is for L1 norm reconstruction loss while $p = 2$ is for L2 norm reconstruction loss.

Since in variational inference we want to use the decoder to induce a generative process, the prior $p(z)$ is conveniently chosen as a zero mean and unit standard deviation Gaussian distribution. Also, the d components of $z \in R^d$ are conveniently chosen to be independent.

The encoder learns a mapping from the observation x to the conditional distribution $q_\theta(z|x)$. Since the sampled $z \sim q_\theta(z|x)$ should seem like they are from the prior distribution $p(z)$, it is convenient to assume that $q_\theta(z|x)$ is Gaussian distribution as well where the dimensions of z are independent of each other. Hence for a given observable x and its corresponding latent variable vector z sampled from $q_\theta(z|x)$, the following will hold true where $z = [z^{(0)}z^{(1)}..z^{(d-1)}]^T$:

$$q_\theta(z|x) = \prod_{j=0}^{d-1} q_\theta^{(j)}\left(z^{(j)}|x\right)$$

$$q_\theta^{(j)}\left(z^{(j)}|x\right) \sim N\left(u^{(j)}, \sigma^{(j)2}\right)$$

Since we have made the independent Gaussian distribution assumption over z, the encoder would output a mean $u^{(j)}$ and variance $\sigma^{(j)2}$ for each dimension of z. We should be able to sample each $z^{(j)}$ from $N(u^{(j)}, \sigma^{(j)2})$ to come up with the z that can be fed to the decoder. However, basic random sampling of $z^{(j)}$ based on $N(u^{(j)}, \sigma^{(j)2})$ would not suffice as we need to ensure that gradient can pass through the predicted $u^{(j)}$ and $\sigma^{(j)2}$ during backpropagation. The trick that we use is to sample a value $\epsilon \sim N(0, 1)$ and take the $z^{(j)}$ realization as follows:

$$z^{(j)} = u^{(j)} + \epsilon * \sigma^{(j)}$$

This is called the reparameterization trick and is a clever way to ensure that gradient can still flow through the predicted $u^{(j)}$ and $\sigma^{(j)2}$.

Now that we have a good understanding as to how variational autoencoders work and how they are based on variational inference, we will work on its implementation in the next section.

Implementation of Variational Autoencoder

In this section, we implement the variation autoencoder using MNIST data. One of the key aspects in variational autoencoder is the predicted posterior distribution $q_\theta(z|x)$ and how we sample z from it before it is passed through the decoder $p_\phi(x|z)$ to get the reconstructed version of x that we denote as $\hat{x}$. The other aspect is the KL divergence loss we minimize between posterior distribution $q_\theta(z|x)$ and prior $p(z)$ to ensure that z sampled from $q_\theta(z|x)$ can be made to look like coming from the prior $p(z)$. This allows us to sample z from $p(z)$ and construct realistic output images $\hat{x}$ through the decoder $p_\phi(x|z)$.

We would have to derive a form for KL divergence $\sum_{i=0}^{M-1} KL\big(q_\theta(z_i|x_i)\|p(z_i)\big)$ as it factors in the loss objective. Since we assume a zero mean unit variance Gaussian per latent dimension $z^{(j)}$, we can leverage the KL divergence formula for two univariate distributions. The KL divergence between two univariate normal distribution is given by the following:

$$KL\Big(N\big(u_1,\sigma_1^2\big)\|N\big(u_2,\sigma_2^2\big)\Big)=\log\frac{\sigma_2}{\sigma_1}+\frac{\sigma_1^2+\big(u_1-u_2\big)^2}{2\sigma_2^2}-\frac{1}{2}$$

Since we want to sample from a known distribution, we conveniently take the following:

$$p(z^{(j)}) = N(0,1)$$

Taking $q_\theta^{(j)}\big(z^{(j)}|x\big)=N\big(u_1,\sigma_1^2\big)$ and $p(z^{(j)}) = N(0,1)$, we have the following:

$$KL\Big(q_\theta^{(j)}\big(z^{(j)}|x\big)\|p\big(z^{(j)}\big)\Big)=-\log\sigma_1+\frac{\sigma_1^2+u_1^2}{2}-\frac{1}{2}$$

This expression for KL divergence would be used in the loss objective for the variational autoencoder training for each latent dimension $z^{(j)}$.

The detail implementation of the variational autoencoder implementation is outlined in Listing 5-6.

Listing 5-6. Variational Autoencoder Implementation

```python
import tensorflow as tf
import tensorflow.keras import layers, Model
import pandas a pd

# Variational AutoEncoder(VAE) Encoder
class encoder(Model):
    def __init__(self,n_hidden):
        super(encoder,self).__init__()
        self.n_hidden = n_hidden
        self.conv1 = layers.Conv2D(filters=64,kernel_size=4,strides=(2,2),
        padding='SAME')
        self.batchnorm1 = layers.BatchNormalization()
        self.act1 = layers.LeakyReLU(0.3)
        self.conv2 = layers.Conv2D(filters=64,kernel_size=4,strides=(2,2),
        padding='SAME')
        self.batchnorm2 = layers.BatchNormalization()
        self.act2 = layers.LeakyReLU(0.3)
        self.conv3 = layers.Conv2D(filters=64,kernel_size=4,strides=(2,2),
        padding='SAME')
        self.batchnorm3 = layers.BatchNormalization()
        self.act3 = layers.LeakyReLU(0.3)
        self.flatten = layers.Flatten()
        self.mean_layer = layers.Dense(n_hidden)
        self.log_var_layer  = layers.Dense(n_hidden)

    def call(self,x):
        x = self.conv1(x)
        x = self.batchnorm1(x)
        x = self.act1(x)
        x = self.conv2(x)
        x = self.batchnorm2(x)
        x = self.act2(x)
        x = self.conv3(x)
        x = self.batchnorm3(x)
        x = self.act3(x)
```

```python
        x = self.flatten(x)
        mean = self.mean_layer(x)
        log_var = self.log_var_layer(x)
        epsilon = tf.random.normal(tf.stack([tf.shape(x)[0],
        self.n_hidden]))
        z   = mean + tf.multiply(epsilon,tf.exp(0.5*log_var))
        return z, mean, log_var

# Variational AutoEncoder(VAE) Decoder
class decoder(Model):
    def __init__(self):
        super(decoder,self).__init__(0.3)
        self.fc1 = layers.Dense(32)
        self.act1 = layers.LeakyReLU(0.3)
        self.fc2 = layers.Dense(64)
        self.act2 = layers.LeakyReLU()
        self.conv1 = layers.Conv2DTranspose(filters=64,kernel_size=4,
        strides=(2,2),padding='SAME',activation='relu')
        self.conv2 = layers.Conv2DTranspose(filters=64,kernel_size=4,
        strides=(1,1),padding='SAME',activation='relu')
        self.conv3 = layers.Conv2DTranspose(filters=64,kernel_size=4,
        strides=(1,1),padding='SAME', activation='relu')
        self.flatten = layers.Flatten()
        self.fc3 = layers.Dense(32*32,activation='sigmoid')

    def call(self,x):
        x = self.fc1(x)
        x = self.act1(x)
        x = self.fc2(x)
        x = self.act2(x)
        x = tf.reshape(x,[-1,8,8,1])
        x = self.conv1(x)
        x = self.conv2(x)
        x = self.conv3(x)
        x = self.flatten(x)
```

```python
        x = self.fc3(x)
        x = tf.reshape(x,[-1,32,32,1])
        return x

# VAE model class
class encoder_decoder(Model):

    def __init__(self,encoder,decoder):
        super(encoder_decoder,self).__init__()
        self.encoder = encoder
        self.decoder = decoder

    def call(self,x):
        z, mean, log_var = self.encoder(x)
        x = self.decoder(z)
        return x, mean, log_var

# KL divergence loss
def KL_Div(mean,log_var):

    latent_loss = - 0.5 * tf.reduce_mean(log_var - tf.square(mean) -
    tf.exp(log_var) + 1)
    return latent_loss

def data_load():
    (train_X, train_Y), (test_X, test_Y) = tf.keras.datasets.mnist.
    load_data()
    train_X, test_X , = train_X.reshape(-1,28*28), test_X.reshape(-1,28*28)
    train_X, test_X = train_X/255.0, test_X/255.0
    return np.float32(train_X), train_Y, np.float32(test_X), test_Y

def resize_batch(imgs):
    imgs = imgs.reshape((-1, 28, 28, 1))
    resized_imgs = np.zeros((imgs.shape[0], 32, 32, 1))
    for i in range(imgs.shape[0]):
        resized_imgs[i,..., 0] = transform.resize(imgs[i, ..., 0], (32, 32))
    return resized_imgs
```

```python
# Train routine for the VAE
def train(n_hidden,lr,num_epochs=100,batch_size=128):

    enc = encoder(n_hidden)
    dec = decoder()
    model = encoder_decoder(encoder=enc, decoder=dec)
    model_graph = tf.function(model)

    train_X,train_Y, test_X, test_Y = data_load()
    num_train_recs = train_X.shape[0]
    order = np.arange(num_train_recs)
    num_batches = num_train_recs//batch_size

    optimizer = tf.keras.optimizers.Adam(lr)
    loss_fn = tf.keras.losses.MeanSquaredError(reduction=tf.keras.losses.
    Reduction.NONE)

    for i in range(num_epochs):
        epoch_loss = 0
        np.random.shuffle(order)
        train_X = train_X[order]
        for batch in range(num_batches):
            X_batch = train_X[batch*batch_size:(batch+1)*batch_size]
            X_batch = resize_batch(X_batch)
            bsize = X_batch.shape[0]

            with tf.GradientTape() as tape:
                X_pred_batch, mean, log_sigma = model_graph(X_batch)
                X_batch = tf.constant(X_batch)
                loss_reconst = tf.reduce_mean(tf.reduce_sum(loss_fn
                (X_batch,X_pred_batch),(1,2)))
                loss_KL = KL_Div(mean,log_sigma)
                loss = loss_reconst + loss_KL

            gradients = tape.gradient(loss, model.trainable_variables)
            optimizer.apply_gradients(zip(gradients, model.trainable_
            variables))
            epoch_loss += loss.numpy()
```

```
    if i % 5 == 0:
        print(f"Training Loss at epoch {i}: {epoch_loss/num_
        train_recs}")

# Check the quality of the 1st 50 test image reconstruction
X_batch = test_X[:50,:]
X_batch_resize = resize_batch(X_batch)
X_pred_test,_,_ = model_graph(tf.constant(X_batch_resize))

# plot the reconstructed images and their ground truths (inputs)
plt.figure(1)
plt.title('Reconstructed Images')
for i in range(50):
    plt.subplot(5, 10, i+1)
    plt.imshow(X_pred_test[i, ..., 0], cmap='gray')

plt.figure(2)
plt.title('Grounf Truth Images')
for i in range(50):
    plt.subplot(5, 10, i+1)
    plt.imshow(X_batch_resize[i, ..., 0], cmap='gray')

noise_z = tf.constant(np.random.normal(size=[50,n_hidden]))
out_img = model.decoder(noise_z)
plt.figure(3)
plt.title('Input Images')
for i in range(50):
    plt.subplot(5, 10, i+1)
    plt.imshow(out_img[i, ..., 0], cmap='gray')
plt.show()
return model, model_graph

model, model_graph = train(n_hidden=8,lr=0.001,num_epochs=10,
batch_size=256)

--output--
Training Loss at epoch 0: 0.17983988234202067
Training Loss at epoch 5: 0.06145528294245402
```

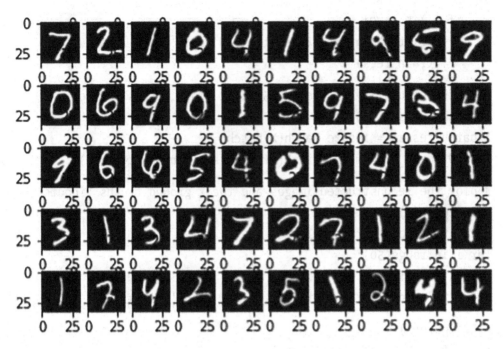

Figure 5-26. *Original images fed to variational autoencoder*

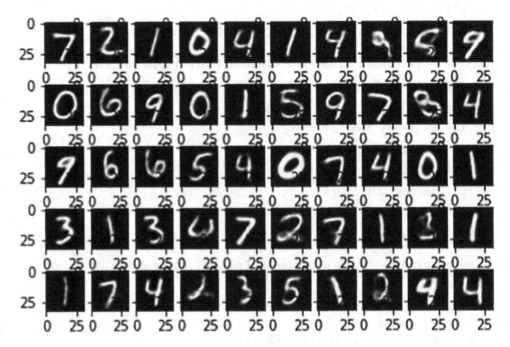

Figure 5-27. *Reconstructed images by the variational autoencoder*

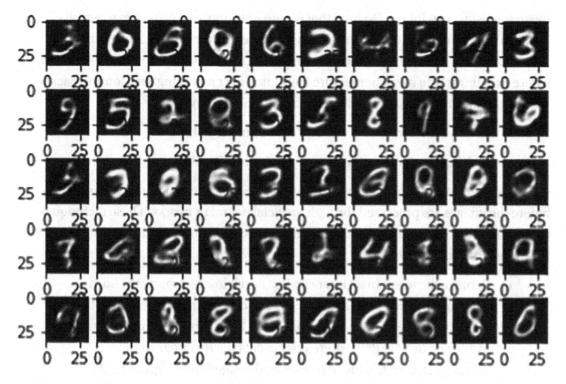

Figure 5-28. *Images generated through the variational autoencoder's decoder with noise z sampled from prior p(z) as input*

We can see from Figure 5-26 and 5-27 that the variational autoencoder has done a good job of reconstructing the input images.

Also, we can see from Figure 5-28 that the decoder of the variational autoencoder has learned to act as a generative model and is able to construct realistic MNIST images by taking in noise input z sampled from prior $p(z)$.

PCA and ZCA Whitening

Generally, images contain pixels whose intensities are highly correlated in any neighborhood of the image, and hence such correlation is highly redundant to a learning algorithm. These dependencies in the form of correlation between nearby pixels are generally of little use to any algorithm. Thus, it makes sense to remove this two-way correlation so that the algorithm puts more emphasis on higher-order correlations. Similarly, the mean intensity of an image might not be of any use to a learning algorithm in cases where the image is a natural image. Therefore, it makes sense to remove the

505

mean intensity of an image. Do note that we are not subtracting the mean per-pixel location but rather the mean of pixel intensities of each image. This kind of mean normalization is different from the other mean normalization we do in machine learning where we subtract the mean per feature computed over a training set. Coming back to the concept of whitening, the advantages of whitening are twofold:

- Remove the correlation among features in the data.

- Make the variance equal along each feature direction.

PCA and ZCA whitening are two techniques generally used to preprocess images before the images are processed through artificial neural networks. These techniques are almost the same, with a subtle difference. The steps involved in PCA whitening are illustrated first, followed by ZCA.

- Remove the mean pixel intensity from each image separately. So, if a 2D image is converted into a vector, one can subtract the mean of the elements in the image vector from itself. If each image is represented by the vector $x^{(i)} \in R^{nX1}$, where i represents the *ith* image in the training set, then the mean normalized image for $x^{(i)}$ is given by the following:

$$x^{(i)} = x^{(i)} - \frac{1}{n}\sum_{j=1}^{n}x_{j}^{(i)}$$

- Once we have the mean normalized images, we can compute the covariance matrix as follows:

$$C = \frac{1}{m}\sum_{i=1}^{m}x^{(i)}x^{(i)T}$$

- Next, we need to decompose the covariance matrix through singular value decomposition (SVD) as follows:

$$C = UDU^{T}$$

- In general, SVD decomposes as $C = UDV^T$, but since C is a symmetric matrix, $U = V$ in this case. U gives the Eigen vectors of the covariance matrix. The Eigen vectors are aligned in a column-wise fashion in U. The variances of the data along the direction of the Eigen vectors are given by the Eigen values housed along the diagonals of D, while the rest of the entries in D are zero since D is the covariance matrix for the uncorrelated directions given by the Eigen vectors.

- In PCA whitening, we project the data along the Eigen vectors, or one may say principal components, and then divide the projection value in each direction by the square root of the Eigen value—i.e., the standard deviation along that direction on which the data is projected. So, the PCA whitening transformation is as follows:

$$T = D^{-\frac{1}{2}} U^T$$

- Once this transformation is applied to the data, the transformed data has zero correlation and unit variance along the newly transformed components. The transformed data for original mean-corrected image $x^{(i)}$ is as follows:

$$x^{(i)}{}_{PW} = Tx^{(i)}$$

The problem with PCA whitening is that although it decorrelates the data and makes the new feature variances unity, the features are no longer in the original space but rather are in a transformed rotated space. This makes the structure of objects such as images lose a lot of information in terms of their spatial orientation, because in the transformed feature space, each feature is the linear combination of all the features. For algorithms that make use of the spatial structure of the image, such as convolutional neural networks, this is not a good thing. So, we need some way to whiten the data such that the data is decorrelated and of unit variances along its features, but the features are still in the original feature space and not in some transformed rotated feature space. The transformation that provides all these relevant properties is called ZCA transform.

Mathematically, any orthogonal matrix R (the column vectors of which are orthogonal to each other) when multiplied by the PCA whitening transform T produces another whitening transform. If one chooses, $R = U$, and the transform

$$Z = UT = UD^{-\frac{1}{2}}U^T$$

is called the ZCA transform. The advantage of ZCA transform is that the image data still resides in the same feature space of pixels, and hence, unlike in PCA whitening, the original pixel doesn't get obscured by the creation of new features. At the same time, the data is whitened—i.e., decorrelated—and of unit variance for each of the features. The unit variance and decorrelated features may help several machine-learning or deep-learning algorithms achieve faster convergence. At the same time, since the features are still in the original space, they retain their topological or spatial orientation, and algorithms such as convolutional neural networks that make use of the spatial structure of the image can make use of the information.

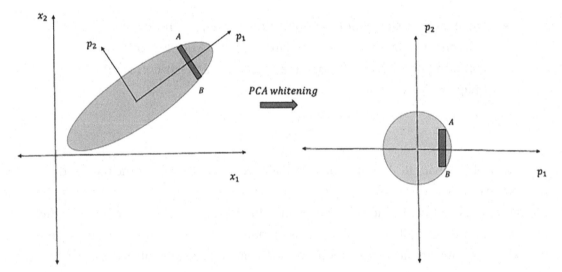

Figure 5-29. *PCA whitening illustration*

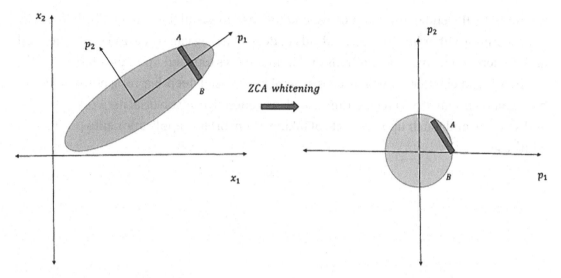

Figure 5-30. *ZCA whitening illustration*

The key difference between PCA whitening and ZCA whitening is illustrated in Figure 5-29 and Figure 5-30. As we can see, in both case, the 2D correlated data is transformed into uncorrelated new data. However, there is a major difference. While in PCA whitening, the new axes have changed from the original axes based on the principal components given by p_1 and p_2, the axes remain same as those of the original with ZCA whitening. The p_1 and p_2 are the Eigen vectors of the covariance matrix for the data. Also, we see the orientation of the marker AB has changed in the new axes for PCA whitening, while it remains intact for ZCA whitening. In both cases, the idea is to get rid of the not-so-useful two-way covariances between the input variables so that the model can concentrate on learning about higher-order correlations.

Summary

In this chapter, we went through the most popular unsupervised techniques in deep learning, namely, restricted Boltzmann machines and autoencoders. Also, we discussed the different applications of using these methods and the training process related to each of these algorithms. Finally, we ended with PCA and ZCA whitening techniques, which are relevant preprocessing techniques used in several supervised deep-learning methods. By the end of this chapter, we had touched upon all the core methodologies in deep learning. Other improvised methods in deep learning can be easily comprehended and implemented given the methods and mathematics touched upon thus far.

In the next chapter, we will discuss several improvised deep-learning networks that have gained popularity in recent times, such as generative adversarial networks, R-CNN, and so forth, and touch upon aspects of taking a TensorFlow application into production with ease.

CHAPTER 6

Advanced Neural Networks

In this chapter, we will look at some of the advanced concepts and models in deep learning. Image segmentation and object localization and detection are some of the key areas that have garnered a lot of importance lately. Image segmentation plays a crucial role in detecting diseases and abnormalities through the processing of medical images. At the same time, it is equally crucial in industries such as aviation, manufacturing, and other domains to detect anomalies such as cracks or other unwanted conditions in machinery. Alternately, images of the night sky can be segmented to detect previously unknown galaxies, stars, and planets. Object detection and localization has profound use in places requiring constant automated monitoring of activities, such as in shopping malls, local stores, industrial plants, and so on. Also, it can be used to count objects and people in an area of interest and estimate various densities, such as traffic conditions at various signals. We will begin this chapter by going through a few of the traditional methods of image segmentation so that we can appreciate how neural networks are different from their traditional counterparts. Then, we will look at object-detection and localization techniques, followed by generative adversarial networks, which have gained lot of popularity recently because of their use and potential as a generative model to create synthetic data. This synthetic data can be used for training and inference in case there is not much data available or the data is expensive to procure. Alternatively, generative models can be used for style transfer from one domain to another.

In the latter part of the chapter, we introduce readers to the field of geometric deep learning—an upcoming research area that aims to replicate deep-learning success stories such as CNN on Euclidean domain data structures(such as images, audio, and text) to the non-Euclidean domain areas of graphs and manifold. In this context, we discuss about convolution operation on graphs and the common graph convolution

© Santanu Pattanayak 2023
S. Pattanayak, *Pro Deep Learning with TensorFlow 2.0*, https://doi.org/10.1007/978-1-4842-8931-0_6

model architectures that are currently being used. Also, we closely look at methods to characterize graphs and manifolds using traditional methods such as spectral embedding as well as deep-learning methods such as Node2Vec.

Aside from image segmentation and generative adversarial network, we will also look at geometric deep learning and graph neural networks in this chapter.

Image Segmentation

Image segmentation is a computer vision task that partitions an image into pertinent segments, such as pixels within the same segment sharing some common attributes such as pixel intensity, texture, and color. The attributes can differ from domain to domain and from task to task. In this section, we will go through some of the basic segmentation techniques, such as thresholding methods based on a histogram of pixel intensities and watershedding thresholding techniques to get some insights about image segmentation before we start with the deep learning–based image segmentation methods.

Binary Thresholding Method Based on Histogram of Pixel Intensities

Often in an image there are only two significant regions of interest—the object and the background. In such a scenario, a histogram of pixel intensities would represent a probability distribution that is bimodal, i.e., has high density around two-pixel intensity values. It would be easy to segment the object and the background by choosing a threshold intensity and setting all pixel intensities below the threshold as 255 and those above the threshold as 0. Such a thresholding scheme would give us a background and an object. If an image is represented as $I(x, y)$ and a threshold t is selected based on the histogram of pixel intensities, then the new segmented image $I'(x, y)$ can be represented as follows:

$$I'(x,y)=0 \ \ when \ \ I(x,y)>t$$
$$=255 \ \ when \ \ I(x,y)\leq t$$

When the bimodal histogram is not distinctly separated by a region of zero density in between, then a good strategy to choose a threshold t is to take the average of the pixel intensities at which the bimodal regions peak. If those peak intensities are represented by p_1 and p_2, then the threshold t can be chosen as follows:

$$t = \frac{(p_1 + p_2)}{2}$$

Alternately, one may use the pixel intensity between p_1 and p_2 at which histogram density is minimum as the thresholding pixel intensity. If the histogram density function is represented by $H(p)$, where $p \in \{0, 1, 2. ., 255\}$ represents the pixel intensities, then

$$t = \underbrace{Arg \ Min}_{p \in [p_1, p_2]} H(p)$$

This idea of binary thresholding can be extended to multiple thresholding based on the histogram of pixel intensities.

Otsu's Method

Otsu's method for image segmentation determines the thresholds by maximizing the variance between the different segments of the images. If using binary thresholding via Otsu's method, here are the steps to be followed:

- Compute the probability of each pixel intensity in the image. Given that N pixel intensities are possible, the normalized histogram would give us the probability distribution of the image.

$$P(i) = \frac{count(i)}{M} \forall i \in \{0, 1, 2, ..., N-1\}$$

- If the image has two segments C_1 and C_2 based on the threshold t, then the set of pixels $\{0, 2 .... t\}$ belong to C_1, while the set of pixels $\{t+1, t+2.....L-1\}$ belong to C_2. The variance between the two segments is determined by the sum of the square deviation of the mean of the clusters with respect to the global mean. The square deviations are weighted by the probability of each cluster.

$$var(C_1, C_2) = P(C_1)(u_1 - u)^2 + P(C_2)(u_2 - u)^2$$

where u_1, u_2 are the mean intensities of cluster 1 and cluster 2 while u is the overall global mean.

$$u_1 = \sum_{i=0}^{t} P(i)i \qquad u_2 = \sum_{i=t+1}^{L-1} P(i)i \qquad u = \sum_{i=0}^{L-1} P(i)i$$

The probability of each of the segments is the number of pixels in the image belonging to that class. The probability of segment C_1 is proportional to the number of pixels that have intensities less than or equal to the threshold intensity t, while that of segment C_2 is proportional to the number of pixels with intensities greater than threshold t. Hence,

$$P(C_1) = \sum_{i=0}^{t} P(i) \qquad P(C_2) = \sum_{i=t+1}^{L-1} P(i)$$

- If we observe the expressions for u_1, u_2, $P(C_1)$, and $P(C_2)$, each of them is a function of the threshold t, while the overall mean u is constant given an image. Hence, the between-segment variance $var(C_1, C_2)$ is a function of the threshold pixel intensity t. The threshold $\hat{t}$ that maximizes the variance would provide us with the optimal threshold to use for segmentation using Otsu's method:

$$\hat{t} = \underbrace{Arg\ Max}_{t}\ var(C_1, C_2)$$

Instead of computing a derivative and setting it to zero to obtain $\hat{t}$, one can evaluate the $var(C_1, C_2)$ at all values of $t = \{0, 1, 2, ..., L - 1\}$ and then choose the $\hat{t}$ at which the $var(C_1, C_2)$ is maximum.

Otsu's method can also be extended to multiple segments where instead of one threshold, one needs to determine $(k - 1)$ thresholds corresponding to k segments for an image.

The logic for both methods just illustrated—i.e., binary thresholding based on histogram of pixel intensities and Otsu's method has been illustrated in Listing 6-1 for reference. Instead of using an image-processing package to implement these algorithms, the core logic has been used for ease of interpretability. Also, one thing to note is that these processes for segmentation are generally applicable on grayscale images or if one is performing segmentation per color channel.

Listing 6-1. Python Implementation of Binary Thresholding Method Based on Histogram of Pixel Intensities and Otsu's Method

```python
"""

 Binary thresholding Method

    From the histogram plotted below it's evident that the distribution is
    bimodal with the
    lowest probability around  at around pixel value of 150. Hence 150
    would be a good threshold
    for binary segmentation
"""

import cv2
import matplotlib.pyplot as plt
%matplotlib inline
import numpy as np
img = cv2.imread("/home/santanu/Downloads/coins.jpg")
gray = cv2.cvtColor(img,cv2.COLOR_BGR2GRAY)
plt.figure(1)
plt.imshow(gray,cmap='gray')
row,col = np.shape(gray)
gray_flat = np.reshape(gray,(row*col,1))[:,0]
plt.figure(2)
plt.hist(list(gray_flat))
gray_const = []
for i in range(len(gray_flat)):
    if gray_flat[i] < 150 :
        gray_const.append(255)
    else:
        gray_const.append(0)
gray_const = np.reshape(np.array(gray_const),(row,col))
plt.figure(3)
plt.imshow(gray_const,cmap='gray')
"""
```

```
  Otsu's thresholding Method  - Determines the threshold by maximizing the
  interclass variance
"""

img = cv2.imread("/home/santanu/Downloads/otsu.jpg")
gray = cv2.cvtColor(img,cv2.COLOR_BGR2GRAY)
plt.figure(1)
plt.imshow(gray,cmap='gray')
row,col = np.shape(gray)
hist_dist = 256*[0]
# Compute the frequency count of each of the pixel in the image
for i in range(row):
    for j in range(col):
        hist_dist[gray[i,j]] += 1
# Normalize the frequencies to produce probabilities
hist_dist = [c/float(row*col) for c in hist_dist]
# Compute the between segment variance
def var_c1_c2_func(hist_dist,t):
    u1,u2,p1,p2,u = 0,0,0,0,0
    for i in range(t+1):
        u1 += hist_dist[i]*i
        p1 += hist_dist[i]
    for i in range(t+1,256):
        u2 += hist_dist[i]*i
        p2 += hist_dist[i]
    for i in range(256):
        u += hist_dist[i]*i
    var_c1_c2 = p1*(u1 - u)**2 + p2*(u2 - u)**2
    return var_c1_c2
# Iteratively run through all the pixel intensities from 0 to 255 and chose
the one that
# maximizes the variance

variance_list = []
for i in range(256):
    var_c1_c2 = var_c1_c2_func(hist_dist,i)
    variance_list.append(var_c1_c2)
```

```
## Fetch the threshold that maximizes the variance
t_hat = np.argmax(variance_list)
## Compute the segmented image based on the threshold t_hat

gray_recons = np.zeros((row,col))

for i in range(row):
    for j in range(col):
        if gray[i,j] <= t_hat :
            gray_recons[i,j] = 255
        else:
            gray_recons[i,j] = 0
plt.figure(2)
plt.imshow(gray_recons,cmap='gray')

--output --
```

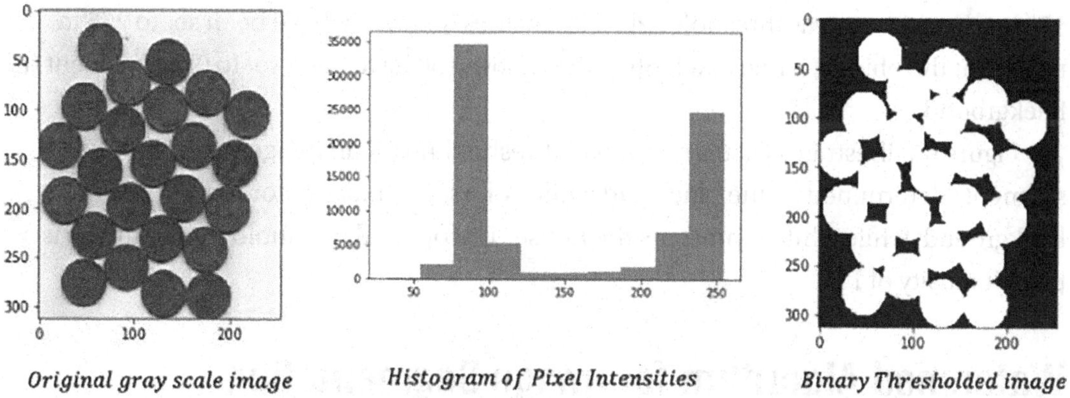

Original gray scale image   Histogram of Pixel Intensities   Binary Thresholded image

Figure 6-1. *Binary thresholding method based on histogram of pixel intensities*

 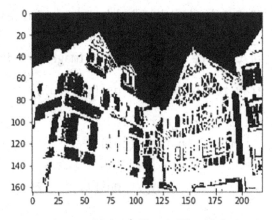

Original Gray scale Image   *Image with Otsu's Thresholding Method*

Figure 6-2. *Otsu's method of thresholding*

In Figure 6-1, the original grayscale image of the coin has been binary thresholded based on the histogram of pixel intensities to separate the objects (i.e., the coins) from the background. Based on the histogram of pixel intensities, the pixel intensity of 150 has been chosen as threshold. Pixel intensities below 150 have been set to 255 to represent the objects, while pixel intensities above 150 have been set to 0 to represent the background.

Figure 6-2 illustrates Otsu's method of thresholding for an image to produce two segments determined by the black and white colors. The black color represents the background, while white represents the house. The optimal threshold for the image is a pixel intensity of 143.

Watershed Algorithm for Image Segmentation

The watershed algorithm aims at segmenting topologically placed local regions around local minima of pixel intensities. If a grayscale image pixel intensity value is considered a function of its horizontal and vertical coordinates, then this algorithm tries to find regions around local minima called basins of attraction or catchment basins. Once these basins are identified, the algorithm tries to separate them by constructing separations or watersheds along high peaks or ridges. To get a better idea of the method, let's look at this algorithm with a simple illustration as represented in Figure 6-3.

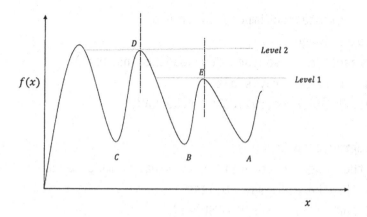

A, B, C — Minima of Catchment Basins
D, E — Peaks or Maximas where watersheds need to be constructed

Figure 6-3. *Watershed algorithm illustration*

If we start filling water in the catchment basin with its minima as *B*, water will keep on filling the basin up to Level 1, at which point an extra drop of water has a chance of spilling over to the catchment basin at *A*. To prevent the spilling of water, one needs to build a dam or watershed at *E*. Once we have built a watershed at *E*, we can continue filling water in the catchment basin *B* till Level 2, at which point an extra drop of water has a chance of spilling over to the catchment basin *C*. To prevent this spilling of water to *C*, one needs to build a watershed at *D*. Using this logic, we can continue to build watersheds to separate such catchment basins. This is the principal idea behind the watershed algorithm. Here, the function is univariate, whereas in the case of a grayscale image, the function representing the pixel intensities would be a function of two variables: the vertical and the horizontal coordinates.

The watershed algorithm is particularly useful in detecting objects when there is overlap between them. Thresholding techniques are unable to determine distinct object boundaries. We will work through illustrating this in Listing 6-2 by applying watershed techniques to an image containing overlapping coins.

Listing 6-2. Image Segmentation Using Watershed Algorithm

```
import numpy as np
import cv2
import matplotlib.pyplot as plt
from scipy import ndimage
from skimage.feature import peak_local_max
```

```python
from skimage.segmentation import watershed
# Load the coins image
img = cv2.imread("/home/santanu/Downloads/coins.jpg")
# Convert the image to gray scale
imgray = cv2.cvtColor(img,cv2.COLOR_BGR2GRAY)
plt.figure(1)
plt.imshow(imgray,cmap='gray')
# Threshold the image to convert it to Binary image based on Otsu's method
thresh = cv2.threshold(imgray, 0, 255,
cv2.THRESH_BINARY | cv2.THRESH_OTSU)[1]
"""

    Detect the contours and display them.
    As we can see in the 2nd image below that the contours are not prominent
    at the regions of
    overlap with normal thresholding method. However with Wateshed
    algorithm the
    the same is possible because of its ability to better separate regions of
    overlap by
    building watersheds at the boundaries of different basins of pixel
    intensity minima
"""
contours, hierarchy = cv2.findContours(thresh,cv2.RETR_TREE,cv2.CHAIN_
APPROX_SIMPLE)
y = cv2.drawContours(imgray, contours, -1, (0,255,0), 3)
plt.figure(2)
plt.imshow(y,cmap='gray')
"""
Hence we will proceed with the Watershed algorithm so that each of the
coin form its own
cluster and hence its possible to have separate contours for each coin.
Relabel the thresholded image to be consisting of only 0 and 1
as the input image to distance_transform_edt should be in this format.
"""
thresh[thresh == 255] = 5
thresh[thresh == 0] = 1
```

```
thresh[thresh == 5] = 0
"""

    The distance_transform_edt and the peak_local_max functions helps building
    the markers by detecting
    points near the centre points of the coins. One can skip these steps and
    create a marker
    manually by setting one pixel within each coin with a random number
    represneting its cluster
"""
D = ndimage.distance_transform_edt(thresh)
localMax = peak_local_max(D, indices=False, min_distance=10,
labels=thresh)
markers = ndimage.label(localMax, structure=np.ones((3, 3)))[0]
"""

    Provide the EDT distance matrix and the markers to the watershed algorithm
    to detect the clusters
    labels for each pixel. For each coin, the pixels corresponding to it will
    be filled with the cluster number
"""
labels = watershed(-D, markers, mask=thresh)
print("[INFO] {} unique segments found".format(len(np.unique(labels)) - 1))
# Create the contours for each label(each coin and append to the plot)
for k in np.unique(labels):
    if k != 0 :
        labels_new = labels.copy()
        labels_new[labels == k] = 255
        labels_new[labels != k] = 0
        labels_new = np.array(labels_new,dtype='uint8')
        contours, hierarchy = cv2.findContours(labels_new,cv2.RETR_
        TREE,cv2.CHAIN_APPROX_SIMPLE)
        z = cv2.drawContours(imgray,contours, -1, (0,255,0), 3)
        plt.figure(3)
        plt.imshow(z,cmap='gray')
```

```
--output --
```

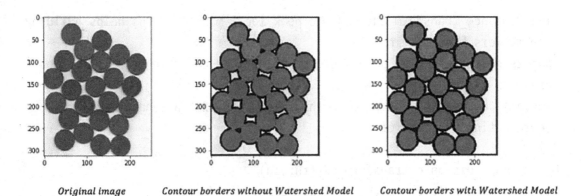

Original image   Contour borders without Watershed Model   Contour borders with Watershed Model

Figure 6-4. *Illustration of watershed algorithm for image segmentation*

As we can see from Figure 6-4, the borders for overlapping coins are distinct after applying the watershed algorithm, whereas the other thresholding methods are not able to provide a distinct border to each of the coins.

Image Segmentation Using K-means Clustering

The famous K-means algorithm can also be used to segment images, especially medical images. The term K is a parameter of the algorithm which determines the number of distinct clusters to be formed. The algorithm works by forming clusters, and each such cluster is represented by its cluster centroids based on specific input features. Image segmentation through K means is generally based on input features such as pixel intensity and its three spatial dimensions, i.e., horizontal and vertical coordinates and the color channel. So, the input feature vector $u \in \mathrm{R}^{4 \times 1}$ $u \in R^{4 \times 1}$ can be represented as follows:

$$u = \left[I(x,y,z), x, y, z \right]^{T}$$

Similarly, one can ignore the spatial coordinates and take the pixel intensities along the three-color channels as the input feature vector, i.e.,

$$u = \left[I_{R}(x,y), I_{G}(x,y), I_{B}(x,y) \right]^{T}$$

where $I_R(x, y)$, $I_G(x, y)$, and $I_B(x, y)$ represent the pixel intensities along the red, green, and blue channels, respectively, at the spatial coordinates (x, y).

The algorithm uses a distance measure such as an L^2 or L^1 norm, as shown here:

$$D(u^{(i)}, u^{(j)}|L^2) = \|u^{(i)} - u^{(j)}\|_2^2 = \sqrt{\left(u^{(i)} - u^{(j)}\right)^T \left(u^{(i)} - u^{(j)}\right)}$$

$$D(u^{(i)}, u^{(j)}|L^1) = \|u^{(i)} - u^{(j)}\|_1^1 = \sum_{k=0}^{m-1} \left|u_k^{(i)} - u_k^{(j)}\right|$$

In the preceding equation, m denotes the number of features used for clustering. The following are the working details of the K-means algorithm:

- *Step 1:* Start with K randomly selected cluster centroids $C_1, C_2 \ldots C_k$ corresponding to the K clusters $S_1, S_2 \ldots S_k$.

- *Step 2:* Compute the distance of each pixel-feature vector $u^{(i)}$ from the cluster centroids and tag it to the cluster S_j if the pixel has a minimum distance from its cluster centroid C_j:

$$j = \underbrace{Arg\ Min}_{j} \|u^{(i)} - C_j\|_2$$

- This process needs to be repeated for all the pixel-feature vectors so that in one iteration of K means, all the pixels are tagged to one of the K clusters.

- *Step 3:* Once the new centroids clusters have been assigned for all the pixels, the centroids are recomputed by taking the mean of the pixel-feature vectors in each cluster:

$$C_j = \sum_{u^{(i)} \in S_j} u^{(i)}$$

- Repeat *Step 2* and *Step 3* for several iterations until the centroids no longer change. Through this iterative process, we are reducing the sum of the intra-cluster distances, as represented here:

$$L = \sum_{j=1}^{K} \sum_{u^{(i)} \in S_j} \|u^{(i)} - C_j\|_2$$

A simple implementation of the *K*-means algorithm is shown in Listing 6-3, taking the pixel intensities in the three-color channels as features. The image segmentation is implemented with $K = 3$. The output is shown in grayscale and hence may not reveal the actual quality of the segmentation. However, if the same segmented image as produced in Listing 6-3 is displayed in a color format, it would reveal the finer details of the segmentation. One more thing to add: the cost or loss functions minimized—i.e., the sum of the intra-cluster distances—are a non-convex function and hence might suffer from local minima problems. One can trigger the segmentation several times with different initial values for the cluster centroids and then take the one that minimizes the cost function the most or produces a reasonable segmentation.

Listing 6-3. Image Segmentation Using K means

```
import cv2
import numpy as np
import matplotlib.pyplot as plt
np.random.seed(0)
"""

  K means that one has used in Machine learning
  clustering also provides good segmentation as we see below
"""

img = cv2.imread("/home/santanu/Downloads/kmeans1.jfif")
imgray_ori = cv2.cvtColor(img,cv2.COLOR_BGR2GRAY)
plt.figure(1)
plt.imshow(imgray_ori,cmap='gray')
# Save the dimensions of the image
row,col,depth = img.shape
# Collapse the row and column axis for faster matrix operation.
img_new = np.zeros(shape=(row*col,3))
glob_ind = 0
for i in range(row):
    for j in range(col):
        u = np.array([img[i,j,0],img[i,j,1],img[i,j,2]])
        img_new[glob_ind,:] = u
        glob_ind += 1
```

```python
"""
Set the number of clusters
One can experiment with different values of K and select
the one that provides good clustering. Having said that Image processing
especially image enhancement and segmentation to some extent is
subjective.
"""
K = 5
num_iter = 20
"""
K means suffers from local minima solution and hence
its better to trigger K-means several times with different random
seed value
"""
for g in range(num_iter):
    # Define cluster for storing the cluster number and out_dist to store
    the distances from centroid
    clusters = np.zeros((row*col,1))
    out_dist = np.zeros((row*col,K))
    centroids = np.random.randint(0,255,size=(K,3))

    for k in range(K):
        diff = img_new - centroids[k,:]
        diff_dist = np.linalg.norm(diff,axis=1)
        out_dist[:,k] = diff_dist

# Assign the cluster with minimum distance to a pixel location

    clusters = np.argmin(out_dist,axis=1)

# Recompute the clusters

    for k1 in np.unique(clusters):
        centroids[k1,:] = np.sum(img_new[clusters == k1,:],axis=0)/
np.sum([clusters == k1])

# Reshape the cluster labels in two dimensional image form
clusters = np.reshape(clusters,(row,col))
out_image = np.zeros(img.shape)
```

```
#Form the 3-D image with the labels replaced by their correponding centroid
pixel intensity

for i in range(row):
    for j in range(col):
        out_image[i,j,0] = centroids[clusters[i,j],0]
        out_image[i,j,1] = centroids[clusters[i,j],1]
        out_image[i,j,2] = centroids[clusters[i,j],2]
        out_image = np.array(out_image,dtype="uint8")

# Display the output image after converting into gray scale
# Readers adviced to display the image as it is for better clarity
imgray = cv2.cvtColor(out_image,cv2.COLOR_BGR2GRAY)
plt.figure(2)
plt.imshow(imgray,cmap='gray')

---output ---
```

Original Image   Segmented Image with K = 3

Figure 6-5. *Illustration of segmentation through the K-means algorithm*

We can see from Figure 6-5 that *K*-means clustering has done a good job segmenting the image for *K* = 3.

Semantic Segmentation

Image segmentation through convolutional neural networks has gained a lot of popularity in recent years. One of the things significantly different when segmenting an image through neural networks is the annotation process of assigning each pixel to an object class so that the training of such a segmentation network is totally supervised.

Although the process of annotating images is a costly affair, it simplifies the problem by having a ground truth to compare to. The ground truth would be in the form of an image with the pixels holding a representative color for a specific object. For example, if we are working with a set of "cats and dogs" images that can have a background, then each pixel for an image can belong to one of the three classes—cat, dog, and background. Also, each class of object is generally represented by a representative color so that the ground truth can be displayed as a segmented image. Let's go through some convolutional neural networks that can perform semantic segmentation.

Sliding-Window Approach

One can extract patches of images from the original image by using a sliding window and then feeding those patches to a classification convolutional neural network to predict the class of the central pixel for each of the image patches. Training such a convolutional neural network with this sliding-window approach is going to be computationally intensive, both at training and at test time, since at least N number of patches per image need to be fed to the classification CNN, where N denotes the number of pixels in the image.

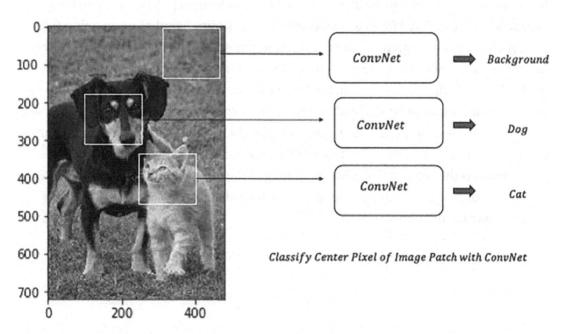

Figure 6-6. *Sliding-window semantic segmentation*

Illustrated in Figure 6-6 is a sliding-window semantic segmentation network for segmenting images of cat, dog, and background. It crops out patches from the original image and feeds it through the classification CNN for classifying the center pixel in the patch. A pretrained network such as AlexNet, VGG19, Inception V3, and so forth can be used as the classification CNN, with the output layer replaced to have only three classes pertaining to the labels for dog, cat, and background. The CNN can then be fine-tuned through backpropagation, with the image patches as input and the class label of the center pixel of the input image patch as output. Such a network is highly inefficient from a convolution of images standpoint since image patches next to each other will have significant overlap and reprocessing them every time independently leads to unwanted computational overhead. To overcome the shortcomings of the above network, one can use a fully convolutional network which is our next topic of discussion.

Fully Convolutional Network (FCN)

A fully convolutional network (FCN) consists of a series of convolution layers without any fully connected layers. The convolutions are chosen such that the input image is transmitted without any change in the spatial dimensions, i.e., the height and width of the image remains the same. Rather than having individual patches from an image independently evaluated for pixel category, as in the sliding-window approach, a fully convolutional network predicts all the pixel categories at once. The output layer of this network consists of C feature maps, where C is the number of categories, including the background, that each pixel can be classified into. If the height and width of the original image are h and w, respectively, then the output consists of C number of $h \times w$ feature maps. Also, for the ground truth, there should be C number of segmented images corresponding to the C classes. At any spatial coordinate (h_1, w_1), each of the feature maps contains the score of that pixel pertaining to the class the feature map is tied to. These scores across the feature maps for each spatial pixel location (h_1, w_1) form a SoftMax over the different classes.

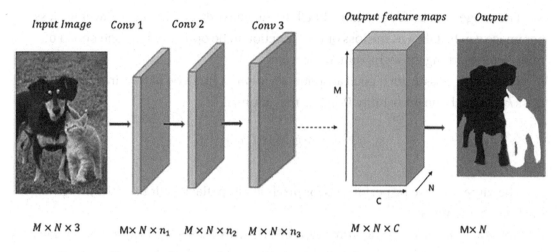

$M \times N \times 3$ $M \times N \times n_1$ $M \times N \times n_2$ $M \times N \times n_3$ $M \times N \times C$ $M \times N$

Figure 6-7. *Fully convolutional network architecture*

Figure 6-7 contains the architectural design of a fully convolutional network. The number of output feature maps as well as ground truth feature maps would be 3, corresponding to the three classes. If the input net activation or score at the spatial coordinate (i, j) for the *kth* class is denoted by $s_k^{(i,j)}$, then the probability of the *kth* class for the pixel at spatial coordinate (i, j) is given by the SoftMax probability, as shown here:

$$P_k(i,j) = \frac{e^{s_k^{(i,j)}}}{\sum_{k'=1}^{C} e^{s_{k'}^{(i,j)}}}$$

Also, if the ground truth labels at the spatial coordinate (i, j) for the *kth* class are given by $y_k(i, j)$, then the cross-entropy loss of the pixel at spatial location (i, j) can be denoted by the following:

$$L(i,j) = -\sum_{k=1}^{C} y_k(i,j) \log P_k(i,j)$$

If the height and width of the images fed to the network are M and N, respectively, then the total loss L for an image is as follows:

$$L = -\sum_{i=0}^{M-1}\sum_{j=0}^{N-1}\sum_{k=1}^{C} y_k(i,j) \log P_k(i,j)$$

The images can be fed as a mini-batch to the network, and hence the average loss per image can be taken as the loss or cost function to be optimized in each epoch of mini-batch learning through gradient descent.

The output class $\hat{k}$ for a pixel at spatial location (i, j) can be determined by taking the class k for which the probability $P_k(i, j)$ is maximum, i.e.,

$$\hat{k} = \underbrace{Arg\ Max}_{k}\ P_k(i, j)$$

The same needs to be performed for pixels at all spatial locations of the image to get the final segmented image.

In Figure 6-8, the output feature maps of a network for segmenting images of cats, dogs, and background are illustrated. As we can see, for each of the three categories or classes, there is a separate feature map. The spatial dimensions of the feature maps are the same as those of the input image. The net input activation, the associated probability, and the corresponding ground label have been shown at the spatial coordinate (i, j) for all the three classes.

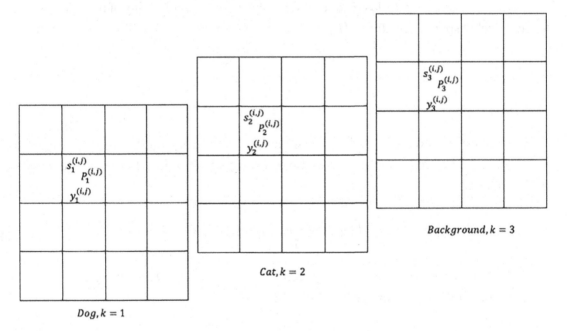

Figure 6-8. *Output feature maps corresponding to each of the three classes for dog, cat, and background*

All convolution layers in the network retain the initial spatial dimensions of the input image. So, for high-resolution images, the network would be computationally intensive, especially if the number of feature maps or channels in each convolution is high. To address this problem, a different variant of a fully convolutional neural network is more widely used that both downsamples the image in the first half of the network and then upsamples the images in the second half of the network. This modified version of the fully convolutional network is going to be our next topic of discussion.

Fully Convolutional Network with Downsampling and Upsampling

Instead of preserving the spatial dimensions of the images in all convolutional layers as in the previous network, this variant of the fully convolutional network uses a combination of convolutions where the image is downsampled in the first half of the network and then upsampled in the final layers to restore the spatial dimensions of the original image. Generally, such a network consists of several layers of downsampling through strided convolutions and/or pooling operations and then a few layers of upsampling. Illustrated in Figure 6-9 is a high-level architectural design of such a network.

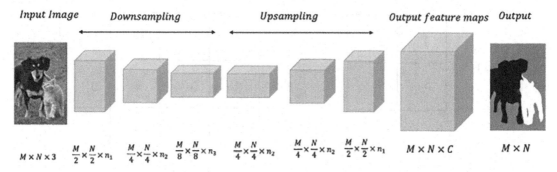

Figure 6-9. *Fully convolutional network with downsampling and upsampling*

The techniques that are commonly used to upsample an image or feature map are as discussed next.

Unpooling

Unpooling can be treated as the reverse operation to pooling. In max pooling or average pooling, we reduce the spatial dimensions of the image by taking either the maximum or the average of the pixel value based on the size of the pooling kernel. So, if we have a 2×2 kernel for pooling, the spatial dimensions of the image get reduced by $\frac{1}{2}$ in each spatial dimension. In unpooling, we generally increase the spatial dimensions of the image by repeating a pixel value in a neighborhood, as shown in Figure 6-10 (A).

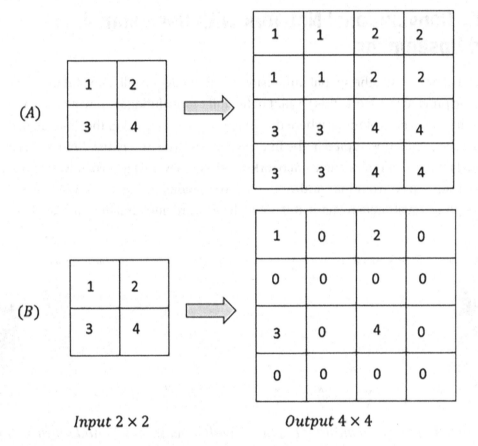

Input 2×2 Output 4×4

Figure 6-10. *Unpooling operation*

Similarly, one may choose to populate only one pixel in the neighborhood and set the rest to zero, as illustrated in Figure 6-10 (B).

Max Unpooling

Many of the fully convolutional layers are symmetric, as a pooling operation in the first half of the network would have a corresponding unpooling in the second half of the network to restore the image size. Whenever pooling is performed, minute spatial information about the input image is lost because of the summarizing of the results of neighboring pixels by one representative element. For instance, when we do max pooling by a 2 × 2 kernel, the maximum pixel value of each neighborhood is passed on to the output to represent the 2 × 2 neighborhood. From the output, it would not be possible to infer the location of the maximum pixel value. So, in this process we are missing the spatial information about the input. In semantic segmentation, we want to classify each pixel as close to its true label as possible. However, because of max pooling, a lot of information about edges and other finer details of the image is lost. While we are trying to rebuild the image through unpooling, one way we can restore a bit of this lost spatial information is to place the value of the input pixel in the output location corresponding to the one where the max pooling output got its input from. To visualize it better, let's look at the illustration in Figure 6-11.

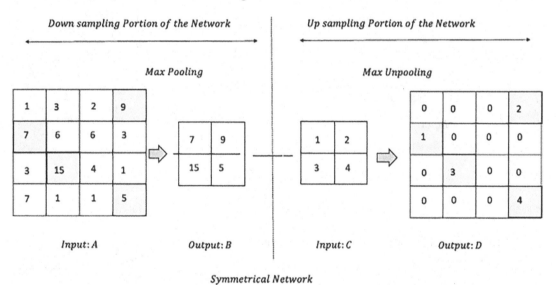

Figure 6-11. *Max unpooling illustration for a symmetric fully connected segmentation network*

As we can see from Figure 6-11, while unpooling only the locations in output map *D* corresponding to the location of the maximal elements in input A with respect to max pooling are populated with values. This method of unpooling is generally called *max unpooling*.

Transpose Convolution

The upsampling done through unpooling or max unpooling is fixed transformations. These transformations don't involve any parameters that the network needs to learn while training the network. A learnable way to do upsampling is to perform upsampling through transpose convolution, which is much like convolution operations that we know of. Since transpose convolution involves parameters that would be learned by the network, the network would learn to do the upsampling in such a way that the overall cost function on which the network is trained on reduces. Now, let's get into the details of how transpose convolution works.

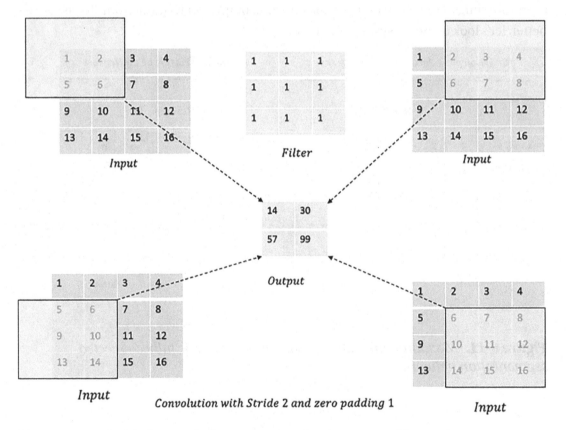

Figure 6-12. *Strided convolution operation for downsampling an image*

In strided convolution, the output dimensions are almost half those of the input for each spatial dimension for a stride of 2. Figure 6-12 illustrates the operation of convolving a 2D input of dimension 5 × 5 with a 4 × 4 kernel with a stride of 2 and 0 padding of 1. We slide the kernel over the input, and at each position the kernel is on, the dot product of the kernel is computed with the portion of the input the kernel is overlapping with.

In transpose convolution, we use the same kind of logic, but instead of downsampling, strides greater than 1 provide upsampling. So, if we use a stride of 2, then the input size is doubled in each spatial dimension. Figures 6-13a, 6-13b, and 6-13c illustrate the operation of transpose convolution for an input of dimension 2 × 2 by a kernel size of 3 × 3 to produce a 4 × 4 output. Unlike the dot product between the filter and the portions of input as in convolution, in transpose convolution, at a specific location, the filter values are weighted by the input value at which the filter is placed, and the weighted filter values are populated in the corresponding locations in the output. The outputs for successive input values along the same spatial dimension are placed at a gap determined by the stride of the transpose convolution. The same is performed for all input values. Finally, the outputs corresponding to each of the input values are added to produce the final output, as shown in Figure 6-13c.

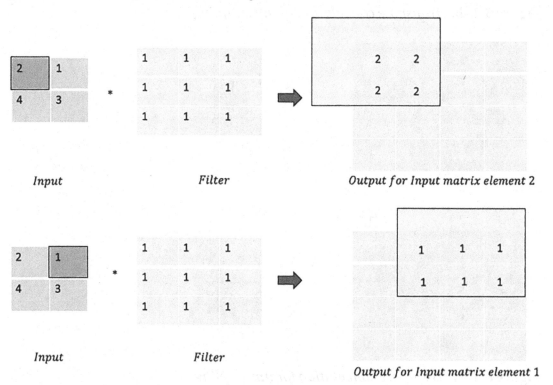

Figure 6-13a. *Transpose convolution for upsampling*

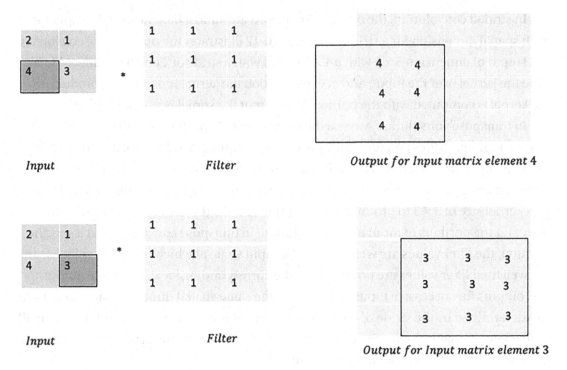

Figure 6-13b. *Transpose convolution for upsampling*

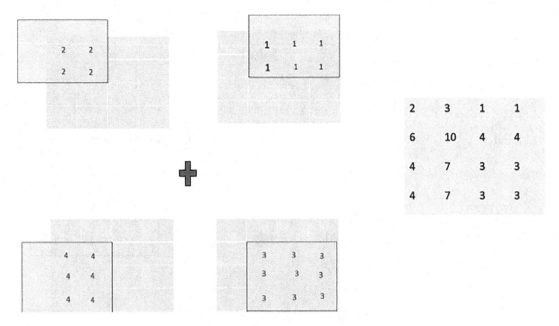

Figure 6-13c. *Transpose convolution for upsampling*

In TensorFlow, the function `tf.nn.conv2d_transpose` can be used to perform upsampling through transpose convolution.

U-Net

The U-Net convolutional neural network is one of the most efficient architectures in recent times for the segmentation of images, especially medical images. This U-Net architecture won the Cell Tracking Challenge at ISBI 2015. The network topology follows a U-shape pattern from the input to the output layers, hence the name U-Net. Olaf Ronneberger, Philipp Fischer, and Thomas Brox came up with this convolutional neural network for segmentation, and the details of the model are illustrated in the white paper "U-Net: Convolutional Networks for Biomedical Image Segmentation." The paper can be located at `https://arxiv.org/abs/1505.04597`.

In the first part of the network, the images undergo downsampling through a combination of convolution and max pooling operations. The convolutions are associated with pixel-wise ReLU activations. Every convolution operation is with 3×3 filter size and without zero padding, which leads to a reduction of two pixels in each spatial dimension for the output feature maps. In the second part of the network, the downsampled images are upsampled till the final layer, where output feature maps correspond to specific classes of objects being segmented. The cost function per image would be pixel-wise categorical cross-entropy or log-loss for classification summed over the whole image, as we saw earlier. One thing to note is that the output feature maps in a U-Net have less spatial dimension than those of the input. For example, an input image having spatial dimensions of 572×572 produces output feature maps of spatial dimension 388×388. One might ask how the pixel-to-pixel class comparison is done for loss computation for training. The idea is simple—the segmented output feature maps are compared to a ground truth segmented image of patch size 388×388 extracted from the center of the input image. The central idea is that if one has images of higher resolution, say 1024×1024, one can randomly create many images of spatial dimensions 572×572 from it for training purposes. Also, the ground truth images are created from these 572×572 subimages by extracting the central 388×388 patch and labeling each pixel with its corresponding class. This helps train the network with a significant amount of data even if there are not many images around for training. Illustrated in Figure 6-14 is the architecture diagram of a U-Net.

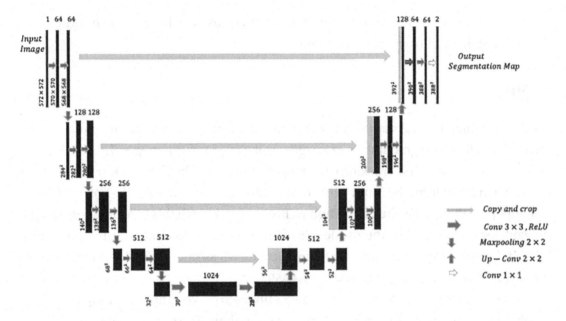

Figure 6-14. *U-Net architecture diagram*

We can see from the architecture diagram that in the first part of the network, the images undergo convolution and max pooling to reduce the spatial dimensions and at the same time increase the channel depth, i.e., to increase the number of feature maps. Every two successive convolutions, along with their associated ReLU activations, is followed by a max pooling operation, which reduces the image size by $\frac{1}{4}$. Every max pooling operation brings the network down to the next set of convolutions and contributes to the U-shape in the first part of the network. Similarly, the upsampling layers increase the spatial dimensions by 2 in each dimension and hence increase the image size by four times. Also, it provides the U structure to the network in the second part. After every upsampling, the image goes through two convolutions and their associated ReLU activations.

The U-Net is a very symmetric network as far as the operations of max pooling and upsampling are concerned. However, for a corresponding pair of max pooling and upsampling, the image size before the max pooling is not the same as the image size after upsampling, unlike with other fully convolutional layers. As discussed earlier, when a max pooling operation is done, a lot of spatial information is lost by having a representative pixel in the output corresponding to a local neighborhood of the image. It becomes difficult to recover this lost spatial information when the image is upsampled back to its original dimensions, and hence the new image lacks a lot of information around the edges and other finer aspects of the image too. This leads to suboptimal segmentation. Had the upsampled image been of the same spatial dimensions as the image before its corresponding max pooling operation, one could have just appended a random number of feature maps before max pooling with the output feature maps after upsampling to help the network recover a bit of the lost spatial information. Since in the case of U-Net these feature-map dimensions don't match, U-Net crops the feature maps before max pooling to be of the same spatial dimensions as the output feature maps from upsampling and concatenates them. This leads to better segmentation of images, as it helps recover some spatial information lost during max pooling. One more thing to note is that upsampling can be done by any of the methods that we have looked at thus far, such as unpooling, max unpooling, and transpose convolution, which is also known as deconvolution.

Few of the big wins with U-Net segmentation are as follows:

- A significant amount of training data can be generated with only a few annotated or hand-labeled segmented images.

- The U-Net does good segmentation even when there are touching objects of the same class that need to be separated. As we saw earlier with traditional image-processing methods, separating touching objects of the same class is tough, and methods such as the watershed algorithm require a lot of input in terms of object markers to come up with reasonable segmentation. The U-Net does good separation between touching segments of the same class by introducing high weights for misclassification of pixels around the borders of touching segments.

Semantic Segmentation in TensorFlow with Fully Connected Neural Networks

In this section, we will go through the working details of a TensorFlow implementation for the segmentation of car images from the background based on the Kaggle competition named Carvana. The input images, along with their ground truth segmentation, are available for training purposes. We train the model on 80% of the training data and validate the performance of the model on the remaining 20% of the data. For training, we use a fully connected convolutional network with a U-Net-like structure in the first half of the network followed by upsampling through transpose convolution. A couple of things different in this network from a U-Net is that the spatial dimensions are kept intact while performing convolution by using padding as *SAME*. The other thing different is that this model doesn't use the skip connections to concatenate feature maps from the downsampling stream to the upsampling stream. The detailed implementation is provided in Listing 6-4.

Listing 6-4. Semantic Segmentation in TensorFlow with Fully Connected Neural Network

```
# Load the different packages
import tensorflow as tf
print(f"Tensorflow version: {tf.__version__}")
from sklearn.model_selection import train_test_split
import matplotlib.pyplot as plt
%matplotlib inline
import os
from subprocess import check_output
import numpy as np
from tensorflow.keras.utils import img_to_array, array_to_img, load_img
from  skimage.transform import resize
from tensorflow.keras import layers, Model
from pathlib import Path
import imageio

class segmentation_model(Model):
    """

    Segmentation Model consisting of downsampling in the 1st half using
    convolution followed by upsampling in the 2nd half using Traspose
    Convolution
```

```
"""
def __init__(self):
    super(segmentation_model,self).__init__()
    self.conv11, self.conv12, self.pool1  = self.conv_
    block(filters=64,kernel_size=3,strides=1,
                                    padding='SAME',activation='relu',
                                    pool_size=2,pool_stride=2)

    self.conv21, self.conv22, self.pool2  = self.conv_
block(filters=128,kernel_size=3,strides=1,
                                    padding='SAME',activation='relu',
                                    pool_size=2,pool_stride=2)

    self.conv31, self.conv32, self.pool3  = self.conv_
block(filters=256,kernel_size=3,strides=1,
                                    padding='SAME',activation='relu',
                                    pool_size=2,pool_stride=2)

    self.conv41, self.conv42, self.pool4  = self.conv_
block(filters=512,kernel_size=3,strides=1,
                                    padding='SAME',activation='relu',
                                    pool_size=2,pool_stride=2)

    self.conv51, self.conv52                 = self.conv_block
(filters=1024,kernel_size=3,strides=1,
                                    padding='SAME',activation='relu',
                                    pool_size=2,pool_stride=2,pool=False)

    self.deconv1 = self.deconv_block(filters=1024,kernel_size=3,
    strides=2,padding='SAME',activation='relu')
    self.deconv2 = self.deconv_block(filters=512,kernel_size=3,
    strides=2,padding='SAME',activation='relu')
    self.deconv3 = self.deconv_block(filters=256,kernel_size=3,
    strides=2,padding='SAME',activation='relu')
    self.deconv4 = self.deconv_block(filters=128,kernel_size=3,
    strides=2,padding='SAME',activation='relu')

    self.convf = layers.Conv2D(filters=1,kernel_size=1,strides=1,
    padding='SAME')
```

541

```
def conv_block(self,filters,kernel_size,strides,padding,activation,
pool_size,pool_stride,pool=True):
    conv11 = layers.Conv2D(filters=filters,kernel_size=kernel_size,
    strides=(strides,strides),padding=padding,activation=activation)
    conv12 = layers.Conv2D(filters=filters,kernel_size=kernel_size,
    strides=(strides,strides),padding=padding,activation=activation)
    if pool:
        pool1 = layers.MaxPool2D(pool_size=(pool_size,pool_size),
        strides=(pool_stride,pool_stride))
        return conv11, conv12, pool1
    return conv11, conv12

def deconv_block(self,filters,kernel_size,strides,padding,activation):
    deconv1 = layers.Conv2DTranspose(filters=filters,kernel_size=
    kernel_size,strides=(strides,strides),padding=padding,activation=
    activation)
    return deconv1

def call(self,x):
    x = self.conv11(x)
    x = self.conv12(x)
    x = self.pool1(x)
    #
    x = self.conv21(x)
    x = self.conv22(x)
    x = self.pool2(x)
    #
    x = self.conv31(x)
    x = self.conv32(x)
    x = self.pool3(x)
    #
    x = self.conv41(x)
    x = self.conv42(x)
    x = self.pool4(x)
    #
```

```python
        x = self.conv51(x)
        x = self.conv52(x)
        #
        x = self.deconv1(x)
        x = self.deconv2(x)
        x = self.deconv3(x)
        x = self.deconv4(x)
        x = self.convf(x)

        return x

def grey2rgb(img):
    """

     utility function to convert greyscale images to rgb
    """

    new_img = []
    for i in range(img.shape[0]):
        for j in range(img.shape[1]):
            new_img.append(list(img[i][j])*3)
    new_img = np.array(new_img).reshape(img.shape[0], img.shape[1], 3)
    return new_img

def data_gen_small(data_dir, mask_dir, images, batch_size, dims):
    """

    Generator that we will use to read the data from the directory
    """

    while True:
        ix = np.random.choice(np.arange(len(images)), batch_size)
        imgs = []
        labels = []
        for i in ix:
            # images
            img_path = f"{Path(data_dir)}/{images[i]}"
            original_img = imageio.imread(img_path)

            resized_img = resize(original_img, dims)
            array_img = img_to_array(resized_img)
            imgs.append(array_img)
```

```python
            # masks
            prefix = images[i].split(".")[0]
            mask_path = f"{Path(mask_dir)}/{prefix}_mask.gif"
            original_mask = imageio.imread(mask_path)
            resized_mask = resize(original_mask, dims)
            array_mask = img_to_array(resized_mask)
            labels.append(array_mask[:, :, 0])
        imgs = np.array(imgs)
        labels = np.array(labels)
        yield imgs, labels.reshape(-1, dims[0], dims[1],1)

def train(data_dir,mask_dir,batch_size=4,train_val_split=[0.8,0.2],
          img_height=128,img_width=128,lr=0.01,num_batches=500):

    model = segmentation_model()
    model_graph = tf.function(model)

    # Get the path to all images for dynamic fetch in each batch generation
    all_images = os.listdir(data_dir)
    # Train val split
    train_images, validation_images = train_test_split(all_images, train_
size=train_val_split[0],
                                                test_size=train_val_split[1])
    # Create train and val generator
    train_gen = data_gen_small(data_dir, mask_dir, train_images,batch_
    size=batch_size,
                            dims=(img_height,img_width))
    validation_gen = data_gen_small(data_dir, mask_dir, validation_
    images,batch_size=batch_size,
                                dims=(img_height,img_width))

    # Setting up the optimizer
    optimizer = tf.keras.optimizers.Adam(lr)
    # setting up the Binary Cross entropy loss for the two
    segmentation class
    loss_fn = tf.keras.losses.BinaryCrossentropy(from_logits=True)
```

```
    for batch in range(num_batches):

        X_batch,y_batch = next(train_gen)
        X_batch, y_batch =  tf.constant(X_batch), tf.constant(y_batch)

        with tf.GradientTape() as tape:
            y_pred_batch = model_graph(X_batch)
            loss_ = loss_fn(y_batch,y_pred_batch)

        # Compute gradient
        gradients = tape.gradient(loss_, model.trainable_variables)
        # Update the parameters
        optimizer.apply_gradients(zip(gradients, model.trainable_
        variables))

        X_val,y_val = next(validation_gen)
        X_val,y_val = tf.constant(X_val), tf.constant(y_val)
        y_val_pred = model_graph(X_val,training=False)
        loss_val = loss_fn(y_val,y_val_pred)
        print(f"Batch : {batch} , train loss:{loss_.numpy()/batch_size},
        val loss: {loss_val/batch_size}")
    return model, model_graph, X_val,y_val, y_val_pred

height,width=128,128
data_dir = "/media/santanu/9eb9b6dc-b380-486e-b4fd-c424a325b976/Kaggle
Competitions/Carvana/train/"
mask_dir = "/media/santanu/9eb9b6dc-b380-486e-b4fd-c424a325b976/Kaggle
Competitions/Carvana/train_masks/"
model, model_graph, X_val,y_val, y_val_pred = train(data_dir,mask_dir,
batch_size=4,train_val_split=[0.8,0.2],img_height=height,img_width=width,
         lr=0.0002,num_batches=500)

--output--
('batch:', 494, 'train loss:', 0.047129884, 'val loss:', 0.046108384)
('batch:', 495, 'train loss:', 0.043634158, 'val loss:', 0.046292961)
('batch:', 496, 'train loss:', 0.04454672,  'val loss:', 0.044108659)
('batch:', 497, 'train loss:', 0.048068151, 'val loss:', 0.044547819)
('batch:', 498, 'train loss:', 0.044967934, 'val loss:', 0.047069982)
('batch:', 499, 'train loss:', 0.041554678, 'val loss:', 0.051807735)
```

Actual Image   *Ground Truth*   *Segmented Image*

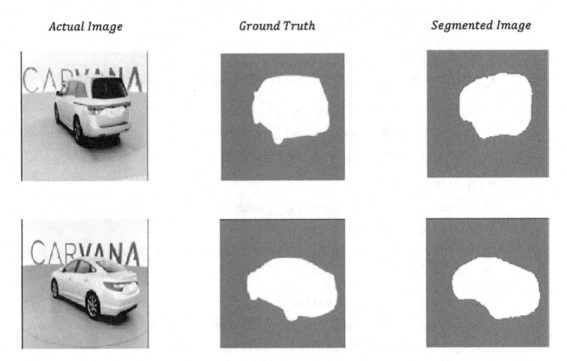

Figure 6-15a. *Segmentation results on the validation dataset with model trained on 128 × 128–size images*

The average training loss and the validation loss are almost the same, which indicates that the model is not overfitting and is generalizing well. As we can see from Figure 6-15a, the results of segmentation look convincing based on the provided ground truths. The spatial dimensions of the images used for this network are 128 × 128. On increasing the spatial dimensions of the input images to 512 × 512, the accuracy and segmentation increase significantly. Since it's a fully convolutional network with no fully connected layers, very few changes in the network are required to handle the new image size. The output of segmentation for a couple of validation dataset images is presented in Figure 6-15b to illustrate the fact that bigger image sizes are most of the time beneficial for image segmentation problems since it helps capture more context.

Actual Image   *Ground Truth*   *Segmented Image*

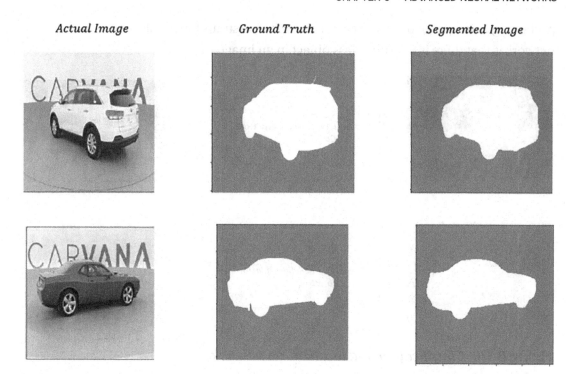

Figure 6-15b. *Segmentation results from the validation dataset with model trained on 512 × 512-size images*

Image Classification and Localization Network

The image classification models predict the class of the object in the image but don't really tell us the location of the object. A bounding box can be used to represent the location of the object in the image. If the images are annotated with bounding boxes and the same is available with the output class, we can train the model to predict these bounding boxes along with the class object. These bounding boxes can be represented by four numbers, two corresponding to the spatial coordinates of the leftmost upper part of the bounding box and the other two to denote the height and width of the bounding box. One can use a convolutional neural network for classification and another for predicting these bounding-box attributes through regression. However, generally the same convolutional neural network with two different prediction head—one for the class of the object and the other for the bounding-box location attributes—is used. This technique of predicting bounding boxes around objects within an image is known as *localization.* Illustrated in Figure 6-16 is an image classification and localization network

547

pertaining to images of dogs and cats. The *a priori* assumption in this type of neural network is that there is only one class object in an image.

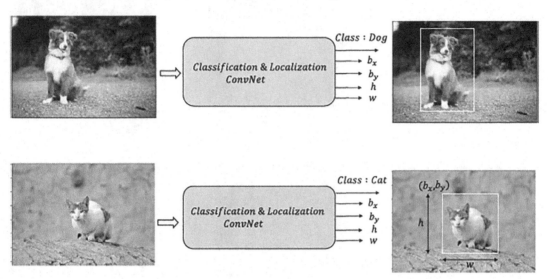

Figure 6-16. *Classification and localization network*

The cost function for this network would be a combination of classification cost over the different object classes and the regression cost associated with the prediction of the bounding-box attributes. Since the cost to optimize is a multitask objective function, one needs to determine how much weight to assign to each task. This is important since the different costs associated with these tasks—let's say categorical cross-entropy for classification and RMSE for regression—would have different scales and hence could drive the optimization haywire if the costs are not weighed properly to form the total cost. The costs need to be normalized to a common scale and then assigned weights based on the complexity of the tasks. Let the parameters of the convolutional neural network that deals with n classes and a bounding box determined by the four numbers be represented by θ. Let the output class be represented by the vector $y = [y_1 y_2 . . y_n]^T \in \{0, 1\}^{n \times 1}$ since each of the $y_j \in \{0, 1\}$. Also, let the bounding-box numbers be represented by the vector $s = [s_1 \ s_2 s_3 \ s_4]^T$ where s_1 and s_2 denote bounding-box coordinates for the upper leftmost pixel while s_3 and s_4 denote the height and width of the bounding box. If the predicted probabilities of the classes are represented by $p = [p_1 p_2 . . p_n]^T$ while the predicted bounding-box attributes are represented by $t = [t_1 t_2 t_3 t_4]^T$, then the loss or cost function associated with an image can be expressed as follows:

$$c(\theta) = -\alpha \sum_{j=1}^{n} y_j \log p_j + \beta \sum_{j=1}^{4} \left(s_j - t_j\right)^2$$

The first term in the preceding expression denotes categorical cross-entropy for the SoftMax over the n classes, while the second term is the regression cost associated with predicting the bounding-box attributes. The parameters α and β are the hyperparameters of the network and should be fine-tuned for obtaining reasonable results. For a mini-batch over m data points, the cost function can be expressed as follows:

$$C(\theta) = \frac{1}{m}\left[-\alpha \sum_{i=1}^{m}\sum_{j=1}^{n} y_j^{(i)} \log p_j^{(i)} + \beta \sum_{i=1}^{m}\sum_{j=1}^{4}\left(s_j^{(i)} - t_j^{(i)}\right)^2\right]$$

where the suffix over i represents different images. The preceding cost function can be minimized through gradient descent. Just as an aside, when comparing the performance of different versions of this network with different hyperparameter values for (α, β), one should not compare the cost associated with these networks as criteria for selecting the best network. Rather, one should use some other metrics such as precision, recall, F1-score, area under the curve, and so on for the classification task and metrics such as overlap area of the predicted and the ground truth bounding boxes and so on for the localization task.

Object Detection

An image in general doesn't contain one object but several objects of interest. There are a lot of applications that benefit from being able to detect multiple objects in images. For example, object detection can be used to count the number of people in several areas of a store for crowd analytics. Also, at an instant, the traffic load on a signal can be detected by getting a rough estimate of the number of cars passing through the signal. Another area in which object detection is being leveraged is in the automated supervision of industrial plants to detect events and generate alarms in case there are safety violations. Continuous images can be captured in critical areas of the plant that are hazardous, and critical events can be captured from those images based on multiple objects detected within the image. For instance, if a worker is working with machinery that requires him to wear safety gloves, eyeglasses, and helmet, a safety violation can be captured based on whether these objects mentioned were detected in the image or not.

The task of detecting multiple objects in images is a classical problem in computer vision. To begin with, we cannot use the classification and localization network described in the preceding section since images can have varying numbers of objects within them. To get ourselves motivated toward solving the problem of object detection, let's get started with a very naïve approach. We can randomly take image patches from the existing image by a brute-force sliding-window technique and then feed it to a pretrained object classification and localization network. Illustrated in Figure 6-17 is a sliding-window approach to detecting multiple objects within an image.

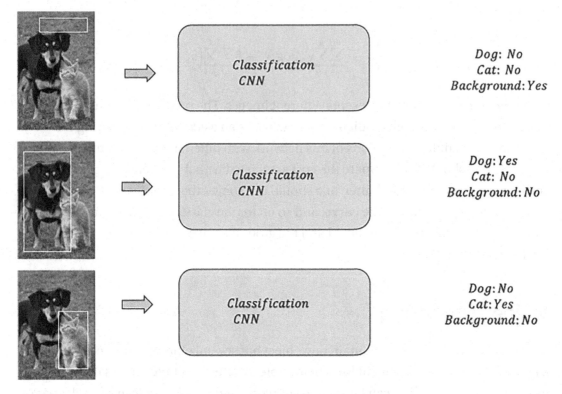

Figure 6-17. *Sliding-window technique to object detection*

Although this method would work, it would be computationally very expensive, or rather computationally intractable, since one would have to try thousands of image patches at different locations and scales in the absence of good region proposals. The current advanced methods in object detection propose several regions in which objects can be located and then feed those image-proposal regions to the classification and localization network. One such object-detection technique is called R-CNN, which we will discuss next.

R-CNN

In an R-CNN, *R* stands for *region* proposals. The region proposals are usually derived in through an algorithm called *selective search*. A selective search on an image generally provides around 2000 region proposals of interest. Selective search usually utilizes traditional image-processing techniques to locate blobby regions in an image as prospective areas likely to contain objects. The following are the processing steps for selective search on a broad level:

- Generate many regions within the image, each of which can belong to only one class.

- Recursively combine smaller regions into larger ones through a greedy approach. At each step, the two regions merged should be most similar. This process needs to be repeated until only one region remains. This process yields a hierarchy of successively larger regions and allows the algorithm to propose a wide variety of likely regions for object detection. These generated regions are used as the candidate region proposals.

These 2000 regions of interest are then fed to the classification and localization network to predict the class of the object along with associated bounding boxes. The classification network is a convolutional neural network followed by a support vector machine for the final classification. Illustrated in Figure 6-18 is a high-level architecture for an R-CNN.

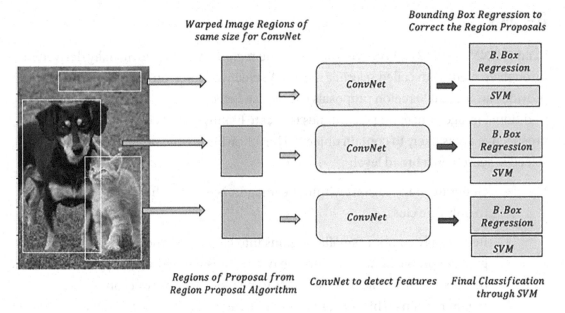

Figure 6-18. R-CNN network

The following are the high-level steps associated with training an R-CNN:

- Take a pretrained ImageNet CNN such as AlexNet, and retrain the last fully connected layer with the objects that need to be detected, along with backgrounds.

- Get all the region proposals per image (2000 per image as per selective search), warp or resize them to match the CNN input size, process them through CNN, and then save the features on disk for further processing. Generally, the pooling layer output maps are saved as features to disk.

- Train SVM to classify either object or background based on the features from CNN. For each class of objects, there should be one SVM that learns to distinguish between the specific object and background.

- Finally, bounding-box regression is done to correct the region proposals.

Although the R-CNN does a good job in at object detection, the following are some of its drawbacks:

- One of the problems with R-CNN is the huge number of proposals, which makes the network very slow since each of these 2000 proposals would have independent flows through convolution neural networks. Also, the region proposals are fixed in a sense they are proposed by a region-proposal algorithm; the R-CNN is not learning these proposals.

- The localization and bounding boxes predicted are from separate models, and hence during model training, we are not learning anything specific to localization of the objects based on the training data.

- For the classification task, the features generated out of the convolutional neural network are used to fine-tune SVMs, leading to a higher processing cost.

Fast and Faster R-CNN

Fast R-CNN overcomes some of the computational challenges of R-CNN by having a common convolution path for the whole image up to a certain number of layers, at which point the region proposals are projected to the output feature maps and relevant regions are extracted for further processing through fully connected layers and then the final classification. The extraction of relevant region proposals from the output feature maps from convolution and resizing them to a fixed size for the fully connected layer are done through a pooling operation known as ROI pooling. Illustrated in Figure 6-19 is an architecture diagram for fast R-CNN.

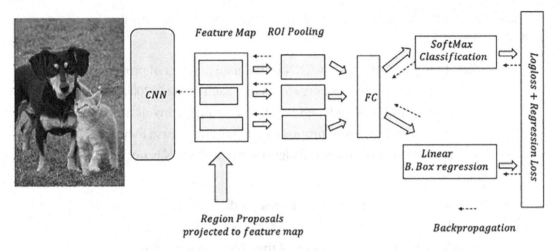

Figure 6-19. *Fast R-CNN schematic diagram*

Fast R-CNN saves a lot of costs associated with multiple convolution operations (2000 per image per selective search) in R-CNN. However, the region proposals are still dependent on the external region-proposal algorithms such as selective search. Because of this dependency on external region-proposal algorithms, fast R-CNN is bottlenecked by the computation of these region proposals. The network must wait for these external proposals to be made before it can move forward. These bottleneck issues are eliminated by faster R-CNN, where the region proposals are done within the network itself instead of depending on external algorithms. The architecture diagram for faster R-CNN is almost like that of fast R-CNN, but with a new addition—a region-proposal network that eliminates the dependency on an external region-proposal scheme such as selective search.

Generative Adversarial Networks

Generative adversarial networks, or GANs, are one of the remarkable advances in deep learning in recent times. Ian Goodfellow and colleagues first introduced this network in 2014 in a NIPS paper titled "Generative Adversarial Networks." The paper can be located at https://arxiv.org/abs/1406.2661. Since then, there has been a lot of interest and development in generative adversarial networks. In fact, Yann LeCun, one of the most prominent deep-learning experts, considers the introduction of generative adversarial networks to be the most important breakthrough in deep learning in recent times. GANs are used as generative models for producing synthetic data like the data produced by a given distribution. GAN has usages and potential in several fields, such as image generation, image inpainting, abstract reasoning, semantic segmentation, video generation, style transfer from one domain to another, and text-to-image generation applications, among others.

Generative adversarial networks are based on the two-agent zero-sum game from game theory. A generative adversarial network has two neural networks, the generator (G) and the discriminator (D), competing against each other. The generator (G) tries to fool the discriminator (D) such that the discriminator is not able to distinguish between real data from a distribution and the fake data generated by the generator (G). Similarly, the discriminator (D) learns to distinguish the real data from the fake data generated by the generator (G). Over a certain period, both the discriminator and the generator improve on their own tasks while competing with each other. The optimal solution to this game-theory problem is given by the Nash equilibrium wherein the generator learns to produce fake data as if they are from the original data distribution, and at the same time the discriminator outputs $\frac{1}{2}$ probability for both real and fake data points.

Now, the most obvious question is how the fake data are constructed. The fake data is constructed through the generative neural network model (G) by sampling noise z from a prior distribution P_z. If the actual data x follows distribution P_x and the fake data $G(z)$ generated by the generator follows distribution P_g, then at equilibrium $P_x(x)$ should equal $P_g(G(z))$; i.e.,

$$P_g\big(G(z)\big) \sim P_x(x)$$

Since at equilibrium the distribution of the fake data would be almost the same as the real data distribution, the generator would learn to sample fake data that would be hard to distinguish from the real data. Also, at equilibrium the discriminator D should output $\frac{1}{2}$ as the probability for both classes—the real data and the fake data. Before we go through the math for a generative adversarial network, it is worthwhile to gain some understanding about the zero-sum game, Nash equilibrium, and minimax formulation.

Illustrated in Figure 6-20 is a generative adversarial network in which there are two neural networks, the generator (G) and the discriminator (D), that compete against each other.

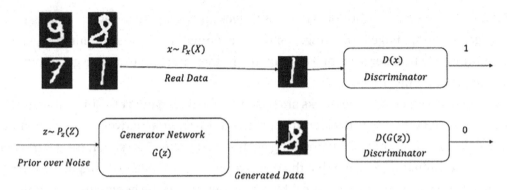

Figure 6-20. *Basic illustration of adversarial network*

Maximin and Minimax Problem

In a game, each participant would try to maximize their payoff and enhance their chance of winning. Considering a game played by *N* competitors, the maximin strategy of candidate *i* is the one that maximizes his or her payoff given the other *N*–1 participants are playing to take candidate *i* down. The payoff of candidate *i* corresponding to the maximin strategy is the maximum value the candidate *i* is sure to get without knowing the others' moves. Hence, the maximin strategy s_i^* and the maximin value L_i^* can be expressed as follows:

$$s_i^* = \underbrace{argmax}_{s_i} \ \underbrace{min}_{s_{-i}} \ L_i\left(s_i, s_{-i}\right)$$

$$L_i^* = \underbrace{max}_{s_i} \ \underbrace{min}_{s_{-1}} \ L_i\left(s_i, s_{-i}\right)$$

An easy way to interpret the maximin strategy for candidate *i* is to consider that *i* already has knowledge of his opponents' moves and that they would try to minimize his maximum payoff possible in each of his moves. Hence, with this assumption, *i* will play a move that would be the maximum of all the minima in each of his moves.

It is easier to explain minimax strategy in this paradigm instead of in its more technical terms. In minimax strategy, the candidate *i* would assume that the other candidates denoted by -*i* would allow the minimum in each of their moves. In such a case, it is logical for *i* to select a move that would provide him the maximum of all the minimum payoffs that other candidates have set for i in each of their moves. Under the minimax strategy, candidate i's payoff is given by the following:

$$L_i^* = \underset{s_{-i}}{\min}\,\underset{s_i}{\max}\ L_i\left(s_i, s_{-i}\right)$$

Do note that the final payoffs or losses when all the players have made their moves might be different from the maximin or minimax values.

Let us try to motivate a maximin problem with an intuitive example in which two agents A and B are competing against each other to maximize their profit from the game. Also, let's assume that A can make three moves, L_1, L_2, L_3, while B can make two moves, M_1 and M_2. The payoff table for this is shown in Figure 6-21. In each cell, the first entry corresponds to A's payoff, while the second entry denotes B's payoff.

B

A		M_1	M_2
	L_1	(6,2)	(4,−41)
	L_2	(10,0.5)	(−20.5,2)
	L_3	(−201,4.1)	(8,8)

Figure 6-21. *Maximin and minimax illustration between two players*

Let us first assume that both A and B are playing maximin strategy; that is, they should make a move to maximize their payoffs with the expectation that the other is going to minimize their payoff as much as possible.

A's maximin strategy would be to select move L_1, in which case the minimum A would get is 4. If he chooses L_2, A runs the risk of ending up with −20, while if he chooses L_3 he may end up with even worse at −201. So, the maximin value for A is the maximum of all the minimums possible in each row, i.e., 4, corresponding to the strategy L_1.

B's maximin strategy would be to select M_1 since along M_1 the minimum B would get is 0.5. If B chooses M_2, then B runs the risk of ending up with −41. So, the minimax value for B is the maximum of all the possible minimums along the columns, i.e., 0.5, corresponding to M_1.

Now, let's say both A and B play their minimax strategy, i.e., (L_1, M_1), in which case A's payoff is 6 and that of B is 2. So, we see that the maximin values are different from the actual payoff values once the players play their maximin strategies.

Now, let's look at the situation where both players wish to play minimax strategy. In minimax strategy, one selects a strategy to arrive at a maximum that is the minimum of all the possible maxima in each of the opponent's moves.

Let's look at A's minimax value and strategy. If B selects M_1, the maximum A can get is 10, while if B selects M_2 the maximum A can get is 88. Obviously, B would allow A to take only the minimum of each of the maxima possible in each of B's moves, and so, thinking of B's mindset, the minimax value that A can expect is 8, corresponding to his move L_2.

Similarly, B's minimax value would be the minimum of all the maxima possible for B in each of A's moves, i.e., minima of 2 and 8. Hence, B's minimax value is 2.

One thing to notice is that the minimax value is always larger than or equal to the maximin value for a candidate solely because of how the maximin and minimax are defined.

Zero-Sum Game

In game theory, a zero-sum game is a mathematical formulation of a situation in which every participant's gain or loss is equally offset by other participants' loss or gain. So, as a system, the net gain or loss of the group of participants is zero. Consider a zero-sum game played by two players A and B against each other. The zero-sum game can be represented by a structure called a payoff matrix, as shown in Figure 6-22.

B

		M_1	M_2	M_3
	L_1	4	−2	−6
A	L_2	0	−10	12
	L_3	8	6	10

Figure 6-22. Payoff matrix for a two-player zero-sum game

Figure 6-22 is an illustration of the two-player payoff matrix where each cell in the matrix represents the payoff of player A's game for every combination of moves for A and B. Since this is a zero-sum game, the payoff of B is not explicitly mentioned; it's just the negative of the payoff of player A. Let's say A plays a maximin game. It would choose the maximum of the minima in each row and hence would select strategy L_3 with its corresponding payoff of the maximum of {-6,-10,6}, i.e., 6. The payoff of 6 corresponds to the move M_2 for B. Similarly, had A played the minimax strategy, A would have been forced to get a payoff equal to the minimum of the maxima of the payoffs along each column, i.e., for each of B's moves. In that case, A's payoff would have been the minimum of {8,6,10}, i.e., 6, corresponding to the minimax strategy of L_3. Again, this payoff of 6 corresponds to move M_2 for B. So, we can see in the case of a zero-sum game the maximin payoff of a participant is equal to the minimax payoff.

Now, let's look at the maximin payoff of player B. The maximin payoff of B is the maximum of the minima of B in each move, i.e., the maximum of $(-8, -6, -12) = -6$, which corresponds to the move M_2. Also, this value corresponds to the move L_3 for A. Similarly, the minimax payoff of B is the minimum of the maxima B can have for each of A's moves, i.e., $(6, 10, -6) = -6$. Again, for B the minimax value is the same as the maximin value, and the corresponding move for B is M_2. The corresponding move for A in this scenario is also L_3.

So, the learnings from a zero-sum game are as follows:

- Irrespective of whether A and B play maximin strategy or minimax strategy, they are going to end up with the moves L_3 and M_2, respectively, corresponding to payoffs of 6 for A and -6 for B. Also, the minimax and maximin values for the players coincided with the actual payoff values the player got when they went with the minimax strategy.

- The preceding point leads to one important fact: in a zero-sum game, the minimax strategy for one player would yield the actual strategies for both players had they both employed pure minimax or maximin strategies. Hence, both the moves can be determined by considering the moves of either A or B. If we consider A's minimax strategy, then both the players' moves are embedded in it. If the payoff utility of A is $U(S_1, S_2)$, then the moves of A and B—i.e.,S_1 and S_2, respectively—can be found out by applying the minimax strategy of A or B alone.

Minimax and Saddle Points

For zero-sum minimax problems involving two players A and B, the payoff $U(x, y)$ of player A can be expressed as follows:

$$\hat{U} = \underset{y}{\min} \underset{x}{\max}\, U(x,y)$$

where x denotes the move of A while y denotes the move of B.

Also, the values of x, y corresponding to $\hat{U}$ are the equilibrium strategies of A and B, respectively, i.e., they will not change the move if they continue to believe in the minimax or the maximin strategy. For a zero-sum two-player game, minimax or maximin would yield the same results, and hence this equilibrium holds true if the players are playing with either the minimax or the maximin strategy. Also, since the minimax value is equal to the maximin value, the order in which the minimax or maximin is defined doesn't matter. We might just as well let A and B independently select their best strategies for each strategy of the other, and we will see that for a zero-sum game, one of the combinations of strategies will overlap. This overlapping condition is the best strategy for both A and B and is identical to their minimax strategy. This is also the Nash equilibrium for the game.

Up to now we have kept the strategies discrete for easy interpretability using the payoff matrix, but they can be of continuous values. As for the GAN, the strategies are the continuous parameter values of the generator and the discriminator neural networks, and so before we go into the details of the GAN utility function, it makes sense to look at the payoff utility function $f(x, y)$ for A, which is a function of two continuous variables in x and y. Further, let x be the move of A and y be the move of B. We need to find the equilibrium point, which is also the minimax or the maximin of the payoff utility function of either player. The payoff corresponding to the minimax of A would provide the strategy of both A and B. Since for the zero-sum two-player game the minimax and maximin are the same, the order of minimax doesn't matter, i.e.,

$$\underset{y}{\min} \underset{x}{\max}\, f(x,y) = \underset{y}{\min} \underset{x}{\max}\, f(x,y) = \underset{\substack{y \\ \underset{x}{\max}}}{\min}\, f(x,y)$$

For a continuous function, this is only possible when the solution to the preceding function is a saddle point. A saddle point is a point at which the gradient with respect to each of the variables is zero; however, it is not a local minima or maxima.

Instead, it tends to a local minimum in some directions of input vector and a local maximum with respect to the other directions of input vector. In fact for the two-player zero-sum game utility equilibrium saddle point would be a local minima with respect to one of the participant's move, while it would be a local maxima with respect to the other participant's move. So, one can just use the methods of finding a saddle point using multivariate calculus. Without loss of generality for a multivariate function $f(x)$ with $x \in R^{n \times 1}$, we can determine the saddle point by the following test:

- Compute the gradient of $f(x)$ with respect to the vector x, i.e., $\nabla_x f(x)$, and set it to zero.

- Evaluate the Hessian $\nabla_x^2 f(x)$ of the function, i.e., the matrix of second-order derivatives at each of the points at which the gradient vector $\nabla_x f(x)$ is zero. If the Hessian matrix has both positive and negative Eigen values at the evaluated point, then the point is a saddle point.

Coming back to the two-variable payoff utility function $f(x, y)$, for A, let us define it as follows to illustrate an example:

$$f(x,y) = x^2 - y^2$$

Hence, the utility function for B would automatically be $-x^2 + y^2$.

We now investigate whether the utility function provides for an equilibrium if both players play a zero-sum minimax or maximin strategy. The game would have an equilibrium beyond which the players won't be able to improve their payoffs due to their strategies being optimal. The equilibrium condition is the Nash equilibrium of the game and is a saddle point for the function $f(x, y)$.

Setting the gradient of $f(x, y)$ to zero, we get the following:

$$\nabla f(x,y) = \begin{bmatrix} \dfrac{\partial f}{\partial x} \\ \dfrac{\partial f}{\partial y} \end{bmatrix} = \begin{bmatrix} 2x \\ -2y \end{bmatrix} = 0 => (x,y) = (0,0)$$

The Hessian of the function is given by the following:

$$\nabla^2 f(x,y) = \begin{bmatrix} \dfrac{\partial^2 f}{\partial x^2} & \dfrac{\partial^2 f}{\partial x \partial y} \\[2ex] \dfrac{\partial^2 f}{\partial y \partial x} & \dfrac{\partial^2 f}{\partial y^2} \end{bmatrix} = \begin{bmatrix} 2 & 0 \\ 0 & -2 \end{bmatrix}$$

The Hessian of the function is $\begin{bmatrix} 2 & 0 \\ 0 & -2 \end{bmatrix}$ for any value of (x, y) including $(x, y) = (0, 0)$. Since the Hessian has both positive and negative Eigen values, i.e., 2 and -2, hence the point $(x, y) = (0, 0)$ is a saddle point. The strategies at equilibrium for A should be to set $x = 0$, while that of y should be to set $y = 0$ in a zero-sum minimax or maximin game.

GAN Cost Function and Training

In generative adversarial networks, both the generator and the discriminator networks try to outdo each other by playing the minimax strategy in a zero-sum game. The moves in this case are parameter values that the networks choose. For ease of notation, let us represent the model parameters by the model notations themselves, i.e., G for the generator and D for the discriminator. Now, let's frame the utility of the payoff function for each of the networks. The discriminator would try to classify both the fake or synthetically generated samples and the real data samples correctly. In other words, it would try to maximize the utility function:

$$U(D,G) = E_{x \sim P_x(x)}\left[\log D(x)\right] + E_{z \sim P_z(z)}\left[\log\left(1 - D(G(z))\right)\right]$$

where x denotes the real data samples drawn from the probability distribution $P_x(x)$ and z is the noise drawn from a prior noise distribution $P_z(z)$. Also, the discriminator is trying to output 1 for real data sample x and 0 for generator-created fake or synthetic data based on the noise samples z. Hence, the discriminator would like to play a strategy that maximizes $D(x)$ to be as close as possible to 1, which would make $\log D(x)$ near a 0 value. The more $D(x)$ is less than 1, the smaller the value for $\log D(x)$ would be and hence the smaller the utility value for the discriminator would be. Similarly, the discriminator would like to catch the fake or synthetic data by setting its probability close to zero; i.e., set $D(G(z))$ as close to zero as possible to identify it as a fake image. When $D(G(z))$ is near zero, the expression $[\log(1 - D(G(z)))]$ tends to zero. As the value of $D(G(z))$ diverges from zero, the payoff for the discriminator becomes smaller since $\log(1 - D(G(z)))$ gets smaller. The discriminator would like to do it over the whole distribution of x and z, and hence the terms for expectation or mean in its payoff function. Of course, the generator G has a say in the payoff function for D in the form of $G(z)$—i.e., the second term—and so it would also try to play a move that minimizes the payoff for D. The more the payoff for D is, the worse the situation is for G. So, we can think of G as having the same utility function as D has, only with a negative sign in it, which makes this a zero-sum game where the payoff for G is given by the following:

$$V(D,G) = -\mathrm{E}_{x \sim P_x(x)}\Big[\log D(x)\Big] - E_{z \sim P_z(z)}\Big[\log\big(1 - D\big(G(z)\big)\big)\Big]$$

The generator G would try to choose its parameters so that $V(D, G)$ is maximized; i.e., it produces fake data samples $G(z)$ such that the discriminator is fooled into classifying them with a 0 label. In other words, it wants the discriminator to think $G(z)$ is real data and assign high probability to them. High values of $D(G(z))$ away from 0 would make $\log(1 - D(G(z)))$ a negative value with a high magnitude, and when multiplied by the negative sign at the beginning of the expression, it would produce a high value of $-E_{z \sim P_z(z)}\Big[\log\big(1 - D\big(G(z)\big)\big)\Big]$, thus increasing the generator's payoff. Unfortunately, the generator would not be able to influence the first term in $V(D, G)$ involving real data since it doesn't involve the parameters in G.

The generator G and the discriminator D models are trained by letting them play the zero-sum game with the minimax strategy. The discriminator would try to maximize its payoff $U(D, G)$ and would try to reach its minimax value.

$$u^* = \min_{D}\max_{G}\, \mathrm{E}_{x \sim P_x(x)}\Big[\log D(x)\Big] + E_{z \sim P_z(z)}\Big[\log\big(1 - D\big(G(z)\big)\big)\Big]$$

Similarly, the generator G would like to maximize its payoff $V(D, G)$ by selecting a strategy.

$$v^* = \min_{D}\max_{G}\, -\mathrm{E}_{x \sim P_x(x)}\Big[\log D(x)\Big] - E_{z \sim P_z(z)}\Big[\log\big(1 - D\big(G(z)\big)\big)\Big]$$

Since the first term is something that is not in the control of G to maximize,

$$v^* = \min_{D}\max_{G}\, -E_{z \sim P_z(z)}\Big[\log\big(1 - D\big(G(z)\big)\big)\Big]$$

As we have seen, in a zero-sum game of two players, one need not consider separate minimax strategies, as both can be derived by considering the minimax strategy of one of the players' payoff utility functions. Considering the minimax formulation of the discriminator, we get the discriminator's payoff at equilibrium (or Nash equilibrium) as follows:

$$u^* = \underset{\substack{G \\ \underbrace{max}_{D}}}{\underbrace{min}} E_{x \sim P_x(x)} \Big[\log D(x) \Big] + E_{z \sim P_z(z)} \Big[\log \big(1 - D(G(z)) \big) \Big]$$

The values of $\hat{G}$ and $\hat{D}$ at u^* would be the optimized parameters for both networks beyond which they can't improve their scores. Also $(\hat{G}, \hat{D})$ gives the saddle point of D's utility function $E_{x \sim P_x(x)} \Big[\log D(x) \Big] + E_{z \sim P_z(z)} \Big[\log \big(1 - D(G(z)) \big) \Big]$.

The preceding formulation can be simplified by breaking down the optimization in two parts, i.e., let D maximize its payoff utility function with respect to its parameters, and let G minimize D's payoff utility function with respect to its parameters in each move.

$$\underbrace{max}_{D} E_{x \sim P_x(x)} \Big[\log D(x) \Big] + E_{z \sim P_z(z)} \Big[\log \big(1 - D(G(z)) \big) \Big]$$

$$\underbrace{min}_{G} E_{z \sim P_z(z)} \Big[\log \big(1 - D(G(z)) \big) \Big]$$

Each would consider the other's move as fixed while optimizing its own cost function. This iterative way of optimization is nothing but the gradient-descent technique for computing the saddle point. Since machine-learning packages are mostly coded to minimize rather than maximize, the discriminator's objective can be multiplied by -1, and then D can minimize it rather than maximizing it.

Presented next is the mini-batch approach generally used for training the GAN based on the preceding heuristics:

- For N number of iterations:

  - For k_D steps:

  - Draw m samples $\{z^{(1)}, z^{(2)}, .. z^{(m)}\}$ from the noise distribution $z \sim P_z(z)$

  - Draw m samples $\{x^{(1)}, x^{(2)}, .. x^{(m)}\}$ from the data distribution $x \sim P_x(x)$

- Update the discriminator D parameters by using stochastic gradient descent. If the parameters of the discriminator D are represented by θ_D, then update θ_D as follows:

$$\theta_D \rightarrow \theta_D - \nabla_{\theta_D}\left[-\frac{1}{m}\sum_{i=1}^{m}\left(\log D\left(x^{(i)}\right)\right) + \log\left(1 - D\left(G\left(z^{(i)}\right)\right)\right)\right]$$

- end

  - For k_G steps:

    - Draw m samples $\{z^{(1)}, z^{(2)}, .. z^{(m)}\}$ from the noise distribution:

$$z \sim P_z(z)$$

    - Update the generator G by stochastic gradient descent. If the parameters of the generator G are represented by θ_G, then update θ_G as follows:

$$\theta_G \rightarrow \theta_G - \nabla_{\theta_G}\left[\frac{1}{m}\sum_{i=1}^{m}\log\left(1 - D\left(G\left(z^{(i)}\right)\right)\right)\right]$$

  - end

- end

The entire pseudo code above would constitute 1 epoch of the GAN training.

Vanishing Gradient for the Generator

Generally, in the initial part of training, the samples produced by the generator are very different from the original data, and hence the discriminator can easily tag them as fake. This leads to close-to-zero values for $D(G(z))$, and so the gradient $\nabla_{\theta_G}\left[\frac{1}{m}\sum_{i=1}^{m}\log\left(1 - D\left(G\left(z^{(i)}\right)\right)\right)\right]$ saturates, leading to a vanishing-gradient problem for parameters of the network of G. To overcome this problem, instead of minimizing $E_{z\sim P_z(z)}\left[\log\left(1 - D(G(z))\right)\right]$, the function $E_{z\sim P_z(z)}\left[\log G(z)\right]$ is maximized, or, to adhere to gradient descent, $E_{z\sim P_z(z)}\left[-\log G(z)\right]$ is minimized. This alteration makes the training method no longer a pure minimax game but seems to be a reasonable approximation that helps overcome saturation in the early phase of training.

GAN Learning from an F-Divergence Perspective

The goal of the generative adversarial network (GAN) generator $G(.)$ is to learn to produce samples $G(z)$ as if they were sampled from the target distribution $P(x)$ that we want to learn. Hence once the GAN has been trained to optimality, the following should hold true:

$$P\big(G(z)\big) \sim P\big(x\big)$$

The discriminator D tries to classify the samples from the target distribution $P(x)$ as real samples and the ones generated by the generator G as fake images. Hence the discriminator tries to minimize the binary cross-entropy loss (see below) associated with classifying the target distribution samples as belonging to class 1 and the generator G samples as belonging to class 0.

$$U\big(G,D\big) = -E_{x \sim P(x)}\Big[\log D\big(x\big)\Big] - E_{z \sim P(z)}\Big[\log\big(1 - D\big(G(z)\big)\big)\Big]$$

Since we are dealing with the generator images for the discriminator loss, we can change the expectation over $P(z)$ to be expectation over the generator samples distribution $P(G(z))$ as shown in the following:

$$U\big(G,D\big) = -E_{x \sim P(x)}\Big[\log D\big(x\big)\Big] - E_{G(z) \sim P(G(z))}\Big[\log\big(1 - D\big(G(z)\big)\big)\Big]$$

The discriminator D tries to minimize the preceding utility $U(G, D)$, while the generator tries to maximize the same. Because of the zero-sum nature of the utility, the optimal generator $\hat{G}$ and discriminator $\hat{D}$ can be found through the following optimization:

$$\min_{D}\max_{G} - E_{x \sim P(x)}\Big[\log D\big(x\big)\Big] - E_{G(z) \sim P(G(z))}\Big[\log\big(1 - D\big(G(z)\big)\big)\Big]$$

We measure the distance between two probability distributions through f-divergence methods such as KL divergence that we have studied in Chapter 5.

To recall, the KL divergence between a distribution $Q(x)$ and a distribution $P(x)$ is given by the following:

$$KL(P\|Q) = E_{x \sim P(x)}\log\frac{P(x)}{Q(x)}$$

As you may recall, KL divergence is not symmetric, i.e., $KL(P \| Q) \neq KL(Q \| P)$.

However, a symmetric divergence measure called **Jensen-Shannon divergence (JSD)** can be defined using KL divergence as in the following:

$$JSD(P \| Q) = \frac{1}{2} KL\left(P \| \frac{P+Q}{2}\right) + \frac{1}{2} KL\left(Q \| \frac{P+Q}{2}\right)$$

$$= E_P \log \frac{P}{\frac{P+Q}{2}} + E_Q \log \frac{Q}{\frac{P+Q}{2}}$$

Coming back to the optimization of the GAN utility $U(G, D)$, we have the following:

$$\min_D \max_G U(G,D) = \min_D \max_G - E_{x \sim P(x)}\left[\log D(x)\right] - E_{G(z) \sim P(G(z))}\left[\log\left(1 - D(G(z))\right)\right]$$

If we denote the generator samples $G(z)$ as x and the probability distribution over the generator samples as G_g, then $P(G(z)) \sim G_g(x)$. Similarly, let's denote the target distribution $P(x)$ by $P_{data}(x)$. Making these substitutions in the utility $U(G, D)$, we have the following:

$$U(G,D) = -E_{x \sim P_{data}(x)}\left[\log D(x)\right] - E_{x \sim G_g(x)}\left[\log\left(1 - D(x)\right)\right]$$

$$= - \int_{x \sim P_{data}(x)} \log D(x) p_{data}(x) dx - \int_{x \sim G_g(x)} \log(1 - D(x)) G_g(x) dx$$

Keeping the generator fixed, the preceding integral would be minimized with respect to discriminator D when the optimal discriminator is as follows:

$$\hat{D}(x) = \frac{P_{data}(x)}{P_{data}(x) + G_g(x)}$$

Substituting the optimal discriminator $\hat{D}$ in the utility function U(G,D), we have the following:

$$U\left(G, \hat{D}\right) = -E_{x \sim P_{data}(x)}\left[\log \frac{P_{data}(x)}{P_{data}(x) + G_g(x)}\right] - E_{x \sim G_g(x)}\left[\log\left(1 - \frac{P_{data}(x)}{P_{data}(x) + G_g(x)}\right)\right]$$

$$= -E_{x \sim P_{data}(x)}\left[\log \frac{P_{data}(x)}{P_{data}(x) + G_g(x)}\right] - E_{x \sim G_g(x)}\left[\log\left(\frac{G_g(x)}{P_{data}(x) + G_g(x)}\right)\right]$$

$$= \log 4 - E_{x \sim P_{data}(x)} \left[\log \frac{P_{data}(x)}{\frac{P_{data}(x) + G_g(x)}{2}} \right] - E_{x \sim G_g(x)} \left[\log \left(\frac{G_g(x)}{\frac{P_{data}(x) + G_g(x)}{2}} \right) \right]$$

$$= \log 4 - JSD\left(P_{data} \| G_g\right)$$

So conditioned on the optimal discriminator, the utility $U\left(G, \hat{D}\right)$ for the generator to maximize turns out to be negative of the Jenson-Shannon divergence of the target data distribution P_{data} and the generator samples distribution G_g. Hence the optimal generator learns to generate samples as if they are coming from the target data distribution P_{data} by minimizing the Jensen-Shannon divergence $JSD(P_{data} \| G_g)$.

TensorFlow Implementation of a GAN Network

In this section, a GAN network trained on MNIST images is illustrated where the generator tries to create fake synthetic images like MNIST, while the discriminator tries to tag those synthetic images as fake while still being able to distinguish the real data as authentic. Once the training is completed, we sample a few synthetic images and see whether they look like the real ones. The generator is a simple feed-forward neural network with three hidden layers followed by the output layer, which consists of 784 units corresponding to the 784 pixels in the MNIST image. The activations of the output unit have been taken to be *tanh* instead of *sigmoid* since *tanh* activation units suffer less from vanishing-gradient problems as compared to sigmoid units. A *tanh* activation function outputs values between -1 and 1, and thus the real MNIST images are normalized to have values between -1 and 1 so that both the synthetic images and the real MNIST images operate in the same range. The discriminator network is also a three-hidden-layer feed-forward neural network with a sigmoid output unit to perform binary classification between the real MNIST images and the synthetic ones produced by the generator. The input to the generator is a 100-dimensional input sampled from a uniform noise distribution operating between -1 and 1 for each dimension. The detailed implementation is illustrated in Listing 6-5.

Listing 6-5. Implementation of a Generative Adversarial Network

```python
import tensorflow as tf
from tensorflow.keras import layers, Model, initializers,activations

## The dimension of the Prior Noise Signal is 100
## The generator would have 150 and 300 hidden units successively before
784 outputs corresponding
## to 28x28 image size

h1_dim = 150
h2_dim = 300
dim = 100
batch_size = 256

class generator(Model):
    """

    Generator class of GAN
    """

    def __init__(self,hidden_units=[500,500]):
        super(generator,self).__init__()
        self.fc1 = layers.Dense(hidden_units[0],activation='relu',kernel_
        initializer=initializers.TruncatedNormal(mean=0.0, stddev=0.1))
        self.fc2 = layers.Dense(hidden_units[1],activation='relu',kernel_
        initializer=initializers.TruncatedNormal(mean=0.0, stddev=0.1))
        self.fc3 = layers.Dense(28*28,activation=activations.tanh,kernel_
        initializer=initializers.TruncatedNormal(mean=0.0, stddev=0.1))

    def call(self,x):
        x = self.fc1(x)
        x = self.fc2(x)
        x = self.fc3(x)
        return x

class discriminator(Model):
    """

    Discriminator Class of the GAN
    """
```

```python
    def __init__(self,hidden_units=[500,500],dropout_rate=0.3):
        super(discriminator,self).__init__()
        self.fc1 = layers.Dense(hidden_units[0],activation='relu',kernel_
        initializer=initializers.TruncatedNormal(mean=0.0, stddev=0.1))
        self.drop1 = layers.Dropout(rate=dropout_rate)
        self.fc2 = layers.Dense(hidden_units[1],activation='relu',kernel_
        initializer=initializers.TruncatedNormal(mean=0.0, stddev=0.1))
        self.drop2 = layers.Dropout(rate=dropout_rate)
        self.fc3 = layers.Dense(1,kernel_initializer=initializers.
        TruncatedNormal(mean=0.0, stddev=0.1))

    def call(self,x):
        x = self.fc1(x)
        x = self.drop1(x)
        x = self.fc2(x)
        x = self.drop2(x)
        x = self.fc3(x)

        return x

class GAN(Model):
    """

    Generator and Discriminator Flow for the fake images
    """

    def __init__(self,G,D):
        super(GAN,self).__init__()
        self.G = G
        self.D = D

    def call(self,z):
        z = self.G(z)
        z = self.D(z)
        return z

def data_load():
    """

    Loading the training MNIST images and normalizing them within the
    range -1 to 1
    """
```

```
(train_X, train_Y), (test_X, test_Y) = tf.keras.datasets.mnist.
load_data()
train_X, test_X , = train_X.reshape(-1,28*28), test_X.reshape(-1,28*28)
train_X, test_X = train_X/255.0, test_X/255.0
train_X, test_X = 2*train_X - 1, 2*test_X - 1
return np.float32(train_X), train_Y, np.float32(test_X), test_Y
def train(lr=0.0001,batch_size=256,hidden_units=[150,130],dim=100,dropout_
rate=0.3,num_epochs=300):

    # Build the GAN model
    G_ = generator(hidden_units=hidden_units)
    D_ = discriminator(hidden_units=hidden_units[::-1],dropout_
    rate=dropout_rate)
    model = GAN(G_,D_)
    G_graph, D_graph, model_graph = tf.function(G_), tf.function(D_),
    tf.function(model)

    # Setup the optimizer
    optimizer = tf.keras.optimizers.Adam(lr)

    # Load the daat
    train_X, train_Y, test_X, test_Y = data_load()
    num_train = train_X.shape[0]
    order = np.arange(num_train)
    num_batches = num_train//batch_size

    # Set up the discriminator loss
    loss_fn = tf.keras.losses.BinaryCrossentropy(from_
    logits=True,reduction=tf.keras.losses.Reduction.NONE)

    # Invoke the training
    for epoch in range(num_epochs):
        np.random.shuffle(order)
        train_X, train_Y = train_X[order], train_Y[order]

        for i in range(num_batches):
            x_ = train_X[batch_size*i:(i+1)*batch_size]
```

```
z_ = np.random.uniform(-1, 1, size=(x_.shape[0],dim)).
astype(np.float32)
y_label_real_dis = np.array([1. for i in range(x_.shape[0])]).
reshape(-1,1)
y_label_fake_dis = np.array([0. for i in range(x_.shape[0])]).
reshape(-1,1)
y_label_gen = np.array([1. for i in range(x_.shape[0])]).
reshape(-1,1)

x_,z_, y_label_real_dis, y_label_fake_dis,y_label_gen =
tf.constant(x_),
                         tf.constant(z_), tf.constant(y_label_
                         real_dis) ,
                         tf.constant(y_label_fake_dis),
                         tf.constant(y_label_gen)

with tf.GradientTape(persistent=True) as tape:
    y_pred_fake = model_graph(z_,training=True)
    y_pred_real = D_graph(x_,training=True)

    loss_discrimator = 0.5*tf.reduce_mean(loss_fn(y_label_fake_
    dis,y_pred_fake)
                                        + loss_fn(y_label_real_
                                        dis,y_pred_real))
    loss_generator = tf.reduce_mean(loss_fn(y_label_gen,y_
    pred_fake))
# Compute gradient
grad_d = tape.gradient(loss_discrimator, D_.trainable_variables)
grad_g = tape.gradient(loss_generator, G_.trainable_variables)
# update the parameters
optimizer.apply_gradients(zip(grad_d, D_.trainable_variables))
optimizer.apply_gradients(zip(grad_g, G_.trainable_variables))
del tape
if (i % 200) == 0:
    print (f"Epoch: {epoch} Iteration : {i}, Discrinator loss:
    {loss_discrimator.numpy()}, Generator loss: {loss_generator.
    numpy()}")
```

```
# Generator some images
z_ = tf.constant(np.random.uniform(-1, 1, size=(batch_size,dim)).
astype(np.float32))
imgs = 0.5*(G_graph(z_,training=False) + 1).numpy()
print(imgs.shape)
for k in range(36):
    plt.subplot(6,6,k+1)
    image = np.reshape(imgs[k],(28,28))
    plt.imshow(image,cmap='gray')

return G_, D_, model, G_graph, D_graph, model_graph, imgs

G_, D_, model, G_graph, D_graph, model_graph,imgs_val   = train()
```

-- output --

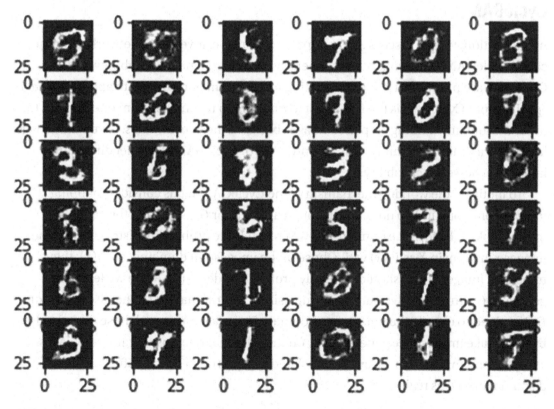

Figure 6-23. *Digits synthetically generated by the GAN network*

From Figure 6-23, we can see that the GAN generator is able to produce images similar to the MNIST dataset digits. The GAN model was trained on 60000 mini-batches of size 256 to achieve this quality of results. The point I want to emphasize is that GANs are relatively hard to train in comparison with other neural networks. Hence, a lot of experimentation and customization is required in order to achieve the desired results.

GAN's Similarity to Variational Autoencoder

Do note that GAN has lot of parallels to the variational autoencoders that we have studied in Chapter 5. In the GAN, the generator $G(.)$ acts a generative model mapping the z sampled from a prior distribution $P(z)$ to actual images x from a known distribution $P(x)$. Similarly, in the variational autoencoder, the decoder $p_\phi(x|z)$ learns to project the samples z from a prior distribution $P(z)$ to images x. Both variational autoencoder and GAN are very popular as generative models in the computer vision space.

CycleGAN

In this section, we will discuss a very popular generative adversarial network technique called CycleGAN. Cycle consistency generative adversarial networks abbreviated as CycleGAN are primarily used for translating images from one domain to another. What is special about CycleGAN is the fact that it learns to map the images from one domain to other without any training on paired images between the two domains. Given that image labeling is a time-consuming affair, this property of CycleGAN of not needing paired images is a powerful capability.

To motivate the working mechanism of the CycleGAN, let us consider mapping images from domain X to domain Y through a generator G_{XY} such that the generated images look realistic in the domain Y. If we consider the probability distribution of the images x in domain X as $P_X(x)$ while those of the images y in domain Y as $P_Y(y)$, then translated images $G_{XY}(x)$ should be highly probable under $P_Y(y)$. Again we learn the generator G_{XY} through adversarial training using a discriminator D_Y. The discriminator D_Y is trained to classify the images from domain Y as real images and those generated by G_{XY} as fake images. The generator G_{XY} on the other hand learns to translate images $x \in X$ to domain Y in such a way the translated images $G_{XY}(x)$ are classified by the discriminator D_Y as real.

574

So, in its minimal form, the training objective for domain translation from X to Y can be represented by the GAN objective as follows:

$$T_{XY} = -E_{y \sim P_Y(y)} \log\big[D_Y(y)\big] - E_{x \sim P_x(X)} \log\big[1 - D_Y\big(G_{XY}(x)\big)\big]$$

Learning the translation $G_{XY}: X \to Y$ with just the translation loss T_{XY} doesn't guarantee any meaningful pairing between individual images x and y from domains X and Y, respectively. This is due the fact that there can be infinitely many translation G_{XY} that induces a distribution similar to G_{XY}. To add more structure to the underconstrained translation objective T_{XY} CycleGAN introduces the cycle consistency loss through another generator G_{YX} that translates images from domain Y to domain X such that the following:

$$G_{YX}G_{XY}(x) \approx x$$

The cycle consistency loss ensures that the translation from domain X to Y through G_{XY} should retain enough structure so that the original image can be reconstructed from the translated images much like in an autoencoder. The cycle consistency reconstruction loss going from domain X to domain Y and then back to domain X can be represented as follows:

$$R_{XYX} = E_{x \sim P_X(x)} \|G_{YX}G_{XY}(x) - x\|_p^p$$

Generally, p is chosen either as 1 for L1 norm reconstruction loss or as 2 for L2 norm reconstruction loss.

Hence the objective L_{XY} associated with translating images from X to Y in a cycle consistent way can be expressed as follows:

$$L_{XY} = T_{XY} - \lambda * R_{XYX}$$

$$= -E_{y \sim P_Y(y)} \log\big[D_Y(y)\big] - E_{x \sim P_x(X)} \log\big[1 - D_Y\big(G_{XY}(x)\big)\big] - \lambda * E_{x \sim P_X(x)} \|G_{YX}G_{XY}(x) - x\|_p^p$$

The term λ controls the importance of the adversarial loss to the cycle consistency loss and is a hyperparameter to the model.

We train the CycleGAN to learn a mapping not only from X to Y but also from Y to X. If we take the discriminator at the domain X side to be D_X, then the associated losses for translating images from Y to X can be expressed as follows:

$$T_{YX} = -E_{x \sim P_X(x)} \log\left[D_X(x)\right] - E_{y \sim P_Y(y)} \log\left[1 - D_X\left(G_{YX}(y)\right)\right]$$

$$R_{YXY} = E_{y \sim P_Y(y)} \left\| G_{XY} G_{YX}(y) - y \right\|_p^p$$

$$L_{YX} = T_{YX} - \lambda * R_{YXY}$$

$$= -E_{x \sim P_X(x)} \log\left[D_X(x)\right] - E_{y \sim P_Y(y)} \log\left[1 - D_X\left(G_{YX}(y)\right)\right] - \lambda * E_{y \sim P_Y(y)} \left\| G_{XY} G_{YX}(y) - y \right\|_p^p$$

Hence the combined objective of the CycleGAN is as follows:

$$L\left(G_{XY}, G_{YX}, D_Y, D_X\right) = L_{XY} + L_{YX}$$

$$= T_{XY} + \lambda * R_{XYX} + T_{YX} + \lambda * R_{YXY}$$

$$= T_{XY} + T_{YX} - \lambda * \left(R_{XYX} + R_{YXY}\right)$$

The optimal generators can be found out by maximizing the overall loss $L(G_{XY}, G_{YX}, D_Y, D_X)$ with respect to the generators G_{XY} and G_{YX} and minimizing the same with respect to the discriminators D_{XY} and D_{YX}.

$$\hat{G}_{XY}, \hat{G}_{YX} = \arg \max_{G_{XY}, G_{YX}} \min_{D_X, D_Y} L\left(G_{XY}, G_{YX}, D_Y, D_X\right)$$

$$= \arg \max_{G_{XY}, G_{YX}} \min_{D_X, D_Y} T_{XY} + T_{YX} - \lambda * \left(R_{XYX} + R_{YXY}\right)$$

To convert the minimization of the cycle consistency reconstruction loss with respect to the generators as a maximization, we have conveniently chosen the negative of the reconstruction loss in the loss objectives.

The CycleGAN architecture is illustrated in Figure 6-24. The readers are advised to go through it to better appreciate the underlying theory about CycleGAN that we have just discussed.

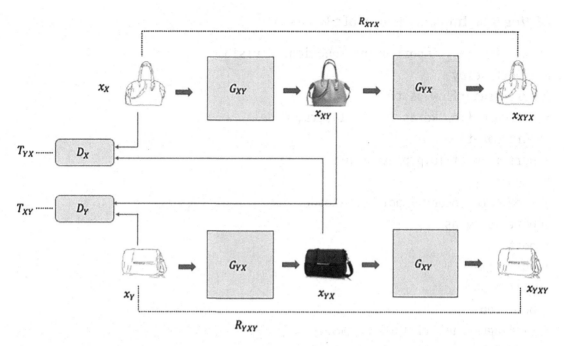

Figure 6-24. *Architectural diagram of a CycleGAN*

CycleGAN Implementation in TensorFlow

In this section, we will implement a CycleGAN in TensorFlow for generating handbag images from sketched outlines while explicitly pairing the sketches to the handbag. We will refer the sketches to belong to the domain A and the handbags to belong to domain B while working through the implementation.

As part of the implementation, we will train two generators of identical structure— one mapping the sketches in domain A to handbags in domain B while the other mapping handbags in domain B to sketches in domain A. The generators in each domain should map the images to the other domain in such a way that the generated images look realistic in the other domain. We will also have two discriminators for the two domains. The discriminator in each domain would learn to distinguish between the real images in the domain from the fake ones. This would enforce the generators to learn to generate better fake images because of the adversarial training.

The dataset for this implementation can be found at `http://efrosgans.eecs.berkeley.edu/pix2pix/datasets/edges2handbags.tar.gz`.

Illustrated in Listing 6-6 is the detailed implementation.

577

Listing 6-6. Implementation of a CycleGAN

```python
from __future__ import print_function, division
# import scipy
import tensorflow as tf
from tensorflow.keras import layers, Model
import datetime
import matplotlib.pyplot as plt
# import sys
# from data_loader import DataLoader
import numpy as np
import os
import time
import glob
import copy
from imageio import imread, imsave
from skimage.transform import resize

#  Load images for creating Training batches
def load_train_data(image_path, dim=64, is_testing=False):
    img_A = imread(image_path[0])
    img_B = imread(image_path[1])
    # Resize
    img_A = resize(img_A, [dim, dim])
    img_B = resize(img_B, [dim, dim])
    if not is_testing:

        if np.random.random() >= 0.5:
            img_A = np.fliplr(img_A)
            img_B = np.fliplr(img_B)

    # Normalize the images pixels to range from -1 to 1
    img_A = img_A / 2 - 1
    img_B = img_B / 2 - 1

    img_AB = np.concatenate((img_A, img_B), axis=2)

    return img_AB
```

```python
def merge(images, size):
    h, w = images.shape[1], images.shape[2]
    img = np.zeros((h * size[0], w * size[1], 3))
    for idx, image in enumerate(images):
        i = idx % size[1]
        j = idx // size[1]
        img[j * h:j * h + h, i * w:i * w + w, :] = image

    return img

# Routines to save images while training
def image_save(images, size, path):
    return imsave(path, merge(images, size))

def save_images(images, size, image_path):
    return image_save(inverse_transform(images), size, image_path)

def inverse_transform(images):
    return (images + 1) * 127.5

# Imagepool to store intermediate generated images
class ImagePool(object):
    def __init__(self, maxsize=50):
        self.maxsize = maxsize
        self.num_img = 0
        self.images = []

    def __call__(self, image):
        if self.maxsize <= 0:
            return image
        if self.num_img < self.maxsize:
            self.images.append(image)
            self.num_img += 1
            return image
        if np.random.rand() > 0.5:
            idx = int(np.random.rand() * self.maxsize)
            tmp1 = copy.copy(self.images[idx])[0]
            self.images[idx][0] = image[0]
```

```
                idx = int(np.random.rand() * self.maxsize)
                tmp2 = copy.copy(self.images[idx])[1]
                self.images[idx][1] = image[1]
                return [tmp1, tmp2]
        else:
            return image

class customConv2D(layers.Layer):
    """

    Custom Convolution layer consisting of convolution, batch normalization
    and relu/leakyrelu activation
    """

    def __init__(self, filters, kernel_size=4, strides=2, padding='SAME',
    norm=True, alpha=0.2, activation='lrelu'):
        super(customConv2D, self).__init__()
        self.norm = norm
        self.conv1 = layers.Conv2D(filters=filters, kernel_size=kernel_
        size, strides=(strides, strides),
                                   padding=padding)
        if self.norm:
            self.bnorm = layers.BatchNormalization()
        if activation == 'lrelu':
            self.activation = layers.LeakyReLU(alpha=alpha)
        elif activation == 'relu':
            self.activation = layers.ReLU()

    def call(self, x):
        x = self.conv1(x)
        if self.norm:
            x = self.bnorm(x)
        x = self.activation(x)
        return x

class customDeConv2D(layers.Layer):
    """

        Custom Transpose Convolution(upsampling) layer
    """
```

```python
    def __init__(self, filters, kernel_size=4, strides=2, padding='SAME',
    dropout_rate=0):
        super(customDeConv2D, self).__init__()
        self.dropout_rate = dropout_rate
        self.deconv1 = layers.Conv2DTranspose(filters=filters, kernel_
        size=kernel_size, strides=(strides, strides),
                                              padding=padding)
        if self.dropout_rate > 0:
            self.drop = layers.Dropout(rate=dropout_rate)

        self.bnorm = layers.BatchNormalization()
        self.activation = layers.ReLU()

    def call(self, x):
        x = self.deconv1(x)
        if self.dropout_rate > 0:
            x = self.drop(x)
        x = self.bnorm(x)
        x = self.activation(x)
        return x

class generator(Model):
    """

    Generator class that can be used for both Domain
    """

    def __init__(self, gf, channels=3):
        super(generator, self).__init__()
        self.down1 = customConv2D(filters=gf, strides=2, norm=False)
        self.down2 = customConv2D(filters=gf * 2, strides=2)
        self.down3 = customConv2D(filters=gf * 4, strides=2)
        self.down4 = customConv2D(filters=gf * 8, strides=2)
        self.down5 = customConv2D(filters=100, strides=1, padding='VALID')

        self.up1 = customDeConv2D(filters=gf * 8, strides=1,
        padding='VALID')
        self.up2 = customDeConv2D(filters=gf * 4, strides=2)
        self.up3 = customDeConv2D(filters=gf * 2, strides=2)
```

```python
        self.up4 = customDeConv2D(filters=gf, strides=2)
        self.convf = layers.Conv2DTranspose(filters=channels, kernel_
        size=4, strides=(2, 2), padding='SAME',
                                            activation=tf.nn.tanh)

    def call(self, x):
        x = self.down1(x)
        x = self.down2(x)
        x = self.down3(x)
        x = self.down4(x)
        x = self.down5(x)

        x = self.up1(x)
        x = self.up2(x)
        x = self.up3(x)
        x = self.up4(x)
        x = self.convf(x)
        return x

class discriminator(Model):
    """

    Discriminator Class that can be used for both Domains
    """

    def __init__(self, df):
        super(discriminator, self).__init__()
        self.down1 = customConv2D(filters=df, strides=2, norm=False)
        self.down2 = customConv2D(filters=df * 2, strides=2)
        self.down3 = customConv2D(filters=df * 4, strides=2)
        self.down4 = customConv2D(filters=df * 8, strides=2)
        self.down5 = layers.Conv2D(filters=1, kernel_size=4, strides=1,
        padding='VALID')

    def call(self, x):
        x = self.down1(x)
        x = self.down2(x)
        x = self.down3(x)
        x = self.down4(x)
```

```python
        x = self.down5(x)

        return x

class GAN_X2Y(Model):
    """
    GAN class for taking an image from one domain to other and evaluating
    the image
    under the other domain discriminator
    """
    def __init__(self, G_XY, D_Y):
        super(GAN_X2Y, self).__init__()
        self.G_XY = G_XY
        self.D_Y = D_Y

    def call(self, x):
        fake_x = self.G_XY(x)
        x = self.D_Y(x)
        return fake_x,x

def process_data(data_dir,skip_preprocess=False):
    """
    Split the images into domain A and domain B images
    Each image contain both Domain A and Domain B images together
    This routines splits it up
    :param data_dir: Input images dir
    :return:
    """

    assert Path(data_dir).exists()

    domain_A_dir = f'{Path(data_dir)}/trainA'
    domain_B_dir = f'{Path(data_dir)}/trainB'
    if skip_preprocess:
        return domain_A_dir, domain_B_dir
    os.makedirs(domain_A_dir,exist_ok=True)
    os.makedirs(domain_B_dir,exist_ok=True)
    files = os.listdir(Path(data_dir))
    print(f'Images to process: {len(files)}')
```

```
    i = 0
    for fl in files:
        i += 1
        try:
            img = imread(f"{Path(data_dir)}/{str(fl)}")
            #print(img.shape)
            w, h, d = img.shape
            img_A = img[:w, :int(h / 2), :d]
            img_B = img[:w, int(h / 2):h, :d]
            imsave(f"{data_dir}/trainA/{fl}_A.jpg", img_A)
            imsave(f"{data_dir}/trainB/{fl}_B.jpg", img_B)
            if (i % 10000) == 0 & (i >= 10000):
                print(f"processed {i+1} images")
        except:
            print(f"Skip processing image {Path(data_dir)}/{str(fl)}")

    return domain_A_dir, domain_B_dir

def train(data_dir,sample_dir,num_epochs=5,lr=0.0002,beta1=0.5,beta2=0.99,
train_size=10000,batch_size=64,epoch_intermediate=10,dim=64,sample_freq=10,
_lambda_=0.5,skip_preprocess=False):

    # Process input data and split to domain A, domain B data
    domain_A_dir, domain_B_dir = process_data(data_dir=data_dir,
    skip_preprocess=skip_preprocess)

    # Build the models
    G_AB, G_BA = generator(gf=64), generator(gf=64)
    D_A, D_B = discriminator(df=64), discriminator(df=64)
    GAN_AB = GAN_X2Y(G_XY=G_AB,D_Y=D_B)
    GAN_BA = GAN_X2Y(G_XY=G_BA,D_Y=D_A)
    G_AB_g, G_BA_g, D_A_g, D_B_g, GAN_AB_g,  GAN_BA_g =  tf.function(G_AB),
    tf.function(G_BA), \
                tf.function(D_A), tf.function(D_B), tf.function(GAN_AB),
                tf.function(GAN_BA)

    # Setup the imagepool
    pool = ImagePool()
```

```
# Set up the Binary Cross Entropy loss to be used for the
Discriminators
loss_fn = tf.keras.losses.BinaryCrossentropy(from_
logits=True,reduction=tf.keras.losses.Reduction.NONE)

# Start the training
for epoch in range(num_epochs):
    data_A = os.listdir(domain_A_dir)
    data_B = os.listdir(domain_B_dir)
    data_A = [f"{domain_A_dir}/{str(x)}" for x in data_A]
    data_B = [f"{domain_B_dir}/{str(x)}" for x in data_B]
    np.random.shuffle(data_A)
    np.random.shuffle(data_B)

    if not train_size:
        train_size = min(len(data_A), len(data_B))
    num_batches = min(len(data_A), len(data_B),train_size)

    # Setup lr based on the schedule
    lr_curr =  lr if epoch < epoch_intermediate else lr* (num_epochs -
    epoch) / (num_epochs - epoch_intermediate)
    # Set the optimizer based on updated learning rate for each epoch
    optimizer = tf.keras.optimizers.Adam(lr_curr,beta_1=beta1,
    beta_2=beta2)

    for i in range(num_batches):
        batch_files = list(zip(data_A[i*batch_size:(i + 1)*
        batch_size],
                               data_B[i*batch_size:(i + 1)*
                               batch_size]))
        batch_images = [load_train_data(batch_file, dim) for batch_file
        in batch_files]
        batch_images = np.array(batch_images).astype(np.float32)
        image_real_A = tf.constant(batch_images[:,:,:,:3])
        image_real_B = tf.constant(batch_images[:,:,:,3:6])
```

```
with tf.GradientTape(persistent=True) as tape:

    fake_AB, logit_fake_AB = GAN_AB_g(image_real_A)
    fake_BA, logit_fake_BA = GAN_BA_g(image_real_B)
    #
    A_reconst = G_BA_g(fake_AB)
    B_reconst = G_AB_g(fake_BA)
    #
    logit_real_D_B = D_B_g(image_real_B)
    logit_real_D_A = D_A_g(image_real_A)
    #
    D_B_loss_fake = tf.reduce_mean(loss_fn(logit_fake_AB,
    tf.zeros_like(logit_fake_AB)))
    D_B_loss_real = tf.reduce_mean(loss_fn(logit_real_D_B,
    tf.ones_like(logit_real_D_B)))
    D_B_loss   = 0.5*(D_B_loss_fake + D_B_loss_real)

    D_A_loss_fake   = tf.reduce_mean(loss_fn(logit_fake_BA,
    tf.zeros_like(logit_fake_BA)))
    D_A_loss_real   = tf.reduce_mean(loss_fn(logit_real_D_A,
    tf.ones_like(logit_real_D_A)))
    D_A_loss   = 0.5*(D_A_loss_fake + D_A_loss_real)
    loss_discriminator = D_B_loss + D_A_loss

    loss_G_ABA = _lambda_*tf.reduce_mean(tf.abs(A_reconst -
    image_real_A))
    loss_G_A_DB  = tf.reduce_mean(loss_fn(logit_fake_AB,
    tf.ones_like(logit_fake_AB)))
    loss_G_AB    =  loss_G_ABA + loss_G_A_DB

    loss_G_BAB = _lambda_*tf.reduce_mean(tf.abs(B_reconst -
    image_real_B))
    loss_G_B_DA  = tf.reduce_mean(loss_fn(logit_fake_BA,
    tf.ones_like(logit_fake_BA)))
    loss_G_BA    =  loss_G_BAB + loss_G_B_DA

    loss_generator = loss_G_AB + loss_G_BA
# Compute gradient
```

```
        grad_D_A = tape.gradient(D_A_loss, D_A.trainable_variables)
        grad_D_B = tape.gradient(D_B_loss, D_B.trainable_variables)
        grad_G_AB = tape.gradient(loss_G_AB,G_AB.trainable_variables)
        grad_G_BA = tape.gradient(loss_G_BA,G_BA.trainable_variables)
        # update the parameters
        optimizer.apply_gradients(zip(grad_D_A, D_A.trainable_variables))
        optimizer.apply_gradients(zip(grad_D_B, D_B.trainable_variables))
        optimizer.apply_gradients(zip(grad_G_AB, G_AB.trainable_
        variables))
        optimizer.apply_gradients(zip(grad_G_BA, G_BA.trainable_
        variables))
        del tape
        print(f"Epoch, iter {epoch,i}:  D_B_loss:{D_B_loss_fake,
        D_B_loss_real},D_A_loss:{D_A_loss_fake,D_A_loss_real},loss_G_
        AB:{loss_G_ABA,loss_G_A_DB},loss_G_BA:{loss_G_BA,loss_G_B_DA}")
        if sample_freq % 200 == 0:
            sample_model(sample_dir,epoch, i)
    return G_AB, G_BA, D_A, D_B, GAN_AB,  GAN_BA

def sample_model(sample_dir,data_dir,epoch, batch_num,batch_size=64,
dim=64):
    assert sample_dir != None
    if not Path(sample_dir).exists():
        os.makedirs(f"{Path(sample_dir)}")
        data_A = os.listdir(data_dir + 'trainA/')
        data_B = os.listdir(data_dir + 'trainB/')
        data_A = [f"{Path(data_dir)}/trainA/{str(file_name)}" for file_name
        in data_A ]
        data_B = [f"{Path(data_dir)}/trainB/{str(file_name)}" for file_name
        in data_B ]

        np.random.shuffle(data_A)
        np.random.shuffle(data_B)
        batch_files = list(zip(data_A[:batch_size], data_B[:batch_size]))
        sample_images = [load_train_data(batch_file, is_
        testing=True,dim=dim) for batch_file in batch_files]
        sample_images = np.array(sample_images).astype(np.float32)
```

```
image_real_A = tf.constant(sample_images[:,:,:,:3])
image_real_B = tf.constant(sample_images[:,:,:,3:6])

fake_AB, logit_fake_AB = GAN_AB_g(image_real_A,training=False)
fake_BA, logit_fake_BA = GAN_BA_g(image_real_B,training=False)

save_images(fake_AB, [batch_size, 1],
        './{}/A_{:02d}_{:04d}.jpg'.format(sample_dir, epoch,
        batch_num))
save_images(fake_BA, [self.batch_size, 1],
        './{}/B_{:02d}_{:04d}.jpg'.format(sample_dir, epoch,
        batch_num))
```

--output—

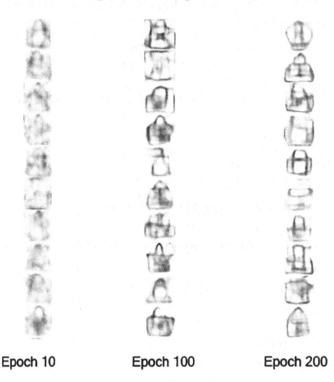

Generated Images of Handbag Sketches (Domain A)

Epoch 10   Epoch 100   Epoch 200

Figure 6-25a. *Handbag sketches generated by trained CycleGAN at different epochs*

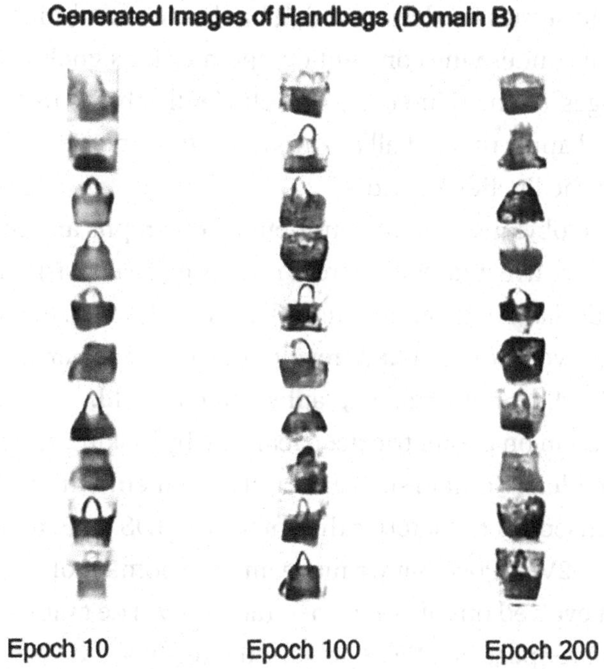

Generated Images of Handbags (Domain B)

Epoch 10   Epoch 100   Epoch 200

Figure 6-25b. *Handbag images generated by trained CycleGAN at different epochs*

We see from Figure 6-25a and 6-25b that the trained CycleGAN has done a great job of converting images in each domain to high-quality realistic images in the other domain.

Geometric Deep Learning and Graph Neural Networks

Deep learning has been very successful on data structures that have an inherent Euclidean nature to them. For instance, RGB images can be thought of as 3D signals on a two-dimensional Euclidean grid. Similarly, audio can be thought of as signals on a one-dimensional Euclidean time axis. Geometric deep learning or more specifically graph neural network is an emerging field that attempts to generalize deep learning to non-Euclidean domain such as graphs and manifolds. Domains such as social networks, 3D modeling, and molecular modeling will benefit immensely from advances in geometric deep learning as they all deal with graph data structures.

One of the success stories for deep learning has been convolution neural network (CNN) and its underlying convolution operator for signals over Euclidean domain such as images. Such architectures are efficiently able to use the local filters with learnable weights and apply them at all the input positions because of the certainty of the neighborhood in the Euclidean grid.

The same is not so obvious for functions defined on graphs and manifolds where the domains can be very irregular and there is no sense of a consistent and ordered neighborhood. The domain of geometric deep learning attempts to take the success of operations such as convolution that have revolutionized deep learning in Euclidean domain to the non-Euclidean domain of graphs and manifold.

We begin this section on geometric deep learning by looking at some of the basic concepts in non-Euclidean domain such as graphs and manifolds and some of the traditional methods used to characterize them such as MDS, LLE, autoencoders, spectral embedding, and Node2Vec. Post that we move into the domain of graph convolution networks and how it evolved out of graph spectral theory. The graph convolution methods we discuss are spectral CNN, K-localized filter-based CNN, ChebNet, GCN, GraphSage, and graph attention network. The convolution method for spectral CNN, K-localized filter-based CNN, ChebNet, and GCN has varying degree of dependency on the graph spectral theory, and hence we look at graph spectral theory in details while working through these model formulations. Alternately we discuss GraphSage and graph attention networks very vividly in the content of some of the shortcoming the graph spectral methods have.

Manifolds

Manifolds can be defined as topological spaces that locally resemble Euclidean spaces. For example, in the spherical surface modeling of the Earth, we take surface to be Euclidean locally.

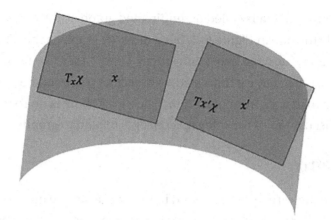

Figure 6-26. *Tangent planes in manifolds*

In Figure 6-26 (see above) is an illustration of a manifold. We can see from the figure that the tangent spaces $T_x\chi$ and $T_{x'}\chi$ of the manifold are Euclidean in nature. Hence the concept of inner product and subsequently the distance is defined on each of the tangent spaces.

Graphs

A graph is the most generic data structure form. A graph G is represented by a set of vertices (also called nodes) V and a set of edges E between the vertices. Hence, we generally denote a graph as $G = (V, E)$. Illustrated in Figure 6-27 are two graphs each with five vertices. The lines that allow connecting two vertices are called edges.

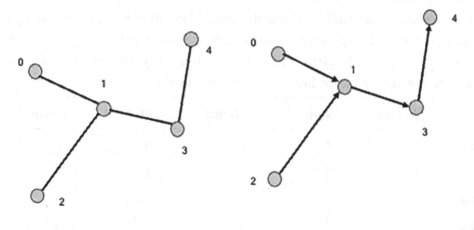

a. Undirected Graph   **b. Directed Graph**

Figure 6-27. *Directed and undirected graph*

The graph in Figure 6-27a is called an undirected graph as the edges are directionless, while the one in Figure 6-27b is a directed graph as the edges have a direction. For instance, the edge between vertices 3 and 2 goes from 3 to 2 in the directed graph. If in the directed graph we think of the vertices as places and the edges as roads, we can think that a road goes from place 3 to 2; however, there is no road from 2 to 3.

We will now discuss few of the relevant concepts related to graphs.

Adjacency Matrix

The connectivity in the vertices is expressed in terms of a square matrix called **adjacency matrix** of the graph. For the five vertices graph, the adjacency matrix would be a 5×5 matrix. If the row indices represent the "from vertex" of the edge while the column indices represent the "to vertex" of the edge, then for an edge with "from vertex" i and "to vertex," the adjacency matrix entry e_{ij} is set to 1. The adjacency matrix for the directed graph in Figure 6-27 is as follows:

	Vertex 0	Vertex 1	Vertex 2	Vertex 3	Vertex 4
Vertex 0	0	1	0	0	0
Vertex 1	0	0	0	1	0
Vertex 2	0	1	0	0	0
Vertex 3	0	0	0	0	1
Vertex 4	0	0	0	0	0

As we would have probably realized, the directed graph would not necessarily be symmetric. For an undirected graph, if there is an edge from vertices i and j, then we mark both e_{ij} and e_{ji} as 1. Hence an undirected graph is always symmetric. The adjacency matrix representation of the undirected graph is as follows:

	Vertex 0	Vertex 1	Vertex 2	Vertex 3	Vertex 4
Vertex 0	0	1	0	0	0
Vertex 1	1	0	1	1	0
Vertex 2	0	1	0	0	0
Vertex 3	0	1	0	0	1
Vertex 4	0	0	0	1	0

Connectedness of a Graph

A graph is said to be **connected** if we can go from any vertex i to any other vertex j either directly or via other vertices. If we think about the undirected graph in Figure 6-27a, it is connected as we can travel from any vertex to the other. However, the directed graph in 6-27b is clearly not connected. For instance, there is no path between vertex 4 and vertex 0 in this directed graph.

Vertex Degree

The **in-degree** of a vertex i is the number of edges that have the "to vertex" as vertex i. Similarly, the **out-degree** of the vertex i is the number of the edges that have the "from vertex" as vertex i. For an undirected graph, the in-degree of a node would always equal the out-degree, and hence for an undirected graph, we can in general refer to in-degree and out-degree by just degree. For an undirected graph with n vertices, the degree matrix would be a diagonal matrix D_{nx} containing the degrees of each vertex.

Laplacian Matrix of a Graph

For an undirected graph $G = (V, E)$ with n vertices $v_0, v_1, \ldots v_{n-1}$, each element of the Laplacian matrix $L_{n \times n}$ can be defined as follows:

$$L_{ij} = degree(v_i) \text{ if } (i = j)$$

$$= -1 \text{ if } (i \neq j) \text{ and edge between } i \text{ and } j$$

$$= 0 \text{ otherwise}$$

In terms of the matrix notation, the Laplacian matrix $L_{n \times n}$ can be expressed in terms of the adjacency matrix $A_{n \times n}$ and the degree matrix $D_{n \times n}$ as follows:

$$L = D - A$$

Function of the Laplacian Matrix

If we consider the vertices of the graph to have features $x = [x_0, x_1, \ldots . x_{n-1}]^T$, then the Laplacian matrix L can be through of an operator that can act on the features. To understand what the Laplacian operator symbolizes, let us apply the Laplacian operator of the graph to feature x:

$$Lx = (D - A)x = Dx - Ax$$

$$Lx = \begin{bmatrix} d_0 & 0 & \ldots & 0 & 0 \\ 0 & d_1 & 0 & 0 & 0 \\ 0 & 0 & \ldots & 0 & .. \\ 0 & 0 & 0 & d_i & 0 \\ 0 & 0 & 0 & 0 & .. \end{bmatrix} \begin{bmatrix} x_0 \\ x_1 \\ .. \\ x_i \\ .. \end{bmatrix} - \begin{bmatrix} 0 & .. & \ldots & .. & 0 \\ 0 & 0 & 0 & .. & 0 \\ 0 & 0 & 0 & .. & .. \\ e_{i0} & e_{i1} & ..e_{ij} & .. & .. \\ .. & .. & .. & .. & \ldots \end{bmatrix} \begin{bmatrix} x_0 \\ x_1 \\ .. \\ x_i \\ .. \end{bmatrix}$$

Let us see what the i-th term would be in the Lx vector of dimension n. From Dx, we would get $d_i x_i$, while from Ax we would get $e_{i0}x_0 + e_{i1}x_1 + \ldots e_{i*(n-1)}x_{n-1}$. Hence the combined term would be as follows:

$$d_i x_i - \left(e_{i0}x_0 + e_{i1}x_1 + \ldots e_{i*(n-1)}x_{n-1} \right)$$

$$= d_i x_i - \sum_{j=0}^{n-1} e_{ij}x_j$$

If we consider only the vertices j that vertex i shares edges with, then we would only have d_i neighbors that i shares edges with. We denote the direct neighbor set of i as $N(i)$. With the above considerations, the above expression can be simplified as follows:

$$d_i x_i - \sum_{j=0}^{n-1} e_{ij}x_j = d_i x_i - \sum_{j \in N(i)} x_j = d_i \left(x_i - \frac{1}{d_i} \sum_{j \in N(i)} x_j \right)$$

The term $\frac{1}{d_i} \sum_{j \in N(i)} x_j$ in the preceding expression can be thought of as the local feature average around the vertex i, and hence $\left(x_i - \frac{1}{d_i} \sum_{j \in N(i)} x_j \right)$ can be thought of as a measure of how different the feature of vertex i is in comparison with its local average (average feature of its direct neighbors).

Let us now probe if the graph Laplacian shares some similarity with the Laplacian operator in the Euclidean domain.

The Laplacian operator in the Euclidean domain is given by the following:

$$\nabla^2 = \frac{\partial^2}{\partial x^2} + \frac{\partial^2}{\partial y^2} + \frac{\partial^2}{\partial z^2}$$

For a 1-D function $f(x)$, the Laplacian operator $L = \nabla^2 = \frac{d^2}{dx^2}$ applied on the function gives the following:

$$Lf(x) = \frac{d^2 f(x)}{dx^2}$$

The finite approximation of the preceding second derivative can be expressed as follows:

$$\left(f(x_{i+1}) - f(x_i)\right) - \left(f(x_i) - f(x_{i-1})\right)$$

$$= -2\left(f(x_i) - 0.5\left(f(x_{i+1}) + f(x_{i-1})\right)\right)$$

From the preceding expression, it is obvious that the Laplacian operator in the Euclidean domain also gives a measure of how different a function value is at a given point with respect to the average of its neighbors.

Different Versions of Laplacian Matrix

The Laplacian matrix that we have been working with is called the unnormalized Laplacian matrix given by $L = D - A$.

If we perform a symmetric normalization of $(D - A)$ using the degree matrix D, then we get the normalized Laplacian as follows:

Normalized Laplacian $L = I - D^{-\frac{1}{2}} A D^{-\frac{1}{2}}$

The edge weights in the normalized Laplacian can be expressed as $e_{ij} = \frac{1}{\sqrt{d_i d_j}}$.

If we apply the inverse of the degree matrix as a normalizer to $(D - A)$, then we get what is commonly referred as random walk Laplacian as shown in the following:

Random walk Laplacian $L = I - D^{-1} A$

Do note that random walk Laplacian won't be symmetric.

Different Problem Formulations in Geometric Learning

Geometric learning on graphs and manifolds deals with two classes of problems:

- Characterize the structure of the data that is non-Euclidean.

- Analyze functions and signals on the non-Euclidean domain.

The following are some of the traditional methods that are currently used to characterize the non-Euclidean data.

- MDS: Multidimensional scaling

- Autoencoders

- LLE

- Spectral clustering

We will briefly go through each of the methods to move on to geometric deep learning.

Multidimensional Scaling

MDS seeks to construct a Euclidean latent space for data points from a given dissimilarity matrix of those points. Hence given a dissimilarity matrix of $D_{n \times n}$ of n data points where the dissimilarity of data points i and j

is denoted by d_{ij}, MDS tries to find latent presentations $z_0, z_1, \ldots . z_{n-1}$ such that the following holds for each pair of data points:

$$d_{ij} \approx \| z_i - z_j \|_2$$

Do note that if the dissimilarities are Euclidean distances to begin with, MDS is equivalent to principal component analysis. However, in the more general sense, MDS is used to project points in a non-Euclidean space to a Euclidean one using the non-Euclidean pairwise distances.

Autoencoders

Autoencoders project points in a low-dimensional latent space using its encoder. We have extensively looked at autoencoders in Chapter 5. Given an autoencoder with encoder E_θ and decoder D_ϕ, it learns by minimizing the reconstruction loss and some form of regularization loss R as shown in the following:

$$\|D_\phi\big(E_\theta(x)\big)-x\|_p + \lambda R\big(\theta,\phi,\{x\}\big)$$

The output of the encoder $z = F(x)$ belongs to a low-dimensional space of interest to us. The regularizer varies based on the nature of the autoencoder.

For instance, for a variational autoencoder where we want the decoder to behave as a generative model, we want the regularizer to minimize the KL divergence between the generated latent samples $z = E_\theta(x)$ and the Gaussian prior $P(z)$ from which we would like to sample using the decoder.

For a denoising autoencoder, we learn a sparse encoded latent space by having an overrepresented latent space and inducing sparsity on it by enforcing a low probability Bernoulli distribution over the latent representation activations.

Locally Linear Embedding

Locally linear embedding (LLE) attempts to find a lower-dimensional projection of data by preserving local neighborhood distance between points. This local neighborhood over which distances are computed can be thought of as the tangent spaces in a manifold which behave as Euclidean spaces themselves.

Algorithmically, LLE attempts to find the low-dimensional projection using the following steps:

A.   Given N data points, find k-nearest neighbors of each data point such that the given data point $x_i \in R^D$ can be expressed as linear sum of the k-nearest neighbors as shown in the following:

$$\widehat{x_i} = \sum_{j=0}^{k-1} w_{ij} x_j$$

The matrix W for the weights w_{ij} is learned by optimizing the following objective:

$$L(W) = \sum_{i} \| x_i - \sum_{j=0}^{k-1} w_{ij} x_j \|_2^2$$

The weights corresponding to each data point follow the following constraints:

- $w_{ij} = 0$ if x_j is not within the K-nearest neighbors of x_i.
- $\sum_{j=0}^{k-1} w_{ij} = 1$

The learned weights w_{ij} for each data point are invariant to rotations and scaling of the data point and its neighbors as is obvious from the preceding optimization objective $L(W)$. The constraint on the sum of the weights being 1 makes the weights invariant to translation of the data points. Imagine the data point x_i and its k-nearest neighbors are translated by Δx. We can see that the learned weights would still be invariant to the same as follows:

$$L(W;x_i) = \left\| x_i + \Delta x - \sum_{j=0}^{k-1} w_{ij}\left(x_j + \Delta x\right) \right\|_2^2$$

$$= \left\| x_i - \sum_{j=0}^{k-1} w_{ij}x_j + \left(\Delta x - \sum_{j=0}^{k-1} w_{ij}\Delta x \right) \right\|_2^2$$

$$\left\| x_i - \sum_{j=0}^{k-1} w_{ij}x_j + \left(\Delta x - \Delta x\sum_{j=0}^{k-1} w_{ij} \right) \right\|_2^2$$

If $\sum_{j=0}^{k-1} w_{ij} = 1$, then $\left(\Delta x - \Delta x\sum_{j=0}^{k-1} w_{ij} \right) = 0$, and hence the above expression reduces to the same objective prior to the translation, $i.\ e.,\ \left\| x_i - \sum_{j=0}^{k-1} w_{ij}x_j \right\|_2^2$.

B.   If the data is the high D dimensional original space that lies in a smooth low $d \ll D$ dimensional manifold, then there would exists a linear mapping in terms of rotation, scaling, and translation that would map the high D dimensional inputs of each neighborhood to global coordinates of the low d dimensional manifold. And since the weights of the k-nearest neighbors w_{ik} for each data point x_i are invariant to rotation, scaling, and translation, we can represent each data point in the new low d dimensional manifold also with the same set of weights. Hence if we map each original data point $x_i \in R^D$ to their low d dimensional representation $y_i \in R^d$, then the following should hold true:

$$y_i = \sum_{j=0}^{k-1} w_{ij}y_j$$

Also, the low dimensional representation y_i for all the data points represented as a matrix $Y \in R^{d \times N}$ can be obtained by minimizing the following objective:

$$L(Y) = \sum_i \left\| y_i - \sum_{j=0}^{k-1} w_{ij} y_j \right\|_2^2$$

Minimizing the preceding objective $L(Y)$ subject to the constraints that the low-dimensional representations have zero mean ($\frac{1}{N} \sum_{i=0}^{N-1} y_i = 0$) and unit covariance ($\frac{1}{N} \sum_{i=0}^{N-1} y_i y_i^T = I_{d \times d}$) leads to the solution of Y being the Eigen vectors corresponding to the lowest $(d + 1)$ Eigen values of the sparse and symmetric matrix $(I - W)^T (I - W)$. The Eigen vector of all ones $[1, 1, ..]^T$ corresponding to the Eigen value of zero can be ignored, and the remaining Eigen d vectors can be treated as the embedding representation for Y.

Spectral Embedding

Spectral embedding is a method of embedding generation that is based on the spectral graph theory. The following are the steps associated with spectral embedding creation.

A. If the given inputs are data points in a high-dimensional manifold and a graph adjacency is not present, we need to first create an adjacency matrix based on the local affinity of data points. Much like LLE for each data point x_i, we can choose the k-nearest neighbors as its neighbors that it shares edges with. Other common approach is to choose data points within a specified ϵ Euclidean distance of a data point as its neighbors.

The edge weight w_{ij} in the adjacency is generally set to 1 if an edge exists between data point i and j; else the weight is set to zero. Instead of direct 1 weight for edges, sometimes a Gaussian edge weighting scheme is preferred as follows:

$$w_{ij} = e^{-\frac{\|x_i - x_j\|_2^2}{\gamma}}$$

B. Once the graph adjacency matrix A is created where the data points are the vertices of the graph, we create the Laplacian Matrix as follows:

$$L = (D - A)$$

The Laplacian matrix is a symmetric positive definite matrix, and hence it has a spectral decomposition form as shown in the following:

$$L = USU^T$$

The U contains the Eigen vectors of the Laplacian matrix, while S is a diagonal matrix that contains the corresponding Eigen values. If the graph contains n vertices, then

$$U = \left[\phi_0, \phi_1, \phi_2, \ldots \phi_{n-1} \right]$$

$$S = \left[\lambda_0, \lambda_1, \ldots \lambda_{n-1} \right]$$

Here the eigen values are in the increasing order with the first Eigen value $\lambda_0 = 0$. Hence, we have the following:

$$\lambda_0 = 0 \leq \lambda_1 \leq \lambda_2 \ldots \ldots \leq \lambda_{n-1}$$

The Eigen vector ϕ_0 corresponding to the Eigen value $\lambda_0 = 0$ is the vector of all 1s vector, i.e., $[1, 1, ..1]^T$. The fact that all 1s Eigen vector corresponds to the Eigen value of 0 can be proven easily by looking at i-th entry on either of the following equation.

$$(D - A) \begin{bmatrix} 1 \\ 1 \\ .. \\ 1 \\ .. \end{bmatrix} = \lambda_0 \begin{bmatrix} 1 \\ 1 \\ .. \\ 1 \\ .. \end{bmatrix}$$

On the left-hand side of the preceding equation, we have for the i-th entry corresponding to the i-th vertex:

$$d_i - \sum_{j=0}^{n-1} e_{ij}$$

The above expression is zero as $\sum_{j=0}^{n-1} e_{ij}$ is nothing but the degree of the vertex i. Hence the i-th entry on the right-hand size, i.e., λ_o should equal to zero.

The Eigen vectors of the Laplacian act as Fourier basis functions in a graph domain, while the Eigen values act as Fourier frequencies. It would be easy to relate to the Eigen vectors of the Laplacian as Fourier basis functions of a graph by comparing it to the set of Eigen functions and Eigen values corresponding to the Euclidean Laplacian operator ∇^2 or its one-dimensional equivalent $\frac{\partial^2}{\partial x^2}$. If we take the complex exponential functions e^{iwx}, we have the following:

$$\frac{\partial^2 \left(e^{iwx} \right)}{\partial x^2} = \left(iw \right)^2 e^{iwx} = -w^2 e^{iwx}$$

The preceding equations prove that the complex exponentials are Eigen functions of the Laplacian operator in the Euclidean domain. Since these complex exponentials (sum of sines and cosines) form the Fourier basis in the Euclidean domain, the graph Laplacian operator Eigen vectors can be treated as Fourier basis in the graph domain. Similarly, the corresponding Eigen values of the graph Laplacian can be thought of as square of the Fourier frequencies as the Eigen values of the Euclidean Laplacian operator is proportional to the square of the frequency.

C. The Eigen vectors in the increasing order are treated as the spectral embedding. Generally, for a d-dimensional embedding space, the first $(d + 1)$ Eigen vectors are chosen as the embeddings.

As the Eigen vector of all 1s corresponding to $\lambda_o = 0$ doesn't provide any discriminative power over the vertices, hence it is discarded.

If the graph has two connected components, then the Eigen value of λ_1 would be zero. Generalizing the above fact, if the graph has k connected components, then the first k Eigen values $\lambda_o, \lambda_1, \ldots, \lambda_{k-1}$ would all be equal to zero. Hence the magnitude of the Eigen values gives a sense of connectedness of the graph. The corresponding Eigen vectors can be used as embedding features as the vertices that form strongly connected community would have similar values in their Eigen vector embedding space. For instance, if we have two strongly connected communities in a graph where the intra community edges dominates heavily the inter community edges, then the second Eigen value would be close to zero and the second Eigen vector values can be easily split into two clusters using a K means or by applying some convenient threshold. Similarly, if there are three strong communities in the graph, we can bring in the third Eigen vector as an additional embedding dimension and use k means to cluster the graph into three communities using $[\phi_1 \, \phi_2] \in R^{n \times 2}$ as embeddings.

Now that we have gone through some of the traditional methods to characterize non-Euclidean data, let us now look at the Node2Vec—a recent geometric deep learning method that is used to create node (vertex) embeddings.

Node2Vec

Node2vec is a representation learning framework where the graph entities along their relationships are leveraged to learn low-dimensional representation of the vertices (nodes) of the graph. The learning objective is based on biased random walks to determine node neighbors based on various definitions of network neighborhood.

Given a graph $G = (V, E)$ in Node2Vec, we attempt to learn a function f that maps the nodes to a low d-dimensional embedding space as shown in the following:

$$f : V \to R^d$$

One of the primary objectives of learning such feature representation for nodes is to use them for downstream prediction tasks.

Let us look at the learning objective of the Node2Vec more closely. For every source node $u \in V$, we should be able to predict its sampled neighborhood $N_S(u)$ with high probability using its feature representation $f(u)$. Hence the learning objective for all nodes is as follows:

$$L(f) = \sum_{u \in V} \log P\big(N_S(u)| f(u)\big)$$

To make the learning tractable, two key assumptions that Node2Vec makes are as follows:

- Conditional independence: We assume the likelihood of observing a neighborhood node is independent of the other neighborhood nodes given the feature representation of the source node. Hence, we can have the following factorization:

$$P\big(N_S(u)|f(u)\big) = \prod_{i \in N_S(u)} P(n_i|f(u))$$

- Feature space symmetry: The source node and its neighbors have a symmetric effect on the feature space. Hence, we can define the likelihood of all source-neighboring node pairs using a SoftMax over the node representation dot products.

$$P\big(n_i|f(u)\big) = \frac{\exp\big(f(n_i)^T f(u)\big)}{\sum_{v \in V} \exp\big(f(v)^T f(u)\big)}$$

Applying the preceding assumptions, the learning objective of Node2Vec $L(f)$ can be simplified as follows:

$$L(f) = \sum_{u \in V}\left[-\log Z_u + \sum_{i \in N_S(u)} f(n_i)^T f(u) \right]$$

The partition function for each node $Z_u = \sum_{v \in V} \exp\big(f(v)^T f(u)\big)$ is not tractable to compute for large graphs, and hence we generally approximate it with negative sampling.

The desired function $\hat{f}$ can be learned by maximizing the objective $L(f)$:

$$\hat{f} = \underset{f}{\mathrm{argmax}} \sum_{u \in V} \left[-\log Z_u + \sum_{i \in N_S(u)} f(n_i)^T f(u) \right]$$

For each node u, how we sample the neighborhood $N_S(u)$ in Node2vec is of paramount importance as it determines the nature of the node embeddings that would be learned. Typically, the important notions of similarity that we are trying to learn in these embeddings are **structural similarity** and **homophily**.

Under the structural similarity hypothesis, if two nodes are having similar structural roles in the network, their embedding should be similar to each other. Alternately, under the homophily hypothesis, two nodes that are highly interconnected and belong to similar communities or clusters should have similar embedding representation. Most of the time, to do well on the downstream tasks, both the notions of structural similarity and homophily should be captured in the embedding.

Breath first sampling where the neighbors $N_S(u)$ are restricted to be immediate neighbors of u would allow learning structural similarity well but will fail to capture homophily. **Depth first sampling** on the other hand will find strong interconnected neighbors that are multiple hops away and, in the process, would enable learning homophily well but would fail to capture structural similarity. Node2Vec allows one to capture both aspects of structural similarity and homophily by defining two parameters p and q that control the extent of breadth first sampling and depth first sampling in the random walks.

In Node2Vec with each node $u \in V$ as the starting node, we take several random walks (say r) of a specified length l. Hence the total number of random walks in such a setting would be $|V| \times r$. Each of these random walks can be through of sentences of node ids of length l. Now to understand better how these random walk sentences are created starting from a given node u, let us refer to Figure 6-28.

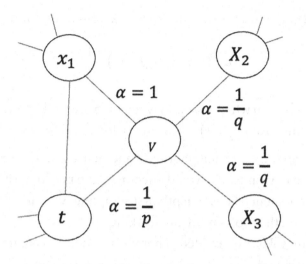

Figure 6-28. *Second-order random walk using parameters p and q*

Let us assume during the walk we have reached node v from the node t. Hence node v is the recent sampled node in the walk, while the node t is the sampled just prior to v. The walk now needs to decide on the new node x based on some transition probability. The unnormalized transition probabilities π_{vx} for choosing the next node x is based on a second-order random walk as follows:

$$\pi_{vx} = \alpha_{pq}(t,x) * w_{vx}$$

In the above equation, w_{vx} denotes the edge weight between nodes v and x. For an unweighted graph $w_{vx} = 1$, if there is an edge between v and x; else $w_{vx} = 0$.

The factor $\alpha_{pq}(t,x)$ depends on parameters p and q as well as the prior node t, and hence this scheme is a second-order random walk. The following is the $\alpha_{pq}(t,x)$ scheme:

$$\alpha_{pq}(t,x) = \begin{aligned} &= \frac{1}{p} \text{ if } d_{tx} = 0 \\ &= 1 \text{ if } d_{tx} = 1 \\ &= \frac{1}{q} \text{ if } d_{tx} = 2 \end{aligned}$$

In the preceding scheme, d_{tx} denotes the shortest path distance between t and x.

In Figure 6-28, node x_1 has a direction connection to node t, and hence the distance $d_{tx} = 1$. For this reason, $\alpha_{pq}(t,x_1) = 1$.

Alternately, nodes x_2 and x_3 are at a distance of 2 from node t and hence

$$\alpha_{pq}\left(t, x_2\right) = \alpha_{pq}\left(t, x_3\right) = \frac{1}{q}.$$

Finally, we look at $d_{tx} = 0$ which essentially refers to the walk returning to the earlier sampled node t. In this case $\alpha_{pq}\left(t, t\right) = \frac{1}{p}$. Hence a lower value of p will encourage the walk to stay around the same neighborhood and hence encourage breadth first sampling. Similarly, a low value of q would encourage the random walk to capture long distance neighbors and hence favor a depth first sampling. We can tune the factor p and q to our liking based on the problem we are working on.

Once the random walks are sampled generally, we look at a context or window size k to define the neighborhood $N_S(u)$ of each sampled node u. To understand the same, let us take a random walk of length 6 starting from node u as follows:

$$u, s_1, s_2, s_3, s_4, s_5$$

From the preceding random walk, we can create the following neighborhoods for the different nodes using a context window of $k = 3$:

$$N_S\left(u\right) = \left\{s_1, s_2, s_3\right\}$$

$$N_S\left(s_1\right) = \left\{s_2, s_3, s_4\right\}$$

$$N_S\left(s_2\right) = \left\{s_3, s_4, s_5\right\}$$

Generalizing the preceding observation, with context size k for each random walk string of length l, we are creating $k * (l - k)$ training samples using the l samples of the random walk. Hence Node2Vec method is sample efficient.

Node2Vec Implementation in TensorFlow

In this section, we will implement the Node2Vec algorithm for creating node embeddings. We will be using the Cora dataset in this exercise where the nodes represent different publications belonging to seven classes—"rule learning," "reinforcement learning," "case based," "probabilistic methods," "genetic algorithms," "theory," and "neural networks."

The citations in the publications create the edges of the Cora dataset graph. Publications belonging to the same class are expected to have several edges among them. With Node2Vec, we expect the learned embeddings with a class to be similar to each other. Also, we expect the learned embeddings across classes to be different from each other.

We will be working with StellarGraph graph neural network capabilities as it works seamlessly with TensorFlow. The graphs that we would be working with will be StellarGraph graph objects, and hence let's look at how we can create a StellarGraph in the following:

Listing 6-7. Creating a StellarGraph-Specific Graph

```
# Please install Stellargraph
from stellargraph import IndexedArray
import pandas as pd

nodes = IndexedArray(np.array([[1,2],[2,1], [5,4],[4,5]]),index=['a','b',
'c','d'])
edges = pd.DataFrame({"source":["a","b","c","d","a"],"target":["b","c","d",
"a","c"]})
GS_example = StellarGraph(nodes,edges)
print("Graph info", GS_example.info())
print(f"Graph directed:{GS_example.is_directed()}")
print(f"Nodes :{GS_example.nodes()}")
print(f"Node features :{GS_example.node_features()}")

--output--

Graph info StellarGraph: Undirected multigraph
 Nodes: 4, Edges: 5

 Node types:
  default: [4]
    Features: int64 vector, length 2
    Edge types: default-default->default

 Edge types:
    default-default->default: [5]
        Weights: all 1 (default)
```

```
        Features: none
Graph directed:False
Nodes :Index(['a', 'b', 'c', 'd'], dtype='object')
Node features :[[1 2]
 [2 1]
 [5 4]
 [4 5]]
```

We can see from Listing 6-7 that we can create a StellarGraph object by feeding in the **nodes** in the form of an index array and edges in the form of a data frame. The node's index array consists of node features and is indexed by the node ids. The edges on the other hand can be feed in the form a data frame with the **source** and **target** columns containing the "from" and "to" node ids, respectively. The **nodes()** and the **node features()** method can be used to get the node ids and the node features corresponding to the graph nodes.

StellarGraph would have methods around common datasets to load them directly into StellarGraph objects as we will see in the implementations to follow.

Now that we have some idea of how a StellarGraph graph works, let us implement the embedding generation of the nodes in the Cora dataset using Node2Vec. The detailed implementation is illustrated in Listing 6-8.

Listing 6-8. Node2Vec on the Cora Dataset

```
import matplotlib.pyplot as plt
from sklearn.manifold import TSNE
import io, os, sys, types
import networkx as nx
import numpy as np
import pandas as pd
from tensorflow import keras
import json
import re
from collections import Counter
from IPython.display import Image, HTML, display
import stellargraph as sg
from IPython.display import display, HTML
from stellargraph import StellarGraph
from stellargraph.data import BiasedRandomWalk
```

```
from stellargraph.data import UnsupervisedSampler
from stellargraph.mapper import Node2VecLinkGenerator,
Node2VecNodeGenerator
from stellargraph.layer import Node2Vec, link_classification
import time
from sklearn.metrics.pairwise import cosine_similarity
%matplotlib inline

import warnings
warnings.filterwarnings('ignore')

# Load the Cora dataset along with the node labels
def load_cora_dataset():
    dataset = sg.datasets.Cora()
    display(HTML(dataset.description))
    GS, node_subjects = dataset.load(largest_connected_component_only=True)
    # GS contains the Stellar Graph
    # node_subjects contains the node_features indexed by the node ids
    return GS, node_subjects

# Define a Baised Random Walker Object
def random_walker(graph_s,num_walks_per_node,walk_length,p=0.5,q=2.0):
    walker = BiasedRandomWalk(
        graph_s,
        n=num_walks_per_node,
        length=walk_length,
        p=p,  # defines probability, 1/p, of returning to source node
        q=q,  # defines probability, 1/q, for moving to a node away from
        the source node
    )
    return walker

# Create sampler for training
def create_unsupervised_sampler(graph_s,walker):
    nodes=list(graph_s.nodes())
    unsupervised_sampler = UnsupervisedSampler(graph_s, nodes=nodes,
    walker=walker)
    return unsupervised_sampler
```

```
# Training Routine
def train(batch_size=50,epochs=2,num_walks_per_node=100,walk_length=5, \
        p=0.5,q=2.0,emb_dim=128,lr=1e-3):

    GS, node_subjects = load_cora_dataset()

    # Create a random walker sampler
    walker = random_walker(graph_s=GS,num_walks_per_node=num_walks_
    per_node, \
                                walk_length=walk_length,p=p,q=q)

    unsupervised_sampler = create_unsupervised_sampler(graph_
    s=GS,walker=walker)

    # Create a batch generator
    generator = Node2VecLinkGenerator(GS, batch_size)

    # Define the Node2Vec model
    node2vec = Node2Vec(emb_dim, generator=generator)
    x_inp, x_out = node2vec.in_out_tensors()
    # link_classification is the output layer that maximizes the dot
    product of the similar nodes
    prediction = link_classification(
    output_dim=1, output_act="sigmoid", edge_embedding_method="dot"
    )(x_out)
    model = keras.Model(inputs=x_inp, outputs=prediction)

    # Compile Model
    model.compile(
    optimizer=keras.optimizers.Adam(learning_rate=lr),
    loss=keras.losses.binary_crossentropy,
    metrics=[keras.metrics.binary_accuracy],
    )

    # Train the model
    history = model.fit(
    generator.flow(unsupervised_sampler),
    epochs=epochs,
    verbose=1,
```

```
            use_multiprocessing=False,
            workers=4,
            shuffle=True,
        )

        # Predict the embedding
        x_inp_src = x_inp[0]
        x_out_src = x_out[0]
        embedding_model = keras.Model(inputs=x_inp_src, outputs=x_out_src)

        node_gen = Node2VecNodeGenerator(GS, batch_size).flow(node_
        subjects.index)
        node_embeddings = embedding_model.predict(node_gen, workers=4,
        verbose=1)
        print(f"Shape of the node Embeddings : {node_embeddings.shape}")
        print('Length of embedded vectors:',len(node_embeddings[0]))
        print('Total embedded vectors:', len(node_embeddings))
        plot_embeddings(node_embeddings,node_subjects)

def plot_embeddings(node_embeddings,node_subjects,n_components=2):

        transform = TSNE
        trans = transform(n_components=2)
        node_embeddings_2d = trans.fit_transform(node_embeddings)
        # draw the embedding points, coloring them by the target label (paper
        subject)
        alpha = 0.7
        label_map = {l: i for i, l in enumerate(np.unique(node_subjects))}
        node_colours = [label_map[target] for target in node_subjects]

        plt.figure(figsize=(7, 7))
        plt.axes().set(aspect="equal")
        plt.scatter(
            node_embeddings_2d[:, 0],
            node_embeddings_2d[:, 1],
            c=node_colours,
            cmap="jet",
            alpha=alpha,
        )
```

```
    plt.title("{} visualization of node embeddings".format(transform._
    name__))
    plt.show()

train()

--output--

link_classification: using 'dot' method to combine node embeddings into
edge embeddings
Epoch 1/2
39760/39760 [==============================] - 140s 4ms/step - loss:
0.3017 - binary_accuracy: 0.8497
Epoch 2/2
39760/39760 [==============================] - 136s 3ms/step - loss:
0.1092 - binary_accuracy: 0.9643
50/50 [==============================] - 0s 3ms/step
Shape of the node Embeddings : (2485, 128)
Length of embedded vectors: 128
Total embedded vectors: 2485
```

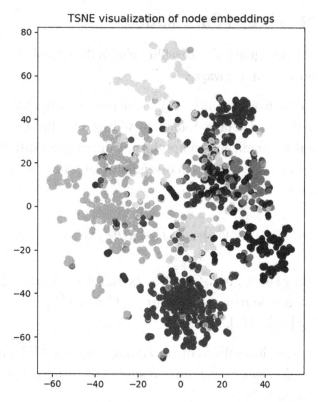

Figure 6-29. *t-SNE plot of Node2Vec embeddings in 2D*

The embedding generated by Node2Vec is projected into the 2D plane using t-SNE (see Figure 6-29). The seven different colors in the plot correspond to the seven different domains to which the publications belong. We can see that the embeddings of the publications with a domain are close to each other while across the domains the embeddings are well separated with very little overlap.

Graph Convolution Networks

Now we will look into graph convolution networks. Since graph neural networks have a strong root in graph spectral theory, we would start with spectral filters in graph convolution and gradually move toward the GCN that we use today. On the way we will discuss about spectral CNN, K-localized filters, and its variant ChevNet and some of their shortcoming that led to simpler formulations such as GCN. Finally we will talk about some other graph convolution methods which are very popular—GraphSage and graph attention network.

Spectral Filters in Graph Convolution

Convolution of a discrete signal $f[n]$ with a filter $g[n]$ in the spatial or time (Euclidean) domain can be performed in two ways:

1. The convolution output $y[n] = f[n](*)g[n]$ can be computed entirely in the spatial/time domain by sliding over the flipped filter over the domain and computing the dot product with $f[n]$ for every value of n. This approach essentially computes the output of the convolution as follows:

$$y[n] = \sum_{m=-\infty}^{m=\infty} f[m]g[n-m]$$

2. We can apply Fourier transform F to the signal and filter to get their frequency domain counterparts $\tilde{f}[k]$ and $\tilde{g}[k]$ where $F\big[f[n]\big] = \tilde{f}[k]$.

    We can then multiply the frequency domain signals $\tilde{f}[k]$ and $\tilde{g}[k]$ to get the frequency domain response of the output of convolution as follows:

$$\tilde{y}[k] = \tilde{f}[k]\tilde{g}[k]$$

Finally, we apply an inverse Fourier transform F^{-1} on $\tilde{y}[k]$ to get the output of convolution in the spatial/time domain as shown in the following:

$$y[n] = F^{-1}\big[\tilde{y}[k]\big]$$

The problem in graphs is that the notions of spatiality are not really well defined as opposed to Euclidean domain where we take spatially for granted because of the nice consistent neighborhood that Euclidean domain presents. One can say the hop distance between nodes gives graphs some essence of spatiality or locality; however, it's not consistent or has a global structure that spatial convolution can benefit from in an obvious way.

We have seen earlier (in the Spectral Embedding section) that the graph Laplacian $L = D - A$ lends itself to a strong foundation for Fourier transform basis in the form of the Eigen vectors $\phi_0, \phi_1, \ldots \phi_{|V|-1}$ of the graph Laplacian L. The square root of the Eigen values $\lambda_0, \lambda_1, \ldots \lambda_{|V|-1}$ corresponding to the Eigen vectors acts as Fourier frequencies in the graph domain. This is analogous to the Euclidean domain where the Eigen functions of the

Laplacian operator ∇^2 are the complex exponential Fourier basis functions e^{jwx}. Hence, we can leverage the graph Fourier transform to convert graph signal to the frequency domain. Let's note down the formula for the spectral decomposition of the symmetric normalized Laplacian, so we can relate to the graph Fourier transform used in spectral ConvNets.

$$L = I - D^{-\frac{1}{2}} A D^{-\frac{1}{2}} = U S U^T$$

$$U = \left[\phi_o, \phi_1, \ldots \phi_{|V|-1} \right]$$

$$S = diag\left(\lambda_o, \lambda_1, \ldots \lambda_i, \lambda_{|v|-1} \right)$$

Because of the symmetric normalization, the matrix U is unitary and hence $UU^T = I$.

We can perform a graph Fourier transform of a function f over the graph nodes by projecting the function into the different graph Fourier basis $U = [\phi_o, \phi_1, \ldots . \phi_{|V|-1}]$. Hence, the graph Fourier transform of the function f can be expressed as follows:

$$\hat{f} = U^T f$$

Alternately a signal in the frequency domain can be transformed to the graph spatial domain by applying the inverse Fourier transform $(U^T)^{-1}$. Since U is a unitary operator, the inverse Fourier transform operator $(U^T)^{-1}$ equals U.

Illustrated in the following (see Figure 6-30) is the mechanism of spectral convolution of a function f with filter g over a graph $G = (V, E)$.

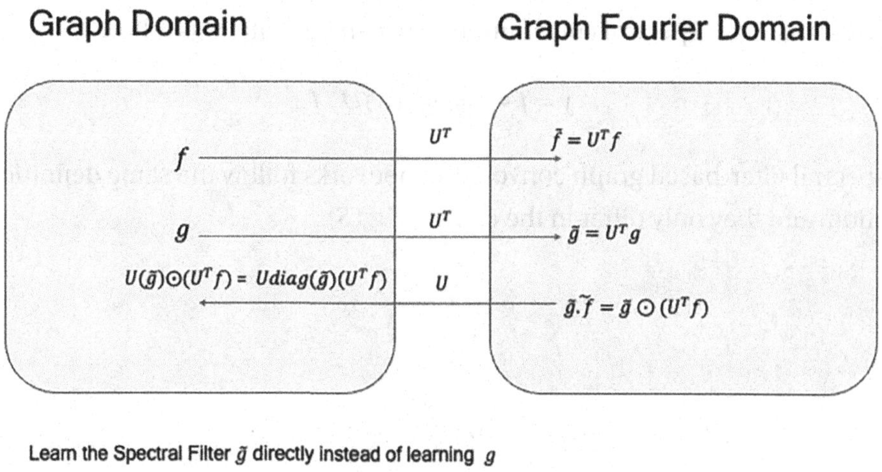

Figure 6-30. *Graph spectral convolution illustration*

As illustrated in the figure, the convolution output of the signal f with g can be expressed as follows:

$$f * g = U\tilde{g} \odot \left(U^T f\right)$$

For spectral ConvNet, we are not interested in learning the filter g but its spectral equivalent $\tilde{g}$, and hence we have retained the same in the expression for the preceding convolution. Also, for mathematical convenience, we can replace the Hadamard product in the preceding expression by a matrix multiplication. This requires replacing the vector $\tilde{g} \in C^{|V|}$ with the diagonal matrix $diag(\tilde{g})$. Hence, with the desired changes, the convolution can be written as follows:

$$f * g = Udiag(\tilde{g})U^T f$$

We parameterize the diagonal matrix $diag(\tilde{g})$ using θ and represent it as $g_\theta(S)$ where S represents the Eigen values. The Eigen values act as graph Fourier frequencies, and since the spectral Filter would contain the coefficients corresponding to the graph frequencies, we can express $g_\theta(S)$ as follows:

$$g_\theta(S) = \begin{bmatrix} \tilde{g}(\lambda_o) & \cdots & .. & 0 \\ . & \tilde{g}(\lambda_1) & . & . \\ . & . & .. & . \\ 0 & . & . & \tilde{g}_{|V|-1}\left(\lambda_{|V|-1}\right) \end{bmatrix}$$

The expression for spectral convolution then can be written as follows:

$$y = f * g = Ug_\theta(S)U^T f$$

All spectral filter-based graph convolution networks follow the same definition of convolution, and they only differ in the choice of $g_\theta(S)$.

Spectral CNN

Spectral CNN keeps the filter unconstrained and has $|V|$ free parameters for the model to learn.

$$g_\theta(S) = \begin{bmatrix} \theta_o & \cdots & \cdot\cdot & 0 \\ \cdot & \theta_1 & \cdot & \cdot \\ \cdot & \cdot & \cdot\cdot & \cdot \\ 0 & \cdot & \cdot & \theta_{|V|-1} \end{bmatrix}$$

There are several major problems with the formulation:

- A filter defined in the spectral Domain is not spatially localized.

- Also, for a graph with $|V|$ nodes, the number of parameters in the spectral filter $g_\theta(S)$ is $|V|$. For a large graph, this can lead to high computation complexity.

- Moreover, the preceding convolution is prohibitory expensive for large graphs as multiplication with the Eigen vector matrix is $O(|V|^2)$. Also performing the Eigen vector decomposition of a large graph might be intractable in the first place.

- Spectral convolution is dependent heavily on the assumption that the graph would not change between training and prediction. If the graph changes, the Fourier basis changes and the filter weights learned with old Fourier basis would not hold good. Hence spectral methods are good for transductive setting where the graph remains fixed between training and prediction.

K-Localized Spectral Filter

Instead of trying to learn all the coefficients of the filter $g_\theta(S)$ as free parameters, we can define $g_\theta(S)$ as a polynomial in S up to order K with only $K+1$ free parameters. Such a filter can be defined as follows:

$$g_\theta(S) = \sum_{k=0}^{K} \theta_k S^K$$

The convolution output of the function f using this spectral filter can be written as follows:

$$y = U g_\theta (S) U^T f$$

$$= U \sum_{k=0}^{K} \theta_k S^K \, U^T f$$

$$= \sum_{k=0}^{K} \theta_k U S^K U^T f$$

Now since USU^T is the spectral decomposition of the Laplacian L, hence $L^k = U S^K U^T$. This allows us to write the output of convolution as follows:

$$y = \sum_{k=0}^{K} \theta_k L^k f$$

The above expression allows us to treat the convolution operator on the function f in terms of the polynomial in the Laplacian, i.e., $\sum_{k=0}^{K} \theta_k L^k$.

A polynomial in the Laplacian of order K can only extract information from the K hop neighborhood, and hence it acts as a K-localized filter.

Although the K-localized filter representation $\sum_{k=0}^{K-1} \theta_k S^K$ involving the polynomial in the Eigen value matrix provided localization and reduction of the free parameters, it is still expensive as it involves multiplication of the Eigen vector matrix which is of $O(|V|^2)$ complexity. Similarly, the convolution involving the polynomial in the Laplacian matrix $\sum_{k=0}^{K-1} \theta_k L^k$ would not be inexpensive either for large graphs as it involves computing the powers of the Laplacian matrix.

ChebNet

A slightly different version of the K-localized spectral filter that we have just discussed in the earlier section ($\sum_{k=0}^{K} \theta_k S^K$) was proposed in the ChebNet paper by Hammond et al. They came up with a well-approximated truncated expansion of $g_\theta(S)$ using Chebyshev polynomials $T_k(x)$ up to order K as shown in the following:

$$g_{\theta'} (S) = \sum_{k=0}^{K} \theta_k' T_k (\tilde{S})$$

In the preceding expression $\tilde{S} = \dfrac{2}{\lambda_{max}} S - I$, λ_{max} denotes the maximum Eigen value of the Laplacian L. The Chebyshev polynomial follows the recurrence given by $T_k(x) = 2xT_{k-1}(x) - T_{k-2}(x)$ with $T_0(x) = 1$ and $T_1(x) = x$.

The output y of the Chebyshev filter-based convolution with function f can be rewritten in terms of the modified Laplacian $\tilde{L} = \dfrac{2}{\lambda_{max}} L - I$ as follows:

$$y = \sum_{k=0}^{K} \theta'_k T_k\left(\tilde{L}\right) f$$

The output of the preceding convolution is a function of the K-th order polynomial of the Laplacian $\tilde{L}$, and hence the ChebNet filter acts as a K-localized filter. In K-localized filters, each node can get information from at most its K hop neighbors. Hence, the complexity of ChebNet-based convolution becomes linear in the number of edges. Also, we don't need to perform the Eigen decomposition of the Laplacian matrix.

Graph Convolution Network (GCN)

GCN uses ChebNet filters but restricts the order of the Chebyshev polynomials in them to order 1. Hence the output y of a GCN-based filter convolution on a function f is as follows:

$$y = \sum_{k=0}^{1} \theta'_k T_k\left(\tilde{L}\right) f$$

$$= \left(\theta'_0 T_0\left(\tilde{L}\right) + \theta'_1 T_1\left(\tilde{L}\right) \right) f$$

$$= \left(\theta'_0 I + \theta'_1 \tilde{L} \right) f$$

Substituting $\tilde{L} = \dfrac{2}{\lambda_{max}} L - I$ in the preceding equation,

$$y = \left(\theta'_0 I + \theta'_1 \left(\frac{2}{\lambda_{max}} L - I \right) \right) f$$

$$= \left((\theta'_0 - \theta'_1) I + \theta'_1 \frac{2}{\lambda_{max}} L \right) f$$

GCN simplifies the preceding expression by assuming $\lambda_{max} = 2$ as the author expects the neural network parameters will adapt to this change. Hence, we have the following:

$$y = \left(\left(\theta_0' - \theta_1'\right)I + \theta_1'L\right)f$$

Substituting the normalized Laplacian expression $L = I - D^{-\frac{1}{2}}AD^{-1/2}$ in the preceding equation, we have the following:

$$y = \left(\left(\theta_0' - \theta_1'\right)I + \theta_1'\left(I - D^{-\frac{1}{2}}AD^{-\frac{1}{2}}\right)\right)f$$

$$= \left(\theta_0'I - \theta_1'\left(D^{-\frac{1}{2}}AD^{-\frac{1}{2}}\right)\right)f$$

To reduce the parameters further, GCN sets $\theta_0' = -\theta_1' = \theta$. This simplifies the output of convolution to the following:

$$y = \theta\left(I + D^{-\frac{1}{2}}AD^{-\frac{1}{2}}\right)f$$

Hence each GCN filter has only one learnable parameter. The GCN filtering operation is linear in L, and it collects information from just the first hop neighbors for each node. To learn a rich set of representation that contains information from multiple hops, GCN can have multiple layers. The advantage with GCN over ChebNet is that it doesn't make the explicit parameterization of k-localization and hence is less likely to overfit. For a fixed computation budget, GCNs can have much deeper layers because of its relatively simpler filters in comparison with other networks.

Implementation of Graph Classification Using GCN

In this section, we perform graph classification using GCN. The graphs are from the MUTAG dataset that contains 188 graphs. Each of the graphs represents a nitroaromatic chemical compound where the nodes stand for the atoms. The input feature for the atoms is "atom type" which is a one hot encoded vector of length 7. Each graph can belong to one of two classes. The goal is to predict each chemical compound mutagenicity on *Salmonella typhimurium* as a binary classification problem.

As far as the model is concerned, we use StellarGraph
GCNSupervisedGraphClassification block that can take in specified number of GCN
layers of specified size followed by a mean pooling over the nodes of the graph. We
follow that up with two dense layers before the final binary class prediction layer. We use
ReLU activation in all the hidden layer. The model architecture is outlined in Figure 6-31.

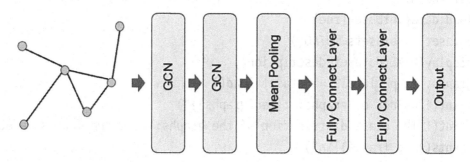

Figure 6-31. *GCN architecture for graph classification*

The implementation of the graph classification model using GCN is illustrated in
detail in Listing 6-9.

Listing 6-9. Graph Classification Using GCN

```
import matplotlib.pyplot as plt
from sklearn.manifold import TSNE
import io, os, sys, types
import pandas as pd
import numpy as np
from stellargraph.mapper import PaddedGraphGenerator
from stellargraph.layer import GCNSupervisedGraphClassification
from stellargraph import StellarGraph
from stellargraph.layer import DeepGraphCNN
from stellargraph import datasets
from sklearn.model_selection import train_test_split
from IPython.display import display, HTML
from tensorflow.keras import Model
from tensorflow.keras.optimizers import Adam
from tensorflow.keras.layers import Dense, Conv1D, MaxPool1D,
Dropout, Flatten
```

```python
from tensorflow.keras.losses import binary_crossentropy
from tensorflow.keras.callbacks import EarlyStopping
import tensorflow as tf
import matplotlib.pyplot as plt

#Import the data
def load_data(verbose=True):
    dataset = datasets.MUTAG()
    display(HTML(dataset.description))
    graphs, graph_labels = dataset.load()
    print(f"Number of graphs: :{len(graphs)}")
    print(f"The class distribution of the graphs:\n {graph_labels.value_
    counts().to_frame()}\n")
    if verbose:
        print(f"Graph 0 info",graphs[0].info())
        print("Graph 0 node features",graphs[0].node_features())
    return graphs, graph_labels

def GCN_model(generator,lr=0.005):
    gc_model = GCNSupervisedGraphClassification(
        layer_sizes=[64, 64],
        activations=["relu", "relu"],
        generator=generator,
        dropout=0.5,
    )
    x_inp, x_out = gc_model.in_out_tensors()
    predictions = Dense(units=32, activation="relu")(x_out)
    predictions = Dense(units=16, activation="relu")(predictions)
    predictions = Dense(units=1, activation="sigmoid")(predictions)

    # Let's create the Keras model and prepare it for training
    model = Model(inputs=x_inp, outputs=predictions)
    model.compile(optimizer=Adam(lr), loss=binary_crossentropy,
    metrics=["acc"])
    return model

def train(epochs=10,lr=0.005,batch_size=8):
```

```python
# Load the graphs
graphs, graph_labels = load_data(verbose=False)

# Convert the graph labels to 0 and 1 instead of -1 and 1
graph_labels = pd.get_dummies(graph_labels, drop_first=True)

# Create train val split
num_graphs = len(graphs)
train_indices, val_indices = train_test_split(np.arange(num_graphs),
test_size=.2, random_state=42)

# Create the data generator for keras training using Stellar's
PaddPaddedGraphGenerator
generator = PaddedGraphGenerator(graphs=graphs)

# Train Generator from the PaddedGraphGenerator Object
train_gen = generator.flow(
    train_indices, targets=graph_labels.iloc[train_indices].values,
    batch_size=batch_size
)
# Test Generator from the PaddedGraphGenerator Object
val_gen = generator.flow(
    val_indices, targets=graph_labels.iloc[val_indices].values, batch_
    size=batch_size
)

# Early Stopping hook
earlystopping = EarlyStopping(
    monitor="val_loss", min_delta=0, patience=25, restore_best_
    weights=True
)

# Define the Model
model = GCN_model(generator,lr=lr)

history = model.fit(
    train_gen, epochs=epochs, validation_data=val_gen, verbose=0,
    callbacks=[earlystopping])
```

```
    # Check validation metrics at the end of training
    val_metrics = model.evaluate(test_gen, verbose=2)
    val_accuracy = val_metrics[model.metrics_names.index("acc")]
    print(f"Training Completed, validation accuracy {val_accuracy}")
    return model

model = train()
```

--output–

```
Number of graphs: :188
The class distribution of the graphs:
    label
1     125
-1     63

2/2 - 0s - loss: 0.5645 - acc: 0.7297 - 28ms/epoch - 14ms/step
Training Completed, validation accuracy 0.7297297120094299
```

The model achieved a reasonable accuracy of 72.9% in classifying the graphs using GCN. One thing to note is that there are only 188 graphs for training and validation and hence one train validation split for training might not give a measure of the robustness of the model. The readers are advised to perform a k-fold cross-validation on this dataset and check for model robustness.

GraphSage

GraphSage is a method for inductive representation learning on graphs. Most of the graph representation methods have focused on generating embeddings on a fixed graph. However, many real-world applications require embeddings to be generated for unseen new nodes as well as for completely new subgraphs without having to retrain a model. In such cases, inductive representation learning comes in handy where a learned model on some nodes can generalize to completely unseen nodes. GraphSage is different from the other embedding methods in the following way:

1. Instead of training for distinct embedding vector for each node, GraphSage trains a set of aggregator functions that learn to aggregate feature representation in the local neighborhood of the node.

2. For each hop, there is a separate aggregator function.

Since GraphSage learns aggregator function, the feature representation of the new nodes can be computed by using the aggregator functions learned.

In factorization method or in approaches such as Node2Vec where the embeddings are learned for each node directly, the embedding ends up being transductive in nature.

GraphSage can be trained to not only generate embeddings in an unsupervised setting but also to learn different tasks under supervision. Hence the embeddings generated from the GraphSage layers are used as features for the task-specific heads of the network.

Outlined in the following is how GraphSage generates embeddings using the aggregator functions during prediction for a graph $G = (V, E)$. We assume there are K GraphSage layers corresponding to which there are K aggregator functions and K weight matrices W^k. The activation we use in each layer is represented by $\sigma(.)$, and we denote the features for node v to be x_v.

1. Initialize $h_v^{(0)} \leftarrow x_v, \forall v \in V$

2. for $k = 1 .. K$ do

3. for $v \in V$ do

4. $\qquad h_{N(v)}^k \leftarrow AGGREGATE_k \left(\left\{ h_u^{(k-1)} \in N(v) \right\} \right);$

5. $\qquad h_v^{(k)} \leftarrow \sigma \left(W^k . concat \left(h_v^{(k)}, h_{N(v)}^k \right) \right)$

6. end

7. $h_v^{(k)} \leftarrow \dfrac{h_v^{(k)}}{{h_v^{(k)}}_2}, v \in V$

8. end

9. $z_v \leftarrow h_v^{(k)}, \forall v \in V$

In the preceding algorithm as we can see in each iteration or search depth, we aggregate information for each node from the local neighbors, and as the iterations proceed, each node gets more and more information from different depths of the graph. The aggregator functions generally used in GraphSage are the mean aggregators and the LSTM aggregators. The neighbor $N(v)$ size for each node v in each iteration or layer of GraphSage is chosen reasonably enough to keep the compute tractable.

In the supervised setting, GraphSage learns the weights W^k based on the task objective. In an unsupervised setting, GraphSage is trained to produce similar embedding for nearby nodes and disparate embedding for highly different nodes.

Hence for a node u, the loss that can be minimized is as follows:

$$L(z_u) = -\log\left(\sigma\left(z_u^T z_v\right)\right) - \lambda E_{v_n \sim P_{n(v)}} \log\left(\sigma\left(-z_u^T z_{vn}\right)\right)$$

In the preceding expression, v is any node that co-occurs with u in random walks for data generation, and hence the model will enforce its representations to be similar. The second part of the loss is to increase the disparity between the node v and other nodes which are dissimilar to node v using negative sampling.

Implementation of Node Classification Using GraphSage

In this section, we implement the node classification on the Cora dataset using GraphSage. The nodes in the Cora dataset graph are publications, while the edges are citations of each other's work. The node features for the publications are bag of word representations based on 1433 keywords in the publications. The keywords are marked 1 if they occur more than once in the publications. Each of these publications falls under the seven subject classes that we would like to predict. Mentioned in the following is the detailed implementation (see Listing 6-10). We use the StellarGraph Node2Vec layer to define a two-layer GraphSage and then build the end-to-end model using tf.keras.

Post the model training, we plot the embeddings coming out of the final GraphSage layer and see if they get clustered based on the subject type of the publication.

Listing 6-10. Node Classification Using GraphSage

```
import networkx as nx
import pandas as pd
import os

import stellargraph as sg
from stellargraph.mapper import GraphSAGENodeGenerator
from stellargraph.layer import GraphSAGE

from tensorflow.keras import layers, optimizers, losses, metrics, Model
from sklearn import preprocessing, feature_extraction, model_selection
```

```python
from stellargraph import datasets
from IPython.display import display, HTML
import matplotlib.pyplot as plt
%matplotlib inline
from collections import Counter
from sklearn.decomposition import PCA
from sklearn.manifold import TSNE
import pandas as pd
import numpy as np

def load_cora_dataset():
    dataset = sg.datasets.Cora()
    display(HTML(dataset.description))
    GS, node_subjects = dataset.load(largest_connected_component_only=True)
    # GS contains the Stellar Graph
    # node_subjects contains the node_features indexed by the node ids
    print(f"The set of classes for nodes:{set(node_subjects)}\n")
    return GS, node_subjects

def train(batch_size = 32,num_samples = [10, 5],lr=0.005,epochs=20,
dropout=0.5):

    # Load Cora Datasets
    GS, node_subjects = load_cora_dataset()

    # Node features dimension to be used for training
    print(f"Training node input features: {GS.node_features().shape}\n")
    train_subjects, val_subjects = model_selection.train_test_split(
        node_subjects, train_size=0.1, test_size=None, stratify=node_
        subjects
    )

    print(f"Class representation in training data: {dict(Counter(node_
    subjects))}\n")

    # Create one hot encoding for node classes
    label_encoder = preprocessing.LabelBinarizer()
```

```
train_y = label_encoder.fit_transform(train_subjects)
val_y = label_encoder.transform(val_subjects)

# Define GraphSAGENodeGenerator object for tf.keras training
generator = GraphSAGENodeGenerator(GS, batch_size, num_samples)
# Create train and val generator using GraphSAGENodeGenerator object
train_gen = generator.flow(train_subjects.index, train_y, shuffle=True)

# Create train and val generator using GraphSAGENodeGenerator object
val_gen = test_gen = generator.flow(val_subjects.index, val_y)

# Define two layer GraphSage Model with 32 units in each layer
GS_model = GraphSAGE(
layer_sizes=[32, 32], generator=generator, bias=True, dropout=0.5,
)

x_inp, x_out = GS_model.in_out_tensors()
prediction = layers.Dense(units=train_y.shape[1],
activation="softmax")(x_out)
# The keras model has the GraphSage layers from Stellar followed by the
Dense prediction layer of tf.keras.layers
model = Model(inputs=x_inp, outputs=prediction)
print(f"Model Summary...\n")
print(model.summary())

# Compile the Model
model.compile(
optimizer=optimizers.Adam(lr=lr),
loss=losses.categorical_crossentropy,
metrics=["acc"],
)

# Train the model
history = model.fit(
train_gen, epochs=epochs, validation_data=test_gen, verbose=2,
shuffle=False
)
```

```python
    # Plot the training loss/metric profile
    sg.utils.plot_history(history)

    val_metrics = model.evaluate(val_gen)
    print("Val Metrics:\n")
    for name, val in zip(model.metrics_names, val_metrics):
        print("\t{}: {:0.4f}".format(name, val))

    # Create embeddings as the output of the Final Graph Sage layer
    # and see if the embeddings of nodes in similar classes are same
    all_nodes = node_subjects.index
    all_gen = generator.flow(all_nodes)
    emb_model = Model(inputs=x_inp, outputs=x_out)
    emb = emb_model.predict(all_gen)
    print(f"Embeddings shape: {emb.shape}\n")
    plot_embeddings(node_embeddings=emb,node_subjects=node_subjects)
    return model

def plot_embeddings(node_embeddings,node_subjects,n_components=2):

    transform = TSNE
    trans = transform(n_components=2)
    node_embeddings_2d = trans.fit_transform(node_embeddings)
    # draw the embedding points, coloring them by the target label (paper
    subject)
    alpha = 0.7
    label_map = {l: i for i, l in enumerate(np.unique(node_subjects))}
    node_colours = [label_map[target] for target in node_subjects]

    plt.figure(figsize=(7, 7))
    plt.axes().set(aspect="equal")
    plt.scatter(
        node_embeddings_2d[:, 0],
        node_embeddings_2d[:, 1],
        c=node_colours,
        cmap="jet",
        alpha=alpha,
    )
```

629

```
    plt.title("{} visualization of node embeddings".format(transform.__
name__))
    plt.show()

model = train()
--output--

Epoch 19/20
8/8 - 1s - loss: 0.2059 - acc: 0.9919 - val_loss: 0.6328 - val_acc:
0.8216 - 1s/epoch - 133ms/step
Epoch 20/20
8/8 - 1s - loss: 0.1877 - acc: 0.9960 - val_loss: 0.6290 - val_acc:
0.8221 - 959ms/epoch - 120ms/step
70/70 [==============================] - 1s 14ms/step - loss: 0.6402 -
acc: 0.8185
Val Metrics:

        loss: 0.6402
        acc: 0.8185
78/78 [==============================] - 1s 16ms/step
Embeddings shape: (2485, 32)
```

Model accuracy on the validation dataset is around ~82% which is reasonable.

Also, the node embeddings coming out of the final GraphSage layer form very distinct clusters (see Figure 6-32) based on the subject of the publications.

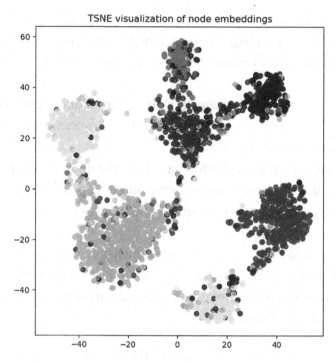

Figure 6-32. *GraphSage-based embedding on supervised task*

Graph Attention Networks

In all the Graph Conv methods we have studied so far, except GraphSage, the learned filters depend on the Laplacian Eigen functions which in turn depend on the graph structure. Thus, a model trained on a specific graph would not work on a new graph with a different structure. The nonspectral approaches on the other hand apply convolution directly to the graph by operating on groups of spatially close neighbors. One of the challenges of applying convolution directly to graphs has been to find an operator which works on all different-sized neighborhoods and maintain the weight sharing property of convolution.

Attention mechanism has really revolutionized the natural language processing domain and has become the standard for sequence-based tasks such as machine translation, text summarization, etc. The advantage with attention is that it provides a direct correlation between two entities. At the same time, it can be highly parallelized for faster processing. Sequential models such as LSTMs are inefficient in this sense as they cannot be a parallelized. This is a big drawback in the age of GPUs which thrive on parallelism.

631

Generally, there are two kinds of attention—self-attention and cross-attention. Self-attention is generally used to attend to the part of the same sequence for representation learning. In a cross-attention, part of one sequence attends to words or other entities in another sequence.

A part of the graph attention network plan is to learn hidden representation of each node by attending to its neighbors using self-attention. The same is done through the graph attention layer that we discuss next.

We take the input to a graph attention layer to be the node feature representations $h = [h_0, h_1, ...h_{|V|-1}]$, $h_i \in R^F$ where $|V|$ denotes the number of nodes in the graph $G = (V, E)$. The layer is expected to produce a new set of node features $h' = \left[h_0', h_1', ..h_{|V|-1}' \right], h_i' \in R^{F'}$ as its output.

In order to be able to transform the feature dimension from F to F' in a given layer, we define a set of weights $W \in R^{F' \times F}$ that is applied to each of the nodes. We then perform self-attention where the attention mechanism a computes attention coefficient between nodes i and j as follows:

$$e_{ij} = a\left(Wh_i, Wh_j\right)$$

In general, we can have each node attend to every other node without considering any relationship between nodes that the graph provides. However, in the graph attention layer for a given node i, we only compute attention weights for all nodes $j \in N(i)$ where $N(i)$ is the neighbors of node i chosen through some scheme. In general, $N(i)$ is mostly chosen as the one-hop neighbors of i and i itself. In order to linearly combine the node embeddings of the nodes $j \in N(i)$ in an expected sense we compute the probability normalized attention weights α_{ij} as

$$\alpha_{ij} = softmax\left(e_{ij}\right) = \frac{\exp\left(e_{ij}\right)}{\sum_{j \in N(i)} \exp\left(e_{ij}\right)}$$

Using the normalized weight α_{ij}, the output representation of the node i can be computed as follows:

$$h_i' = \sigma\left(\sum_{j \in N(i)} \alpha_{ij} Wh_j \right)$$

The σ in the preceding expression is a suitable nonlinear activation.

Let us now revisit the attention function a that gave the initial attention coefficients e_{ij}:

$$e_{ij} = a\left(Wh_i, Wh_j\right)$$

Since e_{ij} is a scalar, hence the attention function can be any function of the form $a : F \times F^1 \to R$:

The simplest attention function a that we can define is the dot product, and hence we could use the following:

$$e_{ij} = \left(Wh_i\right)^T \left(Wh_j\right)$$

We can see the dot product as the attention function doesn't take up any additional parameters that we have to learn. If we want to give some expressive power to the attention function, we can define a parameterized attention function. One such parameterized attention function would be to have dot product of learnable parameter vector $a \in R^{2F'}$ with the concatenated version of the feature representation Wh_i and Wh_j. Also, one can apply a nonlinearity to the output of the dot product. The authors of the graph attention network chose LeakyReLU (with alpha=0.2) as the nonlinearity. The attention function a can in such case be written as follows:

$$a(Wh_i, Wh_j|\, a) = \text{LeakyRELU}[a^T(Wh_i \,||\, Wh_j)]$$

Few of the advantages with the graph attention network formulation are as follows:

1. As discussed, earlier attention mechanism can be highly parallelized, and hence the GAT is computationally efficient.

2. There is no need to perform expensive Eigen decomposition of the Laplacian.

3. GAT allows for implicitly assigning differential importance to nodes of the same neighborhood. The same is not true for GCN. Also, the learned attention weights might help with interpretability.

Summary

With this, we come to the end of both this chapter and this book. The concepts and models illustrated in this chapter, although more advanced, use techniques learned in earlier chapters. After reading this chapter, one should feel confident in implementing the variety of models discussed in the book as well as try implementing other different models and techniques in this ever-evolving deep-learning community. One of the best ways to learn and come up with new innovations in this field is to closely follow the other deep-learning experts and their work. And who better to follow than the likes of Geoffrey Hinton, Yann LeCun, Yoshua Bengio, and Ian Goodfellow, among others. Also, I feel one should be closer to the math and science of deep learning rather than just use it as a black box to reap proper benefits out of it. With this, I end my notes. Thank you.

Index

Printed in the United States
by Baker & Taylor Publisher Services